Exploratory and Multivariate Data Analysis

This is a volume in
STATISTICAL MODELING AND DECISION SCIENCE

Gerald J. Lieberman and Ingram Olkin, editors
Stanford University, Stanford, California

Exploratory and Multivariate Data Analysis

Michel Jambu

National Centre for Telecommunications Studies
Paris, France

ACADEMIC PRESS, INC.
Harcourt Brace Jovanovich, Publishers

Boston   San Diego   New York
London   Sydney   Tokyo   Toronto

This book is printed on acid-free paper. ⊚

English translation copyright © 1991 by Academic Press, Inc.
© BORDAS et C.N.E.T.-E.N.S.T., Paris 1989
All rights reserved.
No part of this publication may be reproduced or
transmitted in any form or by any means, electronic
or mechanical, including photocopy, recording, or
any information storage and retrieval system, without
permission in writing from the publisher.

ACADEMIC PRESS, INC.
1250 Sixth Avenue, San Diego, CA 92101

United Kingdom Edition published by
ACADEMIC PRESS LIMITED
24–28 Oval Road, London NW1 7DX

Library of Congress Cataloging-in-Publication Data:

Jambu, Michel.
   [Exploration informatique et statistique des données. English]
   Exploratory and multivariate data analysis/Michel Jambu.
     p. cm.—(Statistical modeling and decision science)
   Translation of: Exploration informatique et statistique des
données.
   Includes bibliographical references and index.
   ISBN 0-12-380090-0 (alk. paper)
   1. Mathematical statistics—Data processing. I. Title.
II. Series.
   QA276.4.J3613   1991
   519.5′0285—dc20   90-23003
                                              CIP

Printed in the United States of America
91 92 93 94   9 8 7 6 5 4 3 2 1

To Catherine,
Hugo, Sébastien, Thomas

"L'essence de codage des données est de traduire fidèlement les relations observées entre les choses par des relations entre êtres mathématiques, de telle sorte qu'en réduisant par le calcul la structure mathématique choisie pour image du réel, on ait de celui-ci un dessin simplifié accessible à l'intuition et à la réflexion avec la garantie d'une critique mathématique".

J. P. BENZÉCRI
in Les Cahicas de l'Analyse
des Données. Vol. II, 1977, n° 4, 369–406

Contents

Preface　　　　　　　　　　　　　　　　　　　　　　　　　　　　xi

Chapter 1　　General Presentation　　　　　　　　　　　　　　　1
1. Introduction　　　　　　　　　　　　　　　　　　　　　　　　1
2. Examples of Applications　　　　　　　　　　　　　　　　　　5
3. Steps in Data Exploration: Management, Analysis,
   Synthesis　　　　　　　　　　　　　　　　　　　　　　　　　14
4. Computer Aspects　　　　　　　　　　　　　　　　　　　　　17

Chapter 2　　Statistical Data Exploration　　　　　　　　　　　　19
1. Statistics　　　　　　　　　　　　　　　　　　　　　　　　　19
2. Fields of Statistical Data Exploration　　　　　　　　　　　　19
3. Statistics and Experiments　　　　　　　　　　　　　　　　　20
4. Data Analysis, Inductive and Deductive Statistics　　　　　　　21
5. Variables, Statistical Sets, and Data Sets　　　　　　　　　　　21

Chapter 3　　1-D Statistical Data Analysis　　　　　　　　　　　　27
1. Introduction　　　　　　　　　　　　　　　　　　　　　　　27
2. 1-D Analysis of a Quantitative Variable　　　　　　　　　　　27
3. 1-D Analysis of a Categorical Variable　　　　　　　　　　　　50
4. 1-D Analysis of a Categorical Variable with Multiple
   Forms　　　　　　　　　　　　　　　　　　　　　　　　　　53
5. 1-D Analysis of Time Series or Chronological Variables　　　　53
6. Statistical Maps or Cartograms　　　　　　　　　　　　　　　61

Chapter 4　　2-D Statistical Data Analysis　　　　　　　　　　　　63
1. Introduction　　　　　　　　　　　　　　　　　　　　　　　63
2. 2-D Analysis of Two Categorical Variables　　　　　　　　　　64
3. 2-D Analysis of Two Quantitative Variables　　　　　　　　　73
4. 2-D Analysis of a Quantitative Variable and a Categorical
   Variable　　　　　　　　　　　　　　　　　　　　　　　　　90

5. 2-D Analysis of a Quantitative Variable and a Categorical Variable with Multiple Forms	94
6. Conclusion	94

Chapter 5   N-D Statistical Data Analysis   95

1. Introduction	95
2. Joint 3-D Statistical Data Analysis	95
3. Joint N-D Statistical Data Analysis	102
4. Cartograms and N-D Analysis	112

Chapter 6   Factor Analysis of Individuals–Variables Data Sets   113

1. Introduction	113
2. From Linear Adjustment to Factor Analysis	113
3. From the Origin of Factor Analysis to Modern Factor Analysis Techniques	117
4. Mathematical Description of Modern Factor Analysis	117
5. Factor Analysis Formulas	124

Chapter 7   Principal Components Analysis   125

1. Basic Data Sets	125
2. Different Patterns of Principal Components Analysis	125
3. Standardized Principal Components Analysis	127
4. Interpretation of Principal Components Analysis	131
5. Classifying Supplementary Points into Graphics	143
6. Rules for Selecting Significant Axes and Elements	153
7. Standardized Principal Components Analysis Formulas	159
8. Applications and Case Studies	160

Chapter 8   2-D Correspondence Analysis   169

1. Introduction	169
2. Basic Correspondence Data Sets	170
3. Mathematical Description of Correspondence Analysis	171
4. Geometric Representation of the Sets I and J	176
5. Interpretation of the 2-D Correspondence Analysis	189
6. Factor Graphics	200
7. Classifying Supplementary Points into Graphics	203
8. Rules for Selecting Significant Axes and Elements	210
9. 2-D Correspondence Analysis Formulas	216
10. Patterns of Clouds of Points	221
11. Patterns of Acceptable Data Sets	225
12. Case Studies	232

Contents ix

Chapter 9   N-D Correspondence Analysis   241

1. Introduction   241
2. Basic Data Sets   242
3. Equivalence between Analyses of b_{JJ} and k_{IJ} 248
4. Interpretation of N-D Correspondence Analysis   262
5. Factor Graphics   273
6. Classifying Supplementary Points into Graphics   278
7. Rules for Selecting Significant Axes and Points of $N(I)$, $N(J)$, and $N(Q)$ 285
8. N-D Correspondence Analysis Formulas   287
9. Patterns of Acceptable Data Sets   288
10. Case Studies   294

Chapter 10   Classification of Individuals–Variables Data Sets   305

1. Introduction   305
2. Basic Data Sets   306
3. The Mathematical Description of Classifications   306
4. Partitioning Methods   310
5. Hierarchical Classification Methods   331
6. Specific Applications   394
7. Case Studies   398

Chapter 11   Classification and Analysis of Proximities Data Sets   407

1. Introduction   407
2. Proximities Data Sets   407
3. Proximities Data Sets from Individuals–Variables Data Sets   408
4. Elementary Description of Proximities Data Sets   412
5. Factor Analysis of Proximities Data Sets   412
6. Classification of Proximities Data Sets   416
7. Computation of Contributions   417
8. Conclusion   418

Chapter 12   Computer Aspects of Exploratory and Multivariate Data Analysis   419

1. Place of Exploratory and Multivariate Data Analysis in Statistics   419
2. Basic Factors for Exploratory and Multivariate Data Analysis Software   4221
3. Data Analysis Libraries   423
4. Future Prospects   425

Appendix 1	List of Notations	427
Appendix 2	Reference Data Sets	439
REFERENCES		465
AUTHOR INDEX		469
SUBJECT INDEX		471

Preface

Why this book?

After travelling around the world, studying many kinds of data, listening to many lectures on subjects of data analysis, and giving seminars, it became clear that the way data analysis is studied in France, with exploration by Benzécri and his associates, is actually different from data analysis anywhere else in the world.

When I published "Data Analysis and Clustering" in 1983, correspondence analysis and related topics was known world-wide to French-speaking people but not in the English-speaking world. It was one of the first attempts to present correspondence analysis and associated methods of data analysis to readers of English-reading people. Several colleagues then encouraged me to publish a textbook on correspondence analysis and the French method of data analysis. I was not actually satisfied by this proposal, because data analysis is the same around the world, even if the techniques associated with it vary. Finally, I gathered data analysis materials from different sources. There were so many connections and interactions among them that I combined them in order to propose a modern way of thinking and practising data analysis; the point is not only to use techniques but to use interactions and relations between them in view of summarizing data for improving knowledge, drawing valid conclusions, and aiding in decision making. The way was found; it remained to write the book.

What is in this book?

The heart of this book contains methods of exploring data from a statistical data analysis point of view, from the most elementary, associated with univariate and bivariate statistical description, to the most advanced, associated with multivariate statistical description, factor analysis, correspondence analysis and clustering. They are presented in such a manner that they correspond to exploration of data sets, step-by-step, to allow readers to build their own data analysis strategies from their data sets. The titles of the chapters and the general plan of the book are as follows: The first chapter presents a general introduction to the basic principles and steps of statistical data analysis with some case studies. The following chapters are presented in the order of the data analysis process: elaboration of data sets (Chapter 2), 1-D statistical data analysis (Chapter 3), 2-D statistical data analysis (Chapter 4), N-D statistical data analysis (Chapter 5), factor analysis of individuals–variables data sets (Chapter 6), principal components analysis (Chapter 7), 2-D correspondence data analysis (Chapter 8), N-D correspondence data analysis (Chapter 9), classification of individuals–variables data sets (Chapter 10), and analysis and classification of proximities data sets (Chapter 11). Chapter 12 is devoted to the computer aspects of data analysis. A list of notations, an appendix containing the data sets used as examples, and as usual, references, conclude the book.

For whom is the book written?

This book is written for anyone who analyzes data or expects to do so in the future, including students, statisticians, scientists, engineers, managers, and teachers. The material presented here is relevant for applications in various fields, such as physics, chemistry, medecine, business, management, marketing, economics, psychology, sociology, geosciences, biology, astronomy, quality control, engineering, computer science, education, linguistics, and virtually any other field where there are data to be analyzed, synthesized, or explored with the goal of improving knowledge or decision making. This book can also be used as a reference for a supplement to any course in applied statistics, or in applied sciences courses where statistics are taught.

What the prerequisite knowledge needed?

Chapters 1–5 do not assume any previous knowledge. The material can be understood by anyone who wants to learn it and who has some experience or interest in quantitative thinking. Chapters 6–9 assume a knowledge of the previous chapters and an understanding of data in terms of interactions between multiple data sets. These chapters are devoted to methods for solving complex problems involving complex data

Preface

sets. The mathematical background needed is the first level in any linear algebra course. Chapter 10 assumes an interest in taxonomic problems but no specific knowledge, the mathematical background needed is the first level in any university. Chapter 11 assumes a knowledge of Chapters 6–10. It is an introduction to a more general case of data often used in taxonomy and in multidimensional scaling. Chapter 12 assumes an interest in monitoring computer software on real data. It contains some recommendations to users in data analysis. In conclusion, there is no mathematical, statistical, computer knowledge required; just common sense.

Acknowledgments

I would need many pages to thank all the people that have led directly or indirectly to the publication of this book. I have dedicated this book to Professor J. P. Benzecri in acknowledgment of the role he played in my data analysis education. To all those who encouraged me to publish a text-book devoted to data analysis, correspondence analysis, and related topics, I extend my warmest thanks: I. Olkin, C. Hayashi, J. Kruskal, R. Sokal, N. Ohsumi, P. Tukey, J. R. Kettenring, D. Carroll, and D. Merriam, to name a few. Particular thanks are given to H. Teil and F. Murtagh for their critical reading and revising of the manuscript; to G. André, Chief Director of the Centre National d'Etudes des Télécommunications, who controlled efficiently the realization of the manuscript; to the staff of Academic Press for their excellent collaboration in passing the book through the press; last, but not least, to Mrs N. Tissèdre, for her patient work on the pains-taking preparation of the manuscript. Final thanks go to the Centre National d'Etudes des Télécommunications and the Société Francophone de Classification for their generous financial help, and the S.C.C.M. Inc. for its excellent realization of figures.

Paris, 1990

Chapter 1   General Presentation

1. Introduction

1.1.   Aim and Scope of Statistical Data Exploration

The aim of data analysis is to discover the structure of a set of multivariate observations without the assumption of any mathematical hypotheses on the structure of these observations or variables. Because of the size and complexity of the data sets, this structure cannot be discovered directly; specific data processing methods are therefore required to manage, explore, analyze, synthesize, and communicate the results of data processing. These methods are oriented according to the desired goal: improving basic knowledge of a field; diagnosis; forecasting; planning; decision making. Whatever the goal, the statistical features of the observed data sets need to be highlighted. Data analysis methods are the most appropriate ones for doing this.

1.2.   What Does "Data" Mean?

"Data" is a set of organized information of any type, covering all aspects of a domain related to a specific goal (forecasting, improving knowledge, causal analysis, decision making, etc.). It is a quantification of the real world into an image, acceptable to the human brain, and then to the computer. For example, when the quality of cars is studied, the quality is initially defined in terms of certain criteria; the information concerning these criteria observed on a selected set of cars (a sample) is then gathered. For example, criteria such as mileage, number of repairs, headroom, weight, length, turn circle, and gear ratio are collected and recorded in a data file or data base. All the information is stored in a data set that contains heterogeneous data, in general. Examine the data set given in Table 1.1. It is in the form of the rows and columns of a matrix. Each column and each row has a label; at the intersection of a column and a row is the information related to one variable observed on one car model. Naturally, there are many types of data sets. For example, consider the first column of the data set given in Table 1.1. It concerns the price of cars at a given time. This is a simple, or 1-D, data set as only one variable is observed. The whole data set given in Table 1.1 concerns the simultaneous observation of 12 variables on a given set of cars, and so it is a multiple, or *N-D,* data set. The complexity of data depends on the field of study and/or on the initial aim, and/or on the degree of detail associated with the study. Thus, the data sets studied by data analysis involve quantitative information (measurements, ratios, marks, indicators, etc) or qualitative (also called categorical) information (categories, logical attributes, intervals of quantitative information, etc.). A data set can involve homogeneous or heterogeneous information. Finally, depending on the goal, a data set can be divided into explanatory and explainable information. Generally, when the domain is large enough, the reference data sets contain all the different types of information. This is true in information systems or data bases. The problem is how to explore and process the data.

1.3. *What Does "to analyze data" Mean?*

"To analyze data" means to synthesize the content of data in a data base or a data file, by selecting specific data sets on which "data analysis methods" can be applied. Obviously, no method can analyze a disorganized data set. To be described, data must follow specific rules such as homogeneity, exhaustivity, and comparability. Thus, the first step of data analysis is to extract "relevant data sets" that can be analyzed whilst having in mind the objectives, which may vary. In an example about the quality of telephone service, the problem is to study levels of quality and to select statistically determined units from a given range of quality. In medicine, the problem is to study how different variables interact on a group of patients. In marketing, the problem is how to forecast the consumer behavior by observing selected variables on selected users. Basically, to analyze data means to choose data sets on which data analysis methods can be applied, with a view to decision making, selection, planning, forecasting, or understanding. And since data are too complex, too large, and too numerous, specific tools are needed to

1. Introduction

TABLE 1.1. Car models data set (extract). (From *Graphical Methods for Data Analysis*, by J.M. Chambers, W.S. Cleveland, B. Kleiner, and P.A. Tukey. Copyright © 1983 by Bell Telephone Laboratories Incorporated, Murray Hill, NJ. Reprinted by permission of Wadsworth & Brooks/Cole Advanced Books & Software, Pacific Grove, CA 93950.)

Make & Model	Price $	Mileage mpg	Repair Record 1978	Repair Record 1977	Head- room in	Rear Seat in	Trunk Space cu ft	Weight lbs	Length in	Turn Circle ft	Displace- ment cu in	Gear Ratio
Chev. Chevette	3299	29	3	3	2.5	26.0	9	2110	163	34	231	2.93
Chev. Impala	5705	16	4	4	4.0	29.5	20	3690	212	43	250	2.56
Chev. Malibu	4504	22	3	3	3.5	28.5	17	3180	193	41	200	2.73
Chev. Monte Carlo	5104	22	2	3	2.0	28.5	16	3220	200	41	200	2.73
Chev. Monza	3667	24	2	2	2.0	25.0	7	2750	179	40	151	2.73
Chev. Nova	3955	19	3	3	3.5	27.0	13	3430	197	43	250	2.56
Datsun 200-SX	6229	23	4	3	1.5	21.0	6	2370	170	35	119	3.89
Datsun 210	4589	35	5	5	2.0	23.5	8	2020	165	32	85	3.70
Datsun 510	5079	24	4	4	2.5	22.0	8	2280	170	34	119	3.54
Datsun 810	8129	21	4	4	2.5	27.0	8	2750	184	38	146	3.55
Dodge Colt	3984	30	5	4	2.0	24.0	8	2120	163	35	98	3.54
Dodge Diplomat	5010	18	2	2	4.0	29.0	17	3600	206	46	318	2.47
Dodge Magnum XE	5886	16	2	2	3.5	26.0	16	3870	216	48	318	2.71
Dodge St. Regis	6342	17	2	2	4.5	28.0	21	3740	220	46	225	2.94
Fiat Strada	4296	21	3	1	2.5	26.5	16	2130	161	36	105	3.37
Ford Fiesta	4389	28	4	—	1.5	26.0	9	1800	147	33	98	3.15
Ford Mustang	4187	21	3	3	2.0	23.0	10	2650	179	42	140	3.08
Honda Accord	5799	25	5	5	3.0	25.5	10	2240	172	36	107	3.05
Honda Civic	4499	28	4	4	2.5	23.5	5	1760	149	34	91	3.30
Linc. Continental	11497	12	3	4	3.5	30.5	22	4840	233	51	400	2.47
Linc. Cont Mark V	13594	12	3	4	2.5	28.5	18	4720	230	48	400	2.47
Linc. Versailles	13466	14	3	3	3.5	27.0	15	3830	201	41	302	2.47
AMC Concord	4099	22	3	2	2.5	27.5	11	2930	186	40	121	3.58
AMC Pacer	4749	17	3	1	3.0	25.5	11	3350	173	40	258	2.53
AMC Spirit	3799	22	—	—	3.0	18.5	12	2640	168	35	121	3.08
Audi 5000	9690	17	5	2	3.0	27.0	15	2830	189	37	131	3.20
Audi Fox	6295	23	3	3	2.5	28.0	11	2070	174	36	97	3.70
BMW 320i	9735	25	4	4	2.5	26.0	12	2650	177	34	121	3.64
Buick Century	4816	20	3	3	4.5	29.0	16	3250	196	40	196	2.93
Buick Electra	7827	15	4	4	4.0	31.5	20	4080	222	43	350	2.41
Buick Le Sabre	5788	18	3	4	4.0	30.5	21	3670	218	43	231	2.73
Buick Opel	4453	26	—	—	3.0	24.0	10	2230	170	34	304	2.87
Buick Regal	5189	20	3	3	2.0	28.5	16	3280	200	42	196	2.93
Buick Riviera	10372	16	3	4	3.5	30.0	17	3880	207	43	231	2.93
Buick Skylark	4082	19	3	3	3.5	27.0	13	3400	200	42	231	3.08
Cad. Deville	11385	14	3	3	4.0	31.5	20	4330	221	44	425	2.28
Cad. Eldorado	14500	14	2	2	3.5	30.0	16	3900	204	43	350	2.19
Cad. Seville	15906	21	3	3	3.0	30.0	13	4290	204	45	350	2.24

dissect data and to make either numerical or graphical summaries. This specific type of data processing follows a logical process described in Section 3.1 on the different steps of data exploration.

1.4. What Does "to synthesize data" Mean?

In any statistical study, there are two steps: analysis and synthesis. To synthesize data means to gather the most significant or the most telling features within the data. The results are presented in a way that is convenient for the user. Thus, the problem is not only to analyze data in depth, but also to communicate the results in terms of valid conclusions that can be used to make reasonable decisions. When data analysis was first used, analysis meant both analysis and synthesis. But, according to recent developments in methods, size, and complexity of data, analysis and synthesis must be distinguished again. The basic principles of data analysis are presented here in comparison with other scientific trends.

1.5. Basic Principles

Data analysis belongs to Statistics in the following sense:

> "Statistics is concerned with scientific methods for collecting, organizing, summarizing, presenting, analyzing data, as well as drawing valid conclusions and making reasonable decisions on the basis of such analysis."

(*cf.* Spiegel, 1961). It is opposed to experimental methods based on observing the variations of one variable with respect to all the others involved. Statistics and data analysis are based on data as they are collected. All of the possible variations for all of the variables cannot be studied and, most of the time, the control of these variables is impossible, as in economics, marketing, sociology, meteorology, geology, etc. Experimental methods are appropriate for specific classes of measurements. Statistics or data analysis methods can process a larger class of data than those used in experimental methods.

In Statistics, there are two currents: the inductive process and the deductive process. Data analysis is concerned with the deductive process; it means *to deduce* only from gathered data, and not to build a model first. The basic data analysis principles are expressed as follows:

(a) To extract structures from data, and not the reverse.
(b) To process simultaneously information involving multiple variables.

2. Examples of Applications

(c) To elaborate statistical information systems with a view to computer data processing.

(d) To use all the resources of a computer, particularly graphical tools.

Certain remarks can be made:

(a) Often the opposite is done; models smooth out data. Thus, it is taken as real what is purely a mathematical construct. It often happens that data are mutilated because it is thought that they cannot be processed by computer. But, it should be kept in mind that methods and software are now able to process data in depth.

(b) To analyze data variable by variable takes time and does not provide a synthesis. To do so, interactions between pieces of information must be studied globally.

(c) Sometimes data are built in successive layers, producing incoherency. Even if data are elaborated independently from data processing, they must be elaborated with a view to data processing.

(d) Graphics give more information than numerical tables. A histogram highlights the shape of a distribution: factor maps give more information than correlation matrices; dispersion box plots represent more than any statistical measures. In the following, some examples of real applications are given.

2. Examples of Applications

2.1. Economic Data: Car Models

To study the economic quality of cars, 37 cars were selected as a representative set. The variables observed were the price, mileage, repair record, headroom, rear seat, trunk space, weight, length, turn circle, displacement and gear ratio. These variables are assumed to influence both the economic quality and the price of a car (the data are given in Table 1.1). Figures 1.1 and 1.2 give the results of principal components analysis and its hierarchical classification performed on the car data set. The principal components analysis highlights two factors, and the resulting factor map shows the cars and the main criteria as points. To the right of the first axis are found the smaller cars and more generally the Japanese ones (Datsun, Honda) with high gear ratio and mileage; to the left of the first axis occur the larger cars and more generally the American cars, which are comfortable (rear seat, trunk space, headroom)

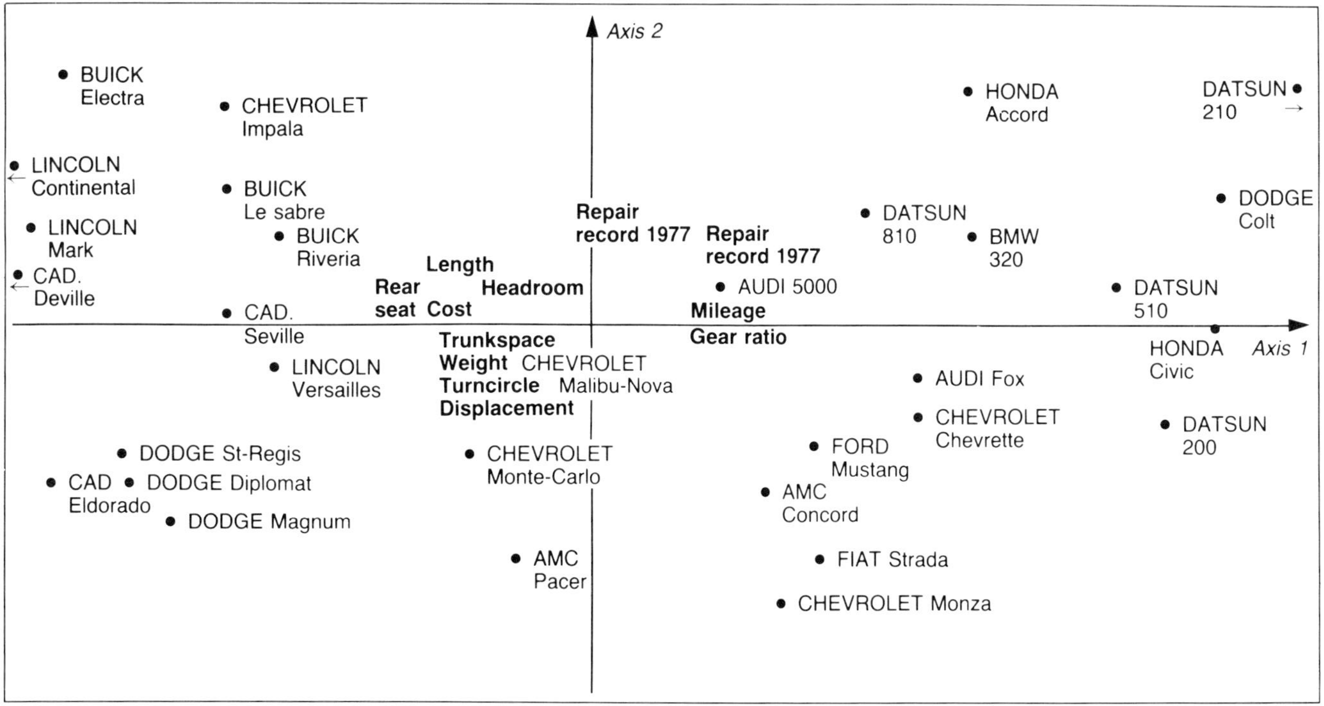

FIGURE 1.1. Analysis of a data set describing models of cars according to 12 criteria expected to have an influence on the quality of models (●). Representation in the two first factors (principal components analysis).

2. Examples of Applications

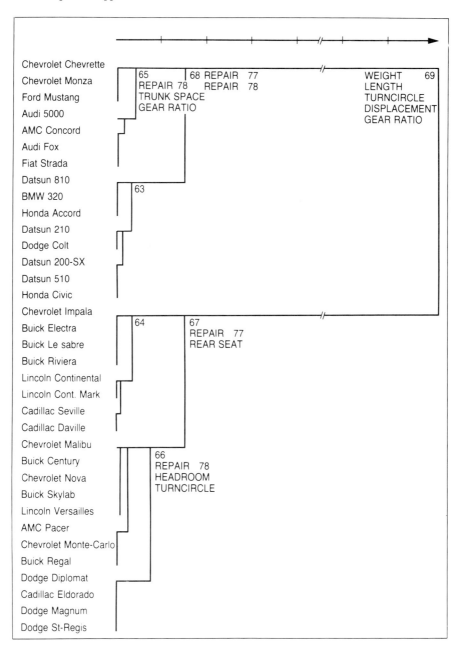

FIGURE 1.2 Car models classification. Hierarchical clustering of a principal components data set.

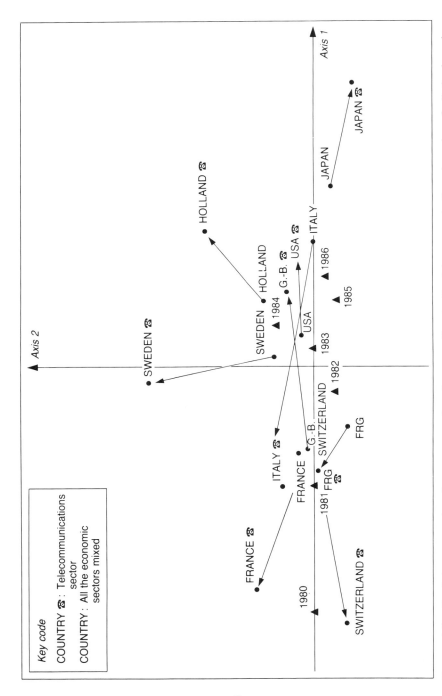

FIGURE 1.3. Statistics concerning the patents registration according to the telecommunications branch and all the branches mixed. Correspondence analysis; representation in the two first factors.

2. Examples of Applications

but heavy and more expensive than the smaller cars. This is confirmed by the hierarchical classification given in Fig. 1.2.

2.2. Industrial Data: International Evolution of Patents Registration

The number of patents is considered a good indicator of industrial activity. Two branches are studied: the telecommunications branch and all of the branches mixed. The data set is organized into two subsets (*cf.* Appendix 2, §3) simultaneously analyzed by correspondence analysis. The factorial map is given in Fig. 1.3. This map shows the relative position of each country for its own telecommunication branch with respect to the total of all branches. During the period 1980–1986, the number of patents registered increased for the USA, Japan, Italy, and Sweden, and was stable or decreased for FRG, Great Britain and France. For the telecommunication branch, the movement is expanded. The telecommunication branches of Japan, The Netherlands, USA, and Great Britain are increasing. But the telecommunication branches of France, FRG, Italy, and Switzerland are decreasing. This map is self-explanatory.

2.3. Marketing Data: Survey Concerning Users of New Services in Telecommunications

To study the behavior and satisfaction (or lack of satisfaction) of users of new services, France Telecom carried out surveys using questionnaires on 1800 people. The new services are the electronic directory and all of the associated distributing services requested using the Minitel, which is a piece of telecommunication equipment resembling to a computer terminal. The questionnaire consists of 70 multiple choice questions. The data set analyzed is a logical data set involving 252 dummy variables and 1800 persons (the dummy variables are the replies to the questions). It was analyzed by N-D correspondence analysis; here, we give two selected graphics, showing a part of the dummy variables. Figure 1.4 represents user satisfaction; Fig. 1.5 represents the usage of the new Minitel services. These two graphics can be superimposed. Data analysis processing of surveys needs more detailed analysis than for contingency data sets. This is will be discussed more in Chapter 9.

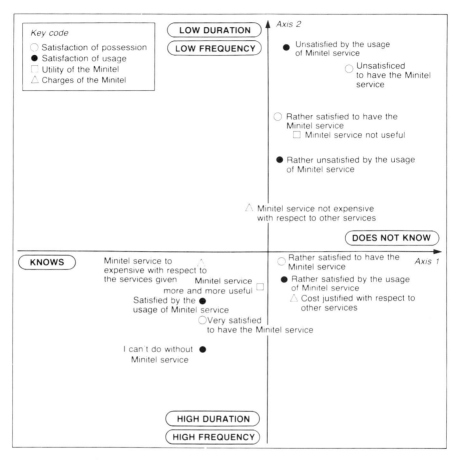

FIGURE 1.4. Questionnaire on Minitel. *N*-D correspondence analysis; representation in the two first factors. Representation of the variables concerning the usage and the price of the Minitel.

2.4. Geological Data: Barataria Grain Size Study

Krumbein and Aberdeen (1937) collected 98 bottom samples from the Kidal lagoon in Barataria Bay at the margin of the Mississippi delta, with the objective of evaluating the depositional environment of the lagoon. Data were recorded on the grain-size distribution of the samples (*cf.* Appendix 2, §13). Only 69 samples (with complete description) are retained for processing by correspondence analysis (*cf.* Fig. 1.6). The first axis clearly represents the evolution from coarse to fine grained sediments (*cf.* Teil, 1985).

2. Examples of Applications

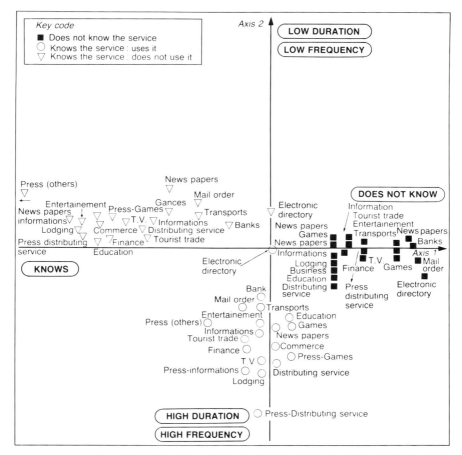

FIGURE 1.5. Questionnaire on Minitel. *N*-D correspondence analysis; representation in the two first factors. Representation of the variables concerning the knowledge and usage of the Minitel services.

2.5. Sociological Data: Family Timetables

In 1965, an international organization wanted to study and compare the lifestyle chosen according to marital status (single or married), sex (male or female), country, and professional activity. In this study, lifestyle was viewed through 10 major activities (professional work, transportation, sleep, household, meals, shopping, children, personal care, TV, leisure). The data set was built by taking into account the number of hours spent by a group on these different activities (*cf.* Appendix 2, §10). A principal components analysis was done (*cf.* Fig. 1.7.); it highlights the relationships between the population groups and variables. For example, the

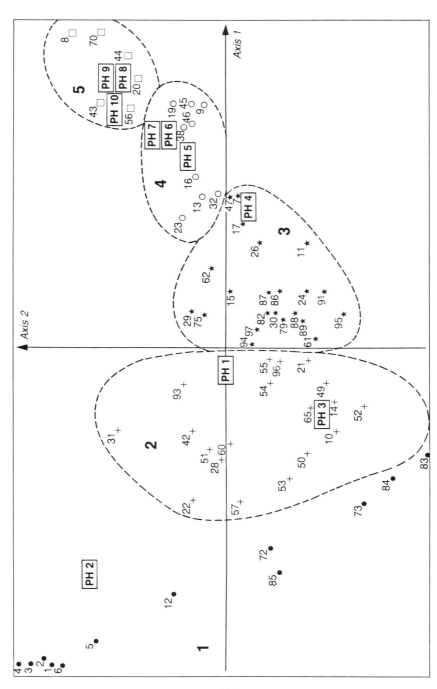

FIGURE 1.6. Correspondence analysis of Barataria raw data—the groups (1–5) correspond to those defined by Krumbein subjectively before the advent of computers. Group 1: ●; 2: +; 3: *; 4: ○; 5: □. PH defines the different grain sizes.

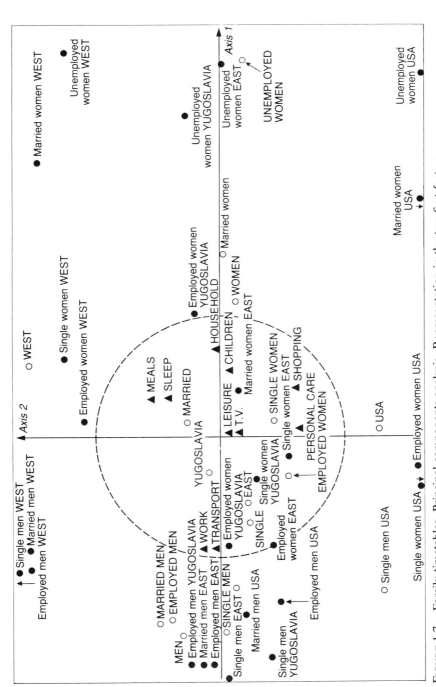

FIGURE 1.7. Family timetables. Principal components analysis. Representation in the two first factors.

second axis opposes the western countries (Europe) to the USA according to two groups of variables: meals and sleep for Europe on the one hand; personal care and shopping on the other hand for USA.

3. Steps in Data Exploration: Management, Analysis, Synthesis

Data analysis involves several steps from data conception to the use of final results in decision making. We present the steps and the relations among them, set in a network where the vertices are the steps and the edges the relations (*cf.* Fig. 1.8). Ten steps are identified and examined in detail. But, keep in mind that data analysis involves interaction with data and steps taken to analyze them.

STEP 1. **Data decision.** At the beginning, there is someone who decides on an action. It could be the manager (in business), the scientist (in fundamental sciences), the physician (in medicine), the agronomist (in studying plants), the decision maker (in marketing), etc. What does he decide? To study a field based on some hypotheses. Therefore, he must define the aim and scope of the study, the boundary of the field, and depending on his knowledge, draw the main features and the orientations of what he wants, and then determine the data expected to be necessary to describe or explain the problem he is trying to solve.

STEP 2. **Data conception, data elaboration (Chapter 2).** This is a hard task. The domain of the study must be already well determined; among all the possible variables that characterize the domain, only those that can be observed with reliability must be selected. Procedures for collecting data must be chosen (automatic collection, samples, surveys, etc.). The acceptable values for the variables must be foreseen so that data input can be validated. For data bases, data must be modeled with a view to carrying out the subsequent steps.

STEP 3. **Data input (Chapter 12).** For data bases, the data manager decides on how to enter the data. Users will then be able to access the data through a data management system. For data files or data matrices, data will be input into standard files acceptable by any statistical software.

STEP 4. **Data management.** We indicate only those data management functions related to data analysis; to manage data means: to create,

3. Steps in Data Exploration: Management, Analysis, Synthesis

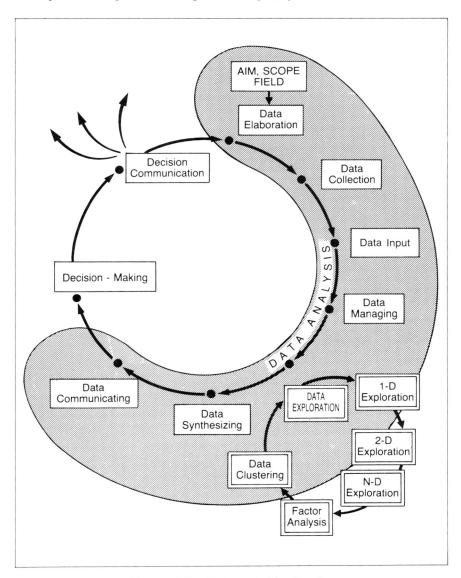

FIGURE 1.8. Data analysis network.

update, check, validate, control, recode, sort, merge, build, extract, aggregate, or question data sets. There are many operations whose aim is to give reliable and easily accessible data. For a data analyst, the most important function is recoding, because data never exist in a form acceptable for the data analysis methods. Chapters 8, 9, and 10 present different patterns of acceptable data sets (acceptable from a statistical data processing point of view).

STEP 5. **1-D analysis. Portraying data (Chapter 3).** The objective of 1-D analysis is to describe the information contained in data sets, variable-by-variable, independently of each other. It only gives a description (a portrait) of each variable. This portrait differs according to the types of variables: qualitative (or categorical), quantitative, categorical with multiple forms, chronological, etc. Most of the time, graphics are used instead of numerical values to present the results: histograms, dispersion box plots, pie charts, bar diagrams, percentage component bar graphs, and cartograms.

STEP 6. **2-D analysis (Chapter 4).** The objective of 2-D analysis is to study the dependency between two variables. This introduces the study of causality and in some sense, exploratory analysis. As in the previous step, different types of analysis may be made according to the type of variables. Graphics have an important place in this step (dispersion scatter plots, colored tables, 2-D histograms, regression curves, etc.).

STEP 7. **N-D analysis—Exploratory data analysis (Chapter 5).** The objective of N-D analysis is to represent *simultaneously* the relationships between N variables without any modeling, just as they are, by the means of graphics. It is also called exploratory data analysis (*cf.* Tukey, 1977). Graphics such as sun ray plots or profile plots are nevertheless limited in the number of variables that can be studied (≤ 10).

STEP 8. **Factor analysis—Advanced exploratory data analysis (Chapters 6, 7, 8, 9).** The objective of factor analysis is to represent observations and variables in a geometric space whose dimension is as small as possible (i.e., containing maximum information for a minimum dimension of the representation space). This involves a mathematical model of representation in which the number of initial variables can be reduced into a smaller number of factors. Based on this model, different factor analyses exist: principal components analysis, 2-D correspondence analysis, N-D correspondence analysis, canonical analysis, etc.). Factor analysis can be applied on a great number of variables or observations (1000 for the variables; as large as possible for the individuals).

STEP 9. **Data classification or clustering (Chapters 10, 11).** The objective of clustering is to build or to recognize classes of observations or variables; classifying methods include clustering as well as pattern recognition methods. Clustering methods can be applied on a great number of variables or observations as in the previous step.

STEP 10. **Data communication, presentation.** The objective is not only to dissect data, but to present results in such a way that decisions can be made. There are results and graphics for the data analyst and graphics for the decision maker. This step is often forgotten by most analysts; but the efficiency of any study is its capability *to present valid conclusions based on understandable graphics.*

4. Computer Aspects

There is much statistical software that involves data analysis methods. After 20 years of processing data, I have found that in order to do data analysis one needs to use two or three different software packages, which are often incompatible in terms of computer specifications. This is a business problem, not a science problem. Whichever software you will use in the future, keep in mind some basic ideas before purchasing one or several of them.

1. All the steps previously described must be considered (not just some of them).
2. The best quality data analysis software is able to do what the statistician wants, not what the program analyst wants.
3. There are as many different types of software as there are needs; the needs are not the same for researchers, teachers, statisticians, engineers, taxonomists, decision makers, managers, students, etc.
4. Do not confuse user-friendliness and data processing capability.

Different computer aspects of data analysis will be studied in depth in Chapter 12.

Chapter 2 — Statistical Data Elaboration

1. Statistics

Statistics has a double meaning. First, Statistics is concerned with scientific methods for collecting, organizing, summarizing, presenting, and analyzing data, as well as drawing valid conclusions and making relevant decisions on the basis of such analysis. In another sense, statistics is used to denote the data themselves. We can speak of economic statistics, geophysical statistics, employment statistics, accident statistics, financial statistics, population statistics, etc. To say that data are statistics, the data sets must be capable of being compared, and must be representative and coherent, and must have been systematically produced so that relevant or significant comparisons or computations can be made. Not all data are statistical data, i.e., able to be analyzed by a statistical method. Keep in mind that a statistical study does not stop, however, at data elaboration; that is made with the objective of future data analysis. Data must be processed to highlight the most significant or most particular features.

2. Fields of Statistical Data Exploration

At the beginning, statistics was employed in economics. Remember the example of nilometers built along the Nile in Egypt, which were used to measure the height of the Nile floods at different points along the river; this allowed the estimation of the harvest size, and therefore the collection of equivalent income taxes. In the 17th century, some applications appeared in different fields: botanics, systematics, natural sciences, taxonomy (Linné, Buffon, Adanson). In the nineteenth century, statistics grew rapidly in importance because of the progress in biology, then psychometry and agronomics (Fisher). Later, statistics was used in physics, astronomy, thermodynamics, and meteorology. Finally, in the twentieth century, statistics has been extended to studying industrial problems such as reliability, quality control, and production control. Statistics has since become an accepted tool in business management control, marketing studies, quality of service, opinion surveys, planning, and forecasting. Thus, statistics is now a decision-making tool as well as a specific method for improving fundamental knowledge.

3. Statistics and Experiments

Statistics and experimental methods are concerned with objective data based on observations. An experimental method is only applied, however, on specific observations, resulting from experiments, whereas statistics uses a larger class of observations; an experiment aims to replace the system of possible causes by a simpler system in which only one cause varies at a time. Consider the study of gas under the action of three variables: temperature, volume, and pressure. At constant temperature, observations are made to highlight the relationship between volume and pressure, and then observations are made at constant pressure to study the relationship between volume and temperature. Generally, an experimental method can be applied any time that the conditions of observations can be fixed by the experimenter, and can be continuously modified, where it is possible to repeat conditions. In some of areas mentioned above, it is obvious that an experimental method cannot be used. An observer of economic facts, or a manager, cannot experiment; he records facts as they are. For instance, to study the consumption of a product with respect to its price, the analyst cannot make the price vary to see how the consumption level varies. The only solution is to observe from time to time the price level and the related consumption. In contrast to experimental methods, data processed by statistics involve many factors, and so factors must be identified, recorded, and processed in a different way. The economist, the sociologist, or the meteorologist cannot experiment. They cannot have any influence on data, which are recorded independently of any action. The agronomist knows that several factors can influence the yield of a plant, but knows that he cannot make these factors vary. Therefore, he must study how combinations of factors vary simultaneously. Hence, statistics and data analysis are scientific methods that are able to draw larger and more valid conclusions than those drawn from only observed facts.

4. Data Analysis, Inductive and Deductive Statistics

Two main processes may be distinguished in statistics: the inductive process and the deductive process. The inductive process is based on mathematical models using the technique of probability calculus. Given mathematical hypotheses, the models are applied to concrete cases. The fundamental question for the scientist (and also the user) is the question of the validity of these mathematical models, which are too often wrongly considered as suitable (they can be correct by construction, but cannot be made relevant to the situation being studied). In the past, models were built because of the impossibility of computing and studying many different situations, but this may be less so now. The second process is the deductive process, which is based on deduction from observed facts only. This may result in building or discovering a mathematical model, derived from the data and not vice versa. Data analysis is concerned the deductive process: from the data to the model. That is the way followed by the well-known modern data analysts such as Benzecri (1972, 1973, 1980, 1981, 1982, 1984), Tukey (1977), and Hayashi (1988). Great progress has been made due to modern computers, which are able to process millions of data and to visualize simultaneously numerous data, allowing a scientific dialogue between the data and the analyzer. The inductive process can be applied to restricted categories of events; the deductive process allows the rapid synthesis of large data sets. This explains the importance of data analysis.

5. Variables, Statistical Sets, and Data Sets

5.1. Variables

Three main types of data can be distinguished: continuous variables, more often called quantitative variables; discrete variables, more often called categorical or qualitative variables; and chronological variables, which in fact involve quantitative or qualitative variables taken at specified times, usually at equal time intervals.

5.1.1. Continuous or Quantitative Variables

A quantitative variable, denoted by X, is a variable capable of having an infinite number of values. Measurements, ratios, and percentages can give quantitative variables. For example, the size, weight, or cranium perimeter of babies at birth are three quantitative variables. The following data sets involve quantitative variables: car models (Appendix 2, §1); measurements of skulls (Appendix 2, §6), samples of steel (Appendix 2, §8), indicators of quality of service in a telephone network (Appendix 2, §14). A counter-example: the number of children of a family is not a quantitative variable as it cannot take all possible values; a family can have 2, 3, or 4 children but not 2.4.

5.1.2. Discrete or Categorical Variables

A discrete or categorical variable, denoted by X, is a variable that takes on a finite number of numerical values, categories, or codes. For example, the number of children in a family is a categorical variable; sex; marital status; class of income taxes; or color of eyes. The responses to a questionnaire generally give categorical variables. Among these categorical variables, different subtypes can be distinguished: variables with multiple forms; logical variables; categorical variables determined as sums of variables from the quantitative ones; or preference variables:

(a) Variables with multiple forms. These often occur in the use of questionnaires. Generally, a question has only one response, which is called a form. But the following situation can happen, however, where several responses can be given to one question; these are called variables with multiple forms. Here is an example of such a situation. In a survey on new services in telecommunications, one question is: What are the reasons for using the Minitel services?

Response 1—To look for precise information (code 1).

Response 2—For curiosity (code 2).

Response 3—For fun (code 3).

Response 4—To learn how to use it (code 4).

Response 5—To show it to people (code 5).

Response 6—For rapidity (code 6).

Response 7—Because it is practical (code 7).

The interviewed can give as many responses as he wants.

5. Variables, Statistical Sets, and Data Sets

(b) *Dummy variables from quantitative variables.* From a quantitative variable a categorical variable can be built as follows: The range is divided into equal sized intervals, each of which is assigned a code. Each original value of the quantitative variable is replaced by its associated code. In this way, the categorical variable is built. Consider the following example: A quantitative variable has a range of [0, 100]; the variable is divided into the subintervals: [0, 25],]25, 50],]50, 75],]75, 100]. The values of the associated categorical variable, called dummy variables, are 1, 2, 3, 4:

$$[0, 25] \quad \text{gives 1;}$$
$$]25, 50] \quad \text{gives 2;}$$
$$]50, 75] \quad \text{gives 3;}$$
$$]75, 100] \quad \text{gives 4.}$$

(c) *Logical variables.* A logical variable is a discrete variable whose only values are one or zero. Generally, they correspond to the presence (one) or absence (zero) of an attribute. They occur in specific domains such as archaeology, psychology, economics, and marketing as dummy variables of a questionnaire or taxonomy. They are also important in N-D correspondence analysis because all of the variables are set in logical forms (*cf.* Chapter 9).

(d) *Preference variables.* Preference variables are specific discrete variables whose values are put into increasing or decreasing order. These occur in the following question:

Would you please rate your main reasons for using the Minitel services? Use 1 for the most appreciated and 8 for the least appreciated.

Code 1—less expensive.
Code 2—equally priced services.
Code 3—price given on the screen.
Code 4—payment for the telephone service only.
Code 5—no charge.
Code 6—fixed price for each requested service.
Code 7—no service rental payment.
Code 8—separation on the bill of the new services and the telephone.

Based on these items, the responses given must be put into order, e.g., code 1 ≤ code 5 ≤ code 2 ≤ code 8 ≤ code 3 ≤ code 4 ≤ code 7 ≤ code 6 (≤ means preferred to).

5.1.3. Chronological Variables

All of the previous variables can be studied during a time range (the quantitative as well as the categorical variables). Sometimes chronological variables require specific analysis such as "times series" processing. But this will not be studied here.

5.2. *Statistical Sets*

Statistical sets are sets of observations on which values of variables are measured or given. For a set of all observations, we speak of an exhaustive population but, if not, we speak of a population. For a subset of units chosen according to some representative criteria, we speak of a sample. The choice of a set depends on the possibility of observations and on the field of study. Statistical sets are often called sets of units of observations, sets of individuals, or statistical units.

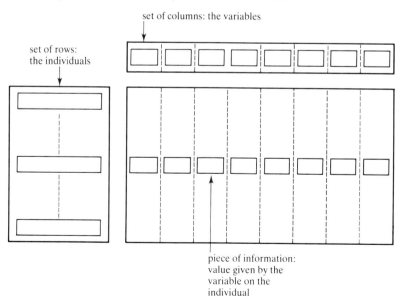

FIGURE 2.1. Model of individuals–variables data set.

5. Variables, Statistical Sets, and Data Sets

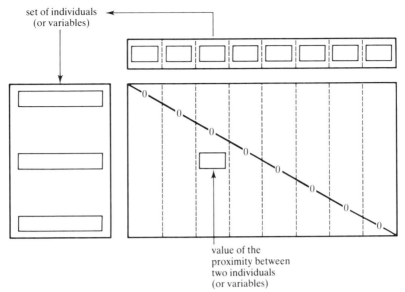

FIGURE 2.2. Model of proximities data set.

5.3. Data Sets

5.3.1. Individuals–Variables Data Sets

On the one hand there is a set of variables; on the other hand, there is a set of units or individuals. A data set is the set of values (or codes, or qualities, or any type of information) associating the set of variables with the set of units. Consider the following representation given in Fig. 2.1. The rows represent the set of variables, the columns represent the set of individuals. At each intersection of a row and a column is a piece of information (i.e., the information given by the specific variable for a specific individual). The set of all the information is called *individuals–variables* data set (*cf.* Fig. 2.1). A 1-D individuals–variables data set is called a statistical series. A 1-D individuals–variables data set, taken at different periods of time, is called a *time series* or a *chronological series*. In Appendix 2, many examples of such data sets are given. The analyses of these data sets are given in Chapters 3 to 10.

5.3.2. Proximities Data Sets

Proximities data sets are sets of *numerical values* corresponding to a distance matrix between two sets of variables or individuals (*cf.* Fig. 2.2). In Appendix 2, some examples of such data sets are given. The analyses of these data sets are given in Chapter 11.

Chapter 3 1-D Statistical Data Analysis

1. Introduction

A statistical series resulting from numerous observations needs to be summarized in a small set of numbers so that one can compare several series and understand them easily.

A summary can be numerical or graphical:

numerical: each numerical summary highlights a specific feature of a series. It is affected by the point of view of the statistician.

graphical: graphics indicate more than numerical summaries, from which they are generally derived.

Most of the time, graphical and numerical summaries are used simultaneously, and they are both studied for the following types of variables: quantitative, qualitative, multiple form qualitative, and chronological.

2. 1-D Analysis of a Quantitative Variable

As an example, consider the series from the data set of prices of cars. (*cf.* Appendix 2, §1). How can one summarize this series? A statistician can study it from three points of view:

1. What are the most representative values of the data set in terms of *local concentration*?
2. What are the most representative values of the data set in terms of *dispersion*?

3. What are the most representative values of the data set in terms of *shape*?

We study these three points of view successively.

2.1. Measures of Central Tendency

Our aim is to characterize a statistical series by a single number (a type-value), representing the order of magnitude of the whole set of numbers, so that we may compare two series by comparing their type-values. The type-value should satisfy the conditions given by Yule (1950):

1. The type-value must be defined independently of the observer, and independently of the conditions under which the observations were taken.
2. The type-value must depend on all the values of the series. In particular, values considered as exceptional or irrelevant must be integrated when computing the type-value.
3. The type-value must have a concrete meaning that is easy to elaborate, because it must be understood by people not versed in statistics.
4. The type-value must be easy to compute.
5. The type-value must not be sensitive to random processes.

We now will study different type-values.

2.1.1. The Median

The median, denoted by M, is the value of the series for which the number of observations smaller than M and the number of observations greater than M are equal.

To compute M, the terms of the series are arranged according to their values, in ascending or descending order (note that the computation of M following does not depend on which order is used):

if there are $n = 2p + 1$ terms in the series, the value whose rank is the $(p + 1)^{th}$ is the median M.

if there are $n = 2p$ terms in the series, any value included between the pth term and the $(p + 1)^{th}$ term could be the median. The interval between these two terms is called the *median range*. By convention, the median M is taken as the arithmetic mean of the pth and the $(p + 1)^{th}$ term of the series.

2. 1-D Analysis of a Quantitative Variable

Properties.

1. The sum of the (absolute values of the) deviations of the terms of the series from the median M is the minimum value possible:

$$\sum_{i \in I} |X_i - M| \quad \text{is minimum for } M.$$

2. The median M divides the frequency histogram into two equal surfaces.

For the series given in Table 3.1, the value of the median M is 5705.

2.1.2. The Arithmetic Mean

The arithmetic mean, denoted by $\bar{X}$, for the series X with the n observations X_i, is the ratio of the sum of the terms of the series to the number of terms n:

$$\bar{X} = \frac{\sum_{i \in I} X_i}{n}.$$

TABLE 3.1. Statistical series associated with car prices. Computation of the median M.

1	Chevrolet Chevette	3299	29	Buick Lesa	5788
5	Chevrolet Monza	3667	17	Honda Accord	5799
11	Chevrolet Nova	3955	13	Dodge Magnum	5886
6	Dodge Colt	3984	7	Datsun 200 S-X	6229
32	Buick Skyline	4082	25	Audi Fox	6295
22	AMC Concord	4089	14	Dodge St Regis	6342
16	Ford Mustang	4187	28	Buick Electra	7827
15	Fiat Strada	4296	10	Datsun 810	8129
18	Honda Civic	4499	24	Audi 5000	9690
3	Chevrolet Malibu	4504	26	BMW 320	9737
8	Datsun 510	4589	31	Buick Riviera	10372
23	AMC Pacer	4749	33	Cadillac Deville	11385
27	Buick Century	4816	19	Lincoln Continental	11447
12	Dodge Diplomat	5010	21	Lincoln Versailles	13866
9	Datsun 510	5079	20	Lincoln Cont. Mark	13594
4	Chev. Monte-Carlo	5104	34	Cadillac Eldorado	14500
30	Buick vega	5189	35	Cadillac Séville	15906
2	Chevrolet Impala	5705			

↑ — median M

The weighted arithmetic mean, also denoted by $\bar{X}$, for the series X with the n observation X_i, each being weighted by a weight p_i, is given by the following ratio:

$$\bar{X} = \frac{\sum_{i \in I} p_i X_i}{\sum_{i \in I} p_i}.$$

Properties.
1. The arithmetic mean satisfies Yule's conditions.
2. For two series X_1 and X_2, with respective arithmetic means $\bar{X}_1$ and $\bar{X}_2$ and respective numbers of observations n_1 and n_2, the arithmetic mean $\bar{X}$ of the series X that is the union of X_1 and X_2 is given by the following formula:

$$\bar{X} = \frac{n_1 \bar{X}_1 + n_2 \bar{X}_2}{n_1 + n_2}.$$

3. The arithmetic mean $\bar{X}$ is the value such that the sum of the deviations of the terms of the series from the arithmetic mean is equal to zero:

$$\sum_{i \in I} (X_i - \bar{X}) = 0.$$

4. The arithmetic mean $\bar{X}$ is the value such that the sum of the deviations of the terms of the series and $\bar{X}$ is the minimum value possible:

$$\sum_{i \in I} (X_i - X_0) \quad \text{is minimum and } X_0 \text{ is equal to } \bar{X}.$$

2.1.3. Mean Values (Generalization)
In general, the mean of a series is a symmetric function of all of the terms of the series such that the following two conditions are satisfied:

If all the terms are equal to the same value X_0, the mean is equal to X_0.

The mean is included between the greatest value and the smallest value of the series.

2. 1-D Analysis of a Quantitative Variable

Let $f(X)$ be any function on real numbers. For a series X, define $f(X)$ by

$$f(X) = \frac{f(X_1) + f(X_2) + \cdots + f(X_i) + \cdots + f(X_n)}{n}.$$

2.1.3.1. The Quadratic Mean Q. Consider $f(X) = X^2$. Then we define Q by the formula

$$Q = \frac{X_1^2 + X_2^2 + \cdots + X_i^2 + \cdots + X_n^2}{n}$$

$$= \frac{1}{n}\sum_{i \in I} X_i^2.$$

2.1.3.2. The Harmonic Mean H. Consider $f(X) = 1/X$. Then, H is defined by the formula

$$\frac{1}{H} = \frac{\dfrac{1}{X_1} + \dfrac{1}{X_2} + \cdots + \dfrac{1}{X_i} + \cdots + \dfrac{1}{X_n}}{n}$$

$$= \frac{\sum_{i \in I} \dfrac{1}{X_i}}{n}.$$

Thus, H is the inverse of the arithmetic mean of the inverses of the observed values.

The weighted harmonic mean is given by the following formula:

$$\frac{1}{H} = \frac{\sum_{i \in I} p_i \dfrac{1}{X_i}}{\sum_{i \in I} p_i}.$$

2.1.3.3. The Geometric Mean G. Consider $f(X) = \log X$. Then, G is defined by the formula

$$\log G = \frac{\log X_1 + \log X_2 + \cdots + \log X_i + \cdots + \log X_n}{n}$$

$$= \frac{\sum_{i \in I} \log X_i}{n};$$

thus,
$$G = \sqrt[n]{X_1 X_2 \cdots X_i \cdots X_n}.$$

The weighted geometric mean is given by the following formula:
$$\log G = \frac{\sum_{i \in I} p_i \log X_i}{\sum_{i \in I} p_i};$$

thus,
$$G = \sqrt[p]{X_1^{p_1} X_2^{p_2} \cdots X_i^{p_i} \cdots X_n^{p_n}} \quad \text{with } p = p_1 + p_2 + \cdots + p_n.$$

G is such that
$$\frac{X_1}{G} \cdot \frac{X_2}{G} \cdots \frac{X_n}{G} = 1,$$

whereas $\bar{X}$ is such that
$$(X_1 - \bar{X}) + (X_2 - \bar{X}) + \cdots + (X_n - \bar{X}) = 0.$$

2.1.3.4. *The α-Mean X_α.* Consider $f(X) = X^\alpha$, with α a real number. We define the α-mean X_α by

$$X_\alpha = \frac{X_1^\alpha + X_2^\alpha + \cdots + X_i^\alpha + \cdots + X_n^\alpha}{n}$$

$$= \frac{\sum_{i \in I} X_i^\alpha}{n}.$$

The weighted α-mean is given by the following formula:
$$X_\alpha = \frac{\sum_{i \in I} p_i X_i^\alpha}{\sum_{i \in I} p_i}.$$

Properties.

1. $X_\alpha \xrightarrow[\alpha \to 0]{} G$.
2. If $\alpha < \beta < \gamma$, then $X_\alpha < X_\beta < X_\gamma$.
3. $H < G < \bar{X} < Q$ since $H = X_{-1}$; $G = X_0$; $\bar{X} = X_1$; $Q = X_2$.

2. 1-D Analysis of a Quantitative Variable

2.1.4. The Mode M_0

The mode, denoted by M_0, for the series X is the value of the maximum of the frequency distribution. The mode M_0 of a series is not easy to compute because a frequency curve must be adjusted to the specific distribution. M_0 does not satisfy the whole set of Yule's conditions.

For a unimodal and a slightly symmetrical distribution the following relations can be used:

$$M_0 = 3M - 2\bar{X}$$

or

$$\bar{X} - M_0 = 3(\bar{X} - M).$$

In this case, the relative positions of the three values, $\bar{X}$, M, and M_0 can be represented as given in Fig. 3.1.

2.1.5. Quantiles, Deciles, and Quartiles

Consider a series X whose values X_i are set in increasing order. It is known that the median M is the value that divides the set of X_i into two equal parts. Other statistical values can be defined as follows:

The *k-quantiles*, denoted by $Q_1, Q_2, \ldots, Q_\alpha, \ldots, Q_{k-1}$ are the values of X that divide the series X into k equal parts; Q_α, the αth quantile is that value such that $\alpha\%$ of the observations are less than Q_α. From that previous definition, the following other values can be defined.

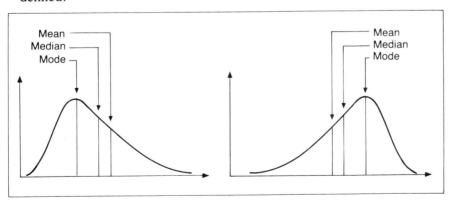

FIGURE 3.1. Relative positions of $\bar{X}$, M, and M_0 according to the shape of the distributions: (a) asymmetrical at left; (b) asymmetrical at right. For asymmetrical distribution, $\bar{X} = M = M_0$.

The *quartiles*, denoted by Q_1, Q_2, Q_3, are the values of X that divide the series X into four equal parts (quantiles = 4-quantiles). Q_1 is the lower quartile; Q_3 is the upper quartile; Q_2 is the median M.

The *deciles*, denoted by $D_1, D_2, \ldots, D_9$ are the values of X which divide the series X into 10 equal parts (deciles = 10-quantiles). D_1 is the first decile; D_9 is the last decile; D_5 is the median.

The *percentiles* (or percentage points) are the values of X which divide the series X into 100 equal parts (percentiles = 100-quantiles).

As with the median, the previous values are not easy to compute.

2.2. Measures of Dispersion

The quantities that characterize the distribution of a series according to the second point of view, the dispersion, are the following:

(a) computation of distances between certain representative values, such as the range, the interdecile range, or the interquartile range;
(b) computations involving the deviations of every observation from the central value, such as the mean deviation from the mean, the mean deviation from the median and the standard deviation;
(c) computations involving the deviations of all the observation among themselves such as the mean difference.

Among these possible values, the most important one is the standard deviation, though they all require some discussion.

2.2.1. The Range R

The range of a series, denoted by R, is the difference between the greatest value and the least value of the series. As a descriptive measure of dispersion, it is little used because it is often unstable; it depends only on two specific values of the series and not on all the terms. The second condition of Yule is not satisfied. The extreme values can result from exceptional observations, but R is useful to show the extent of the series.

2.2.2. The Interquartile Range IQR

In order to reduce the influence of the extreme values of a series, other computations can be used, such as the interquartile range or the interdecile range. The interquartile range, denoted by IQR, is the difference between the upper and the lower quartiles and is thus a range

2. 1-D Analysis of a Quantitative Variable

which contains one half of the total frequency:

$$\text{IQR} = Q_3 - Q_1.$$

The semi-quartile range, denoted by Q, is defined as follows:

$$Q = Q_3 - Q_1/2.$$

Q is also called quartile deviation.

Instead of the difference between quartiles, the difference between two symmetrical percentiles can be used:

$$P_k \text{ and } P_{100-k}.$$

It follows that the interpercentile range is the difference between the 90th percentile and the 10th percentile:

$$\text{IQR} = P_{90} - P_{10}.$$

2.2.3. The Mean Difference Δ

The mean difference, denoted by Δ, is the average of the absolute values of the differences $(X_i - X_j)$:

$$\Delta = \frac{\sum_{i \in I} \sum_{j \in J} |X_i - X_j|}{n^2}.$$

Properties.

1. If X_i is the difference between the ith value and the median, and Y_i is the rank (taken in absolute value) of that element with respect to the median taken as origin, then:

$$\Delta = \frac{\sum_{i \in I} X_i Y_i}{n^2}.$$

2. If the X_i of the series are put into increasing order and their indices are relabelled $(X_1 \leq X_2 \leq \cdots \leq X_i \leq X_{i+1} \leq \cdots \leq X_n)$, then:

$$\Delta = \frac{2 \sum_{i \in I} i(n-i)(X_{i+1} - X_i)}{n^2}.$$

Numerical example:

$X_1 = 6.5$; $X_2 = 8$; $X_3 = 9$; $X_4 = 9.5$; $X_5 = 10$; $X_6 = 11$; $X_7 = 11.5$.

Data table of differences $(X_i - X_j)$

0	1.5	2.5	3	3.5	4.5	5
	0	1	1.5	2	3	3.5
		0	0.5	1	2	2.5
			0	0.5	1.5	2
				0	1	1.5
					0	0.5
						0

$$\Delta = \frac{88}{49} = 1.795.$$

2.2.4. The Mean Deviation MD

The mean deviation, denoted by MD, is the arithmetic mean of the absolute values $|X_i - \bar{X}|$:

$$\mathrm{MD} = \frac{\sum\limits_{i \in I} |X_i - \bar{X}|}{n}.$$

2.2.5. The Standard Deviation s

The standard deviation, denoted by s (sometimes sD, or σ) is the root mean square of the deviations from the arithmetic mean (it is sometimes called the root mean square deviation):

$$s = \sqrt{\frac{\sum\limits_{i \in I} (X_i - \bar{X})^2}{n}}.$$

The quantity s^2 is called the variance of the series.

Sometimes, the standard deviation is defined with $(n - 1)$ instead of n in the denominator of the previous expression; this is because a better estimate of the standard deviation can be obtained for the population from which the sample is taken. For large values of n (≥ 30), there is practically no difference between the two definitions.

2. 1-D Analysis of a Quantitative Variable

Note. The standard deviation of a population is denoted by σ^2 and the standard deviation of a sample is denoted by s^2.

Properties.

1. $s^2 = (\sum_{i \in I} X_i^2) - (\bar{X})^2$.
2. Yule's conditions are satisfied.
3. Consider two samples of X with $\bar{X}_1, \bar{X}_2$ as arithmetic means, s_1, s_2 as standard deviations, and n_1, n_2 as sizes. The $\bar{X}$ and s associated with $(X_1 + X_2)$ are given as follows:

$$\bar{X} = \bar{X}_1 + \bar{X}_2,$$

$$s^2 = \frac{n_1}{n_1 + n_2} s_1^2 + \frac{n_2}{n_1 + n_2} s_2^2 + \frac{n_1 n_2}{n_1 + n_2} (\bar{X}_1 - \bar{X}_2)^2.$$

4. For normal distributions, it turns out that

   (a) 68.27% of the cases are included between $\bar{X} - s$ and $\bar{X} + s$.
   (b) 95.45% of the cases are included between $\bar{X} - 2s$ and $\bar{X} + 2s$.
   (c) 99.73% of the cases are included between $\bar{X} - 3s$ and $\bar{X} + 3s$.

5. Consider two sets consisting of n_1 and n_2 values, with s_1^2 and s_2^2 as variances, and each having the same mean $\bar{X}$. The variance of the union of the two sets is given by the following formula:

$$s^2 = \frac{n_1 s_1^2 + n_2 s_2^2}{n_1 + n_2}.$$

6. Comparison of s^2 and MD (Mean Deviation):

$$\sum_{i \in I} (Z_i - \bar{Z})^2 = \sum_{i \in I} Z_i^2 = n Z^2 = 0.$$

Consider

$$Z_i = |X_i - \bar{X}|;$$

Then,

$$\sum_{i \in I} Z_i^2 = n^2 \quad \text{and} \quad \bar{Z} = \text{MD},$$

$$ns^2 > n \cdot \text{MD},$$

$$s^2 > \text{MD}.$$

7. Comparison of s^2 and the arithmetic mean of the squared differences $(X_i - X_j)^2$:
Consider
$$\sum_{i \in I} \sum_{j \in I} (X_i - X_j)^2.$$
Write
$$(X_i - X_j) = (X_i - \bar{X}) + (\bar{X} - X_j).$$
Then,
$$\sum_{j \in J} (X_i - X_j)^2 = \sum_{j \in I} ((X_i - \bar{X}) + (\bar{X} - X_j))^2$$
$$= n(X_i - \bar{X})^2 + ns^2$$
and
$$\sum_{i \in I} (n(X_i - \bar{X})^2 + ns^2) = 2n^2 s^2.$$
So finally,
$$\sum_{i \in I} \sum_{j \in J} (X_i - X_j)^2 = 2n^2 s^2.$$

8. Numerical examples on particular distributions.
   8.1. Series of the first n natural numbers.
   Let X be such that $X_i = i$ for $i = 1, 2, \ldots, n$. Then,
   $$\bar{X} = \frac{(n+1)}{2}, \qquad s = \frac{(n^2 - 1)}{12}.$$

   8.2. Rectangular distribution.
   Consider N values of X on a segment of length l. This segment is divided into n equal intervals; the abscissa Y_i of the centers of the intervals are:
   $$Y_i = \frac{l}{n}\left(X_i - \frac{1}{2}\right) \qquad \text{for } X_i = 1, 2, \ldots, n.$$
   Then,
   $$\bar{Y} = \frac{l}{n}\left(\bar{X} - \frac{1}{2}\right),$$
   $$s^2 = \frac{l}{N} \sum_{i \in I} N(Y_i - \bar{Y}),$$
   $$= \frac{l^2}{12}\left(\frac{n^2 - 1}{n^2}\right),$$

2. 1-D Analysis of a Quantitative Variable

and
$$\lim_{n \to \infty} s^2 = \frac{l^2}{12}.$$

2.2.6. Median Deviation
The median deviation is the median of a series of deviations taken in absolute values; these deviations are taken from a value of central tendency, in general the median. The median deviation is not commonly used.

2.2.7. Geometric Deviation
The geometric deviation, denoted by sG, is determined from the geometric mean G as follows:

$$\log sG = \frac{\sum_{i \in I} (\log X_i - \log G)^2}{n}.$$

sG is the value whose logarithm is equal to the standard deviation of the series of the logarithms of X_i.

2.2.8. Absolute and Relative Dispersion: Coefficient of Variation
The actual variation or dispersion as determined from the standard deviation or other measures of dispersion being used is called the absolute dispersion. A variation of one meter over a distance of 1000 meters has, however, quite a different effect as the same variation over a distance of 20 meters. A measure of this effect is given by the relative dispersion, which is defined by:

$$\text{Relative dispersion} = \frac{\text{absolute dispersion}}{\text{average}}.$$

If the absolute dispersion is the standard deviation s and the average is the arithmetic mean $\bar{X}$, the relative dispersion is called the coefficient of variation (or coefficient of dispersion), and is given by the following formula:

$$\text{Coefficient of variation} = V = \frac{s}{\bar{X}}.$$

It is expressed as a ratio. Note that the coefficient of variation is independent of the units used (obviously, it cannot be used when $\bar{X} \to 0$).

If Q_1 and Q_3 are given for a set of observations, then $\frac{1}{2}(Q_1 + Q_3)$ is a measure of central tendency, whereas $Q = \frac{1}{2}(Q_3 - Q_1)$, the semi-quartile range, is a measure of the dispersion. Then, a measure of dispersion, denoted by V_Q, can be defined as follows:

$$V_Q = \frac{\frac{1}{2}(Q_3 - Q_1)}{\frac{1}{2}(Q_1 + Q_3)} = \frac{Q_3 - Q_1}{Q_3 + Q_1}.$$

We call V_Q the quartile coefficient of variation or the quartile coefficient of relative dispersion.

2.3. Skewness and Kurtosis Measures

2.3.1. Skewness

Skewness is the degree of asymmetry of a distribution. If the frequency distribution has a longer "tail" to the right of the central maximum than to the left, the distribution is said to be skewed to the right (or to have a positive skewness). If the reverse is true, it is said to be skewed to the left (or to have a negative skewness). For a skewed distribution, the mean tends to lie on the same side of the mode as the longer tail.

Skewness is a dimensionless quantity. One such measure is given by

$$\text{Skewness} = \frac{\text{mean} - \text{mode}}{\text{standard deviation}} = \frac{\bar{X} - M_0}{s}.$$

To avoid having to compute M_0, the following empirical formula is used:

$$\text{Skewness} = \frac{3(\text{mean} - \text{median})}{\text{standard deviation}} = \frac{3(\bar{X} - M)}{s}.$$

These two previous measures are known as Pearson's first and second coefficients of skewness, respectively.

Other measures of skewness can be defined in terms of quartiles or deciles, such as:

The Yule's coefficient:

$$C_Y = \frac{(Q_3 - \bar{X}) - (\bar{X} - Q_1)}{(Q_3 - \bar{X}) + (\bar{X} - Q_1)} = \frac{Q_3 - 2\bar{X} + Q_1}{Q_3 - Q_1}.$$

The Kelley's coefficient:

$$C_K = M - \tfrac{1}{2}(D_1 + D_9).$$

2. 1-D Analysis of a Quantitative Variable

The Pearson's coefficient, denoted by β_1, developed from the moments of a distribution μ_k:

$$\mu_k = \frac{\sum_{i \in I}(X_i - \bar{X})^k}{n};$$

then,

$$\mu_1 = 0, \qquad \mu_2 = s^2;$$

we define

$$\beta_1 = \frac{(\mu_3)^2}{(\mu_2)^3}.$$

The Quartile coefficient:

$$C_Q = \frac{Q_3 - 2Q_2 + Q_1}{Q_3 - Q_1}.$$

The Percentile coefficient:

$$C_P = \frac{(P_{90} - P_{50}) - (P_{50} - P_{10})}{P_{90} - P_{10}}$$

$$= \frac{P_{90} - 2P_{50} - P_{10}}{P_{90} - P_{10}}.$$

The Fisher's coefficient, denoted by γ_1:

$$\gamma_1 = \frac{\mu_3}{\sqrt{\mu_2^3}}.$$

2.3.2. Kurtosis

Kurtosis is the degree of peakedness of a distribution, usually taken relative to a normal distribution. A distribution having a relatively high peak such as the curve of Fig. 3.2(a) is called *leptokurtic*, while the curve of Fig. 3.2(b), which is flat-topped, is called *platykurtic*. The normal distribution, which is not very peaked or very flat topped is called *mesokurtic* (*cf.* Fig. 3.2(c)).

The measures of kurtosis are:

2.3.2.1. The Kelley's Coefficient

$$A_K = \frac{1}{2}\frac{Q_3 - Q_1}{D_9 - D_1}.$$

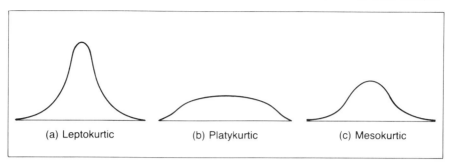

FIGURE 3.2. Different shapes of distribution: (a) leptokurtic; (b) platykurtic; (c) mesokurtic.

2.3.2.2. The Pearson's Coefficient, denoted by β_2

$$\beta_2 = \frac{\mu_4}{\mu_2^2},$$

where

$$\mu_4 = \frac{\sum_{i \in I}(X_i - \bar{X})^4}{n} \quad \text{and} \quad \mu_2 = \frac{\sum_{i \in I}(X_i - \bar{X})^2}{n}.$$

2.3.2.3. The Fisher's Coefficient, denoted by γ_2

$$\gamma_2 = \frac{\mu_4}{\mu_2} - 3 \quad \Rightarrow \quad \gamma_2 = \beta_2 - 3.$$

For a normal distribution, $\gamma_2 = 0$ and $\beta_2 = 3$. Thus, using γ_2, kurtosis is positive for a leptokurtic distribution, negative for a platykurtic distribution, and zero for a normal distribution.

2.4. Frequency Distributions and Histograms

2.4.1. Frequency distributions

When summarizing large masses of data, it is useful to distribute the data into *classes* or *categories,* and to determine the number of observations belonging to each class; these classes are called *class frequencies.* A tabular arrangement of data by classes is called *frequency distribution* or a *frequency table.* Table 3.2 shows a frequency distribution of car prices into 5 *classes* (*cf.* Appendix 2, §1).

Data organized as in the frequency distribution in Table 3.2 are often called *grouped data.* Any grouping process generally destroys much of

2. 1-D Analysis of a Quantitative Variable

TABLE 3.2. Frequency distribution of cars according to their prices (data in Appendix 2, §1).

Class	Lower class limit	Upper class limit	Number of cars
1	3299	5820	20
2	5821	8341	6
3	8342	10862	3
4	10863	13383	2
5	13384	15904	4
			35 = total

the original detail of the data, but highlights the main features contained in the data.

2.4.2. Class Intervals and Class Limits

A symbol defining a class such as "3299–5820" in Table 3.2 is called a *class interval*. The end numbers 3299 and 5820 are called *class limits*. The smaller number, 3299, is the *lower class limit* and the larger, 5820, is the *upper class limit*. The terms class and class interval are often used interchangeably, although the class interval is actually a symbol for the class. A class interval which, at least theoretically, has either no upper or no lower class limit is called an *open class interval*.

2.4.3. Class Boundaries

In the example of cars in Table 3.1, if the prices are computed to the nearest US dollar, the class interval 3299–5820 theoretically includes all measurements from $(3299 - \frac{1}{2})$ to $(5820 + \frac{1}{2})$. These values are the class boundaries of the class 3299–5820. In practice, the class boundaries are obtained by adding the upper limit of one class interval to the lower limit of the next higher class interval and dividing by 2.

2.4.4. The Size or Width of a Class Interval

The size or width of a class interval is the difference between the upper and lower class boundaries, and is also referred to as the class *width*, *class size*, or *class strength*.

2.4.5. General Rules for Forming a Frequency Distribution

(a) Determine the largest and the smallest value in the data and thus find the range (i.e., the difference between the largest and smallest numbers).

(b) Divide the range into a convenient number of class intervals having the same size.
(c) Determine the number of observations falling into each class interval.

2.4.6. Histograms and Frequency Polygons

Histograms and frequency polygons are two graphical representations of frequency distributions:

(a) A *histogram* or *frequency histogram* consists of a set of rectangles having:

   (1) bases on a horizontal axis (the x-axis) with centers at the class midpoint and lengths equal to the class interval sizes;
   (2) areas that are proportional to class frequencies. If the class intervals all have equal size, the heights of the rectangles are proportional to the class frequencies and it is then customary to have the heights numerically equal to the class frequencies. If class intervals do not have equal size, these heights must be adjusted.

(b) A frequency polygon is a line graph of class frequency plotted against class midpoint. It can be obtained by joining the midpoints of the tops of the rectangles in the histogram (*cf.* Fig. 3.3.).

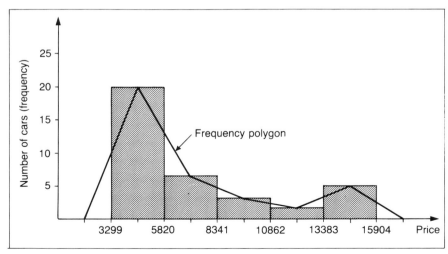

FIGURE 3.3. Frequency histogram of cars according to their prices.

2.4.7. Relative Frequency Distributions

The relative frequency of a class is the frequency of the class divided by the total frequency of all classes. It is generally expressed as a percentage. For example, the relative frequency of the class 5821–8341 is $6/35 = 17\%$. If the frequencies in Table 3.2 are replaced by corresponding relative frequencies, the resulting table is called a *relative frequency table*.

Graphical representations of relative frequency distributions can be obtained from the histogram or frequency polygon by simply changing the vertical scale from frequency to relative frequency, keeping exactly the same diagram. The resulting graphs are called *relative frequency histograms* or *percentage histograms*, and *relative frequency polygons* or *percentage polygons*, respectively.

2.4.8. Cumulative Frequency Distributions. Ogives

The total frequency of all values less than the upper class boundary of a given class interval is called the cumulative frequency up to and including the class interval. For example, the cumulative frequency up to and including the class interval 8341–10862 in the previous example is $20 + 6 + 3 = 29$, signifying that 29 cars have prices less than 10862 US dollars. A table presenting such cumulative frequencies is called a *cumulative frequency distribution,* a *cumulative frequency table* or, briefly, a *cumulative distribution* (*cf.* Table 3.3).

A graph showing the cumulative frequency plotted against the upper boundary is called a *cumulative frequency polygon* or *ogive* (*cf.* Fig. 3.4).

TABLE 3.3. Cumulative frequency distribution of cars according to their prices (data in Appendix 2, §1).

Class	Upper class limit	Number of cars
1	5820	20
2	8341	26
3	10862	29
4	13383	31
5	15904	35

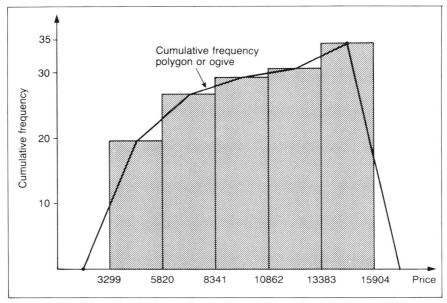

FIGURE 3.4. Cumulative frequency histogram of cars according to their prices (*cf.* data in Appendix 2, §1).

2.4.9. Relative Cumulative Frequency Distributions. Percentage Ogives

The *relative cumulative frequency* is the cumulative frequency divided by the total frequency. For example, the relative cumulative frequency of car prices less than 8341 US dollars is $26/35 = 74\%$; this means that 74% of the cars have prices less than 8341 US dollars. If the frequencies in Table 3.3 are replaced by corresponding relative frequencies, the resulting table is called a *relative cumulative frequency distribution* or *percentage cumulative distribution,* and the resulting graph from that table is called a *relative cumulative frequency polygon* or *percentage ogive.*

2.4.10. Frequency Curves. Smoothed Ogives

Collected data can usually be considered as belonging to a sample from a larger population. Since many observations could be made in the population, it is possible to choose small class intervals and still have sizable numbers of observations falling within each class. Thus, it should be expected that the frequency polygon or relative frequency polygon for a larger population would have so many small broken line segments that they would closely approximate smooth curves; these smooth curves are called *frequency curves* or *relative frequency curves*. It is reasonable to expect that such theoretical curves can be approximated by smoothing the frequency polygons or relative frequency polygons of the sample, the approximation improving as the sample size is increased. For this reason, a frequency curve is sometimes called a *smoothed frequency polygon*. *Smoothed ogives* can be also obtained from the *cumulative frequency polygons* or ogives in the same way. In practice, frequency curves may have characteristic shapes as described in Fig. 3.5.

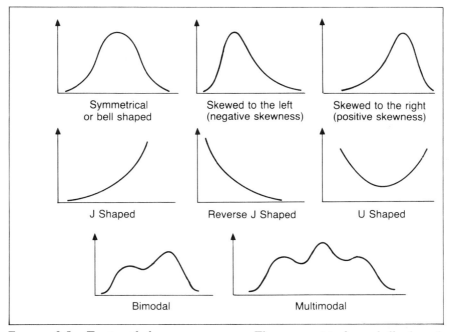

FIGURE 3.5. Types of frequency curves. The *symmetrical* or *bell shaped* frequency curves are characterized by the fact that observations equidistant from the central maximum have the same frequency. An important example is the *normal curve*. In the *moderately asymmetrical* or *skewed* frequency curves the tail of the curve to one side of the central maximum is longer than that to the other. If the longer tail occurs to the right the curve is said to *be skewed to the right* or to have *positive skewness*, while if the reverse is true the curve is said to be *skewed to the left* or to have *negative skewness*. In a *J shaped* or *reverse J shaped* curve a maximum occurs at one end. A *U shaped* frequency curve has maxima at both ends. A *bimodal* frequency curve has two maxima. A *multimodal* frequency curve has more than two maxima. (Reprinted with permission from McGraw-Hill, Inc., Spiegel, M.R., *Schaum's Outline of Theory and Problems of Statistics in SI* Units (1961).)

2.4.11. Labelled Histograms or Class Tables

Consider the case of car prices in Table 3.1. From the series, five class frequencies are built, in which each class contains the labels of the cars relating to that class (*cf.* Table 3.4).

Then, a histogram can be plotted in which each class is represented by a rectangle that contains the labels of the observations falling into that class (*cf.* Fig. 3.6).

2.5. Dispersion Box Plots

To study separately the previous statistical characteristics is not easy, especially if a variable time is involved. To facilitate this study, graphics such as *dispersion box plots* can be used so that different characteristics can be put on the same graphs.

A *dispersion box plot* is a graph (*cf.* Fig. 3.7.) formed as follows:

(a) a rectangle is drawn with a given width and a length equal to the difference $(Q_3 - Q_1)$ (measure of the interquartile range IQR);
(b) the values $Q_3 - 1.5\text{IQR}$ and $Q_1 - 1.5\text{IQR}$ are computed;
(c) the observation, denoted by X_0, with $X_0 = \max_{i \in I} \{X_i; X_i \leq Q_3 + 1.5\text{IQR}\}$ is sought;
(d) the observation, denoted by X_1, with $X_1 = \max_{i \in I} \{X_i; X_i \geq Q_1 - 1.5\text{IQR}\}$ is sought;
(e) two straight lines are drawn from the centers of the top and bottom of the rectangle to the values X_0 and X_1, respectively (a line on each opposite side);

TABLE 3.4. Class table according to the price for the car models data set.

Class	Class interval	Car labels
1	3299–5820	1, 5, 11, 6, 32, 22, 16, 15, 3, 8, 23, 27, 12, 9, 4, 30, 2, 29, 17
2	5821–8341	13, 7, 25, 14, 28, 10
3	8342–10,862	24, 26, 31
4	10,863–13,383	33, 19
5	13,384–15,904	21, 20, 34, 35

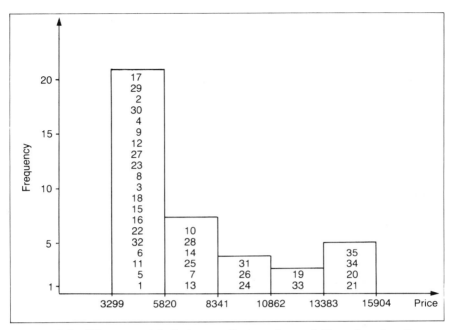

FIGURE 3.6. Histogram labelled according to the variable price, for the car models data set.

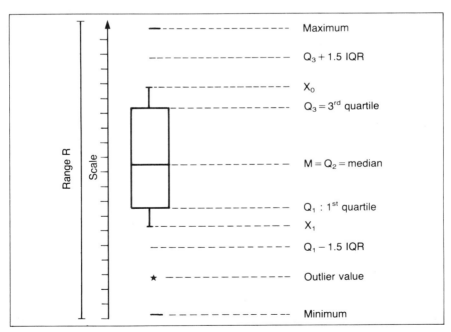

FIGURE 3.7. Dispersion box plot and its parameters.

(f) the median M is indicated by a straight horizontal line drawn in the rectangle;
(g) all the values X_i (if they exist) such that $X_i \geq Q_3 + 1.5\text{IQR}$ or $X_i \leq Q_1 - 1.5\text{IQR}$ are called *outlier values*, and are plotted with a specific symbol (star, dot, etc.);
(h) the scale of the graph is given by the extreme values of the distribution ($R = \text{Maximum} - \text{Minimum}$).

Different types of dispersion box plots can be plotted; for example, the values $(Q_1 - 1.5\text{IQR})$ and $(Q_3 + 1.5\text{IQR})$ can be replaced by D_1 and D_9, respectively. The box plots are the only graphs capable of synthesizing statistical characteristics. An example is given in Fig. 3.15.

3. 1-D Analysis of a Categorical Variable

3.1. Frequency Distributions

A categorical variable (or qualitative variable) is a variable that only *takes a finite number* of distinct values; these values are called *modalities* or *forms* or *categories*. The typical example is given by the responses to a questionnaire (where only one unique response is given to each question).

Let X be the categorical variable; let $p_1, p_2, \ldots, p_k$ be the forms associated with X; and let $n_1, n_2, \ldots, n_k$ be the respective associated frequencies. Then $N = n_1 + n_2 + \cdots + n_k$ is the population size (or the sample) studied. The relative frequencies, denoted by $f_1, f_2, \ldots, f_k$, are computed:

$$f_i = \frac{n_i}{n_1 + n_2 \cdots n_k} = \frac{n_i}{N} \qquad \forall i \in I.$$

An example of such a relative frequency data table is given in Table 3.5.

3.2. Graphics

Three types of graphics are commonly used to represent frequency distributions: *the pie chart*; the *component part bar chart*; and the *bar chart* (or *bar graph* or *bar diagram*).

3. 1-D Analysis of a Categorical Variable

TABLE 3.5. Distribution of responses to the question: How many times a week do you use the Minitel telephone service?

Forms	Frequencies	Relative frequencies in %
every day	180	10.07
>7 times a week	16	0.90
=7 times a week	14	0.78
6 times a week	22	1.23
5 times a week	62	3.47
4 times a week	124	6.94
3 times a week	252	14.10
2 times a week	260	14.55
once a week	360	20.15
3 times a month	153	8.56
2 times a month	128	7.16
once a month	121	6.77
<once a month	98	5.20
		100%

3.2.1. The Pie Chart
The pie chart is formed as follows:

(a) the base is a circle which represents 100% of the relative frequency distribution;
(b) an angle is associated with each form i that is proportional to the relative frequency $f_i = n_i/N$.

Figure 3.8 shows an example of a pie chart, based on Table 3.5.

Advantages: the percentage of each response is clearly represented.

Disadvantages: the graphic is not very precise when a percentage is too small. It cannot be used for comparing chronological variables.

3.2.2. The Component Part Bar Chart
The component part bar chart is formed as follows:

(a) the base is a rectangle, which represents 100% of the relative frequency distribution;

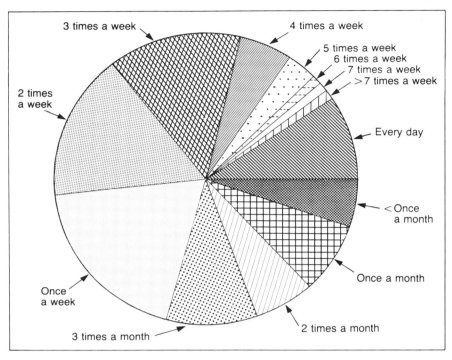

FIGURE 3.8. Pie chart for the responses to the question: How many times a week do you use the Minitel telephone service?

(b) a subrectangle is associated with each form i; the lengths of the subrectangles are proportional to the relative frequencies $f_i = n_i/N$.

Figure 3.9 shows an example of a component part bar chart, based on Table 3.5.

3.2.3. The Bar Chart

The bar chart is formed as follows:

(a) a rectangle is associated with each form i; the lengths of the rectangles are proportional to the relative frequencies $f_i = n_i/N$;
(b) the rectangles have equal widths (the common width has no significance), but are disjoint from each other (they do not overlap);
(c) when the forms i are ordered by a mathematical relation, a cumulative bar chart can be drawn.

Figure 3.10 shows an example of a bar chart, based on Table 3.5.

5. 1-D Analysis of Time Series or Chronological Variables

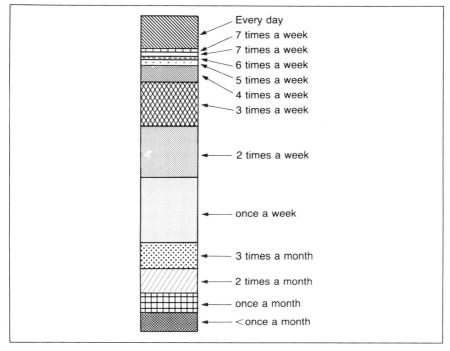

FIGURE 3.9. Component part bar chart for the responses to the question: How many times a week do you use the Minitel telephone service?

4. 1-D Analysis of a Categorical Variable with Multiple Forms

A categorical variable with multiple forms is a categorical variable where several responses may be given to each question. Obviously, the total of the distribution is not equal to 100%. Thus, the pie chart and the component part bar chart cannot be used; a typical example is given by the response to a questionnaire presented in Table 3.6. Figure 3.11 shows an example of graphical representation based on Table 3.6. It is similar to a disjoint bar chart.

5. 1-D Analysis of Time Series or Chronological Variables

5.1. 1-D Analysis of a Chronological Quantitative Variable

A number of the statistical characteristics described in Section 4 can be studied from a temporal point of view. A Cartesian plot is not sufficient

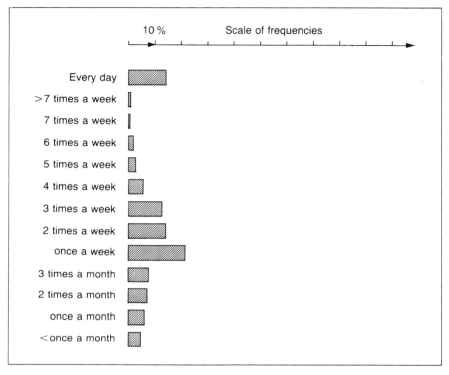

FIGURE 3.10. Bar chart of the responses to the question: How many times a week do you use the Minitel telephone service?

TABLE 3.6. Relative frequency distribution of the responses to the question: What are your reasons for using the Minitel telephone service?

Forms	Frequencies	Relative frequencies
To seek information	1210	67.71
For curiosity	153	8.56
For fun	144	8.06
To learn to use it	16	0.90
To show it to people	5	0.28
For rapidity	315	17.63
For practical aspects	667	37.33
Other reasons	224	12.53
		Total = 153 ≥ 100

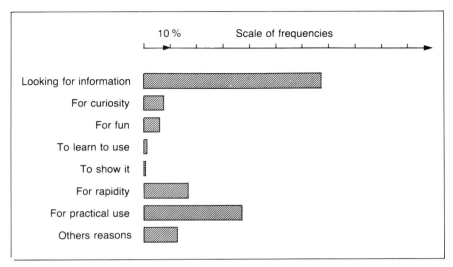

FIGURE 3.11. Bar chart of the relative frequency distribution of the responses to the question: What are the reasons for using the Minitel telephone service?

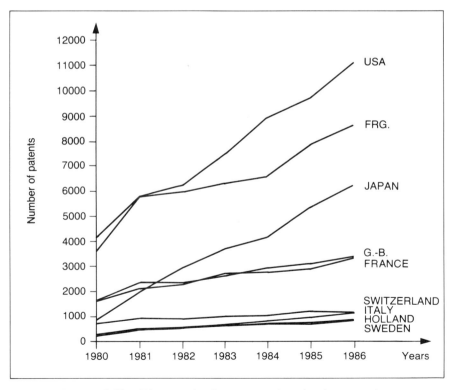

FIGURE 3.12. Line graph of patents registration by countries.

for a real understanding of how the variable evolves with time. The example of patents registration (cf. Appendix 2, §3) is used to present several different kinds of graphics.

5.1.1. Line Graphs
Consider a Cartesian diagram with two axes: x and y. The x-coordinates represent the years; the y-coordinates represent the number of patents. A line joining each point (x, y) of the same country is plotted. There are as many lines as there are countries. Figure 3.12 shows such a representation based on data given in Appendix 2, §3.

5.1.2. Line Graphs of Statistical Characteristics
Statistical characteristics such as median, mode, mean, quartiles, variance, standard deviation, range, and other measures of dispersion, kurtosis, or skewness are computed for each time value. A line graph is then drawn with time as the x-axis and characteristics as the y-axis. Figure 3.13 shows an example of such a representation based on the example of patents registration.

5.1.3. Sun Ray Plots of Statistical Characteristics
Consider a circle where each equally scaled radius is assigned a point of time (day, week, month, year). The angles formed by any two consecutive radii are equal. The value of each statistical characteristic is put on each radius successively. A straight line is then drawn joining the values of the same characteristic ordered by time. Figure 3.14 shows an example of such a representation based on the example of patents registration.

5.1.4. Dispersion Box Plots
For each unit of time, a dispersion box plot is drawn (cf. Section 2.5). The units are ordered according to time, and plotted successively. Figure 3.15. shows an example of such a representation based on the example of patents registration.

5.1.5. Bar Charts
The line graph can be replaced by any bar chart in which a bar corresponds to the values of the variable for one unit of time. If there are several variables, there are as many bars as there are variables; they are disjoint for the same unit of time. Figure 3.16 shows an example of such a representation based on the example of patents registration.

5. 1-D Analysis of Time Series or Chronological Variables

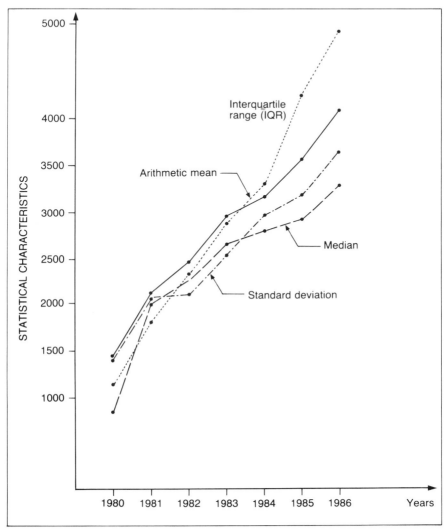

FIGURE 3.13. Line graph of statistical characteristics based on the patents registration data set.

5.1.6. Concluding Remarks

All the previous graphs are very useful, but are not used enough. It is important to study simultaneously the evolution of statistical characteristics. Studying only one of them (for example, the mean $\bar{X}$) does not provide enough understanding of the time series. The standard deviation,

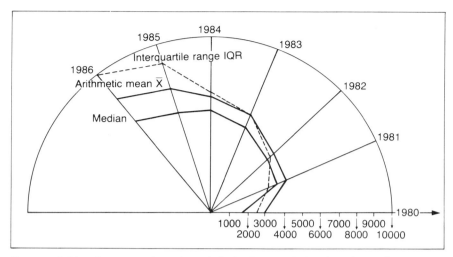

FIGURE 3.14. Sun ray plot of statistical characteristics based on the patents registration data set.

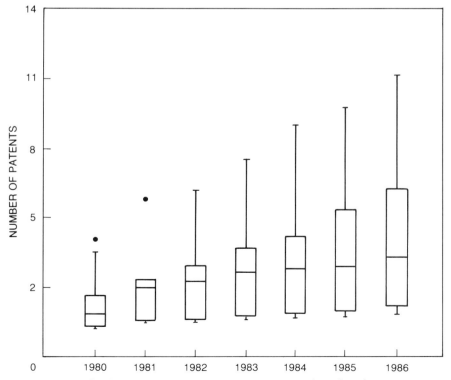

FIGURE 3.15. Box plots based on the patents registration data set.

5. 1-D Analysis of Time Series or Chronological Variables

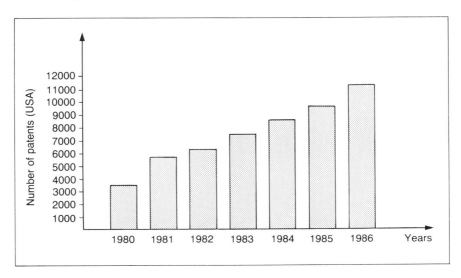

FIGURE 3.16. Bar chart based on the patents registration data set.

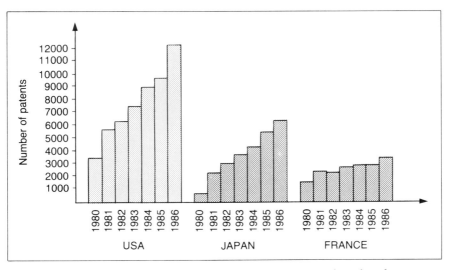

FIGURE 3.17. Multiple bar chart based on the patents registration data set.

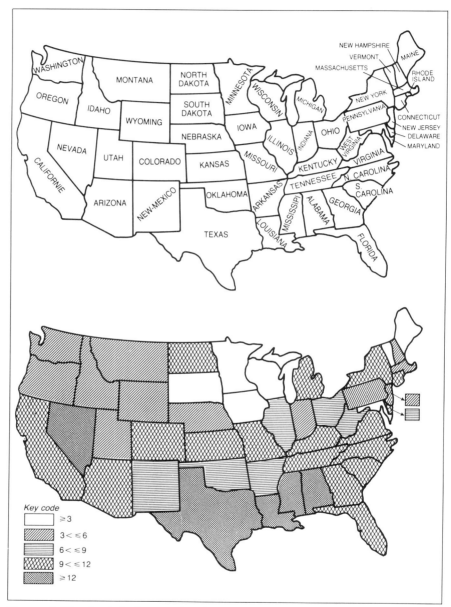

FIGURE 3.18. Statistical map related to the variable murder based on the crime data set (data in Appendix 2, §15). The numbers are expressed in murders per 100,000 population.

the interquartile range, and the range, for example, are as important as the mean.

5.2. 1-D Analysis of a Chronological Categorical Variable

The basic data set for a chronological categorical variable is formed from frequency distributions based on the time variable. Consider the usual graphics for a categorical variable: the pie chart; the component part bar chart; the bar chart. The first two are based on percentages; that reduces their interest a little. Plotting several pie charts does not give more information than the equivalent frequency data table. The bar chart can, however, be used as follows: for each of the categorical variables, as many bars as there are units of time are plotted; these bars are then grouped into a set of bars, so that there is one group of bars for each form. Another way to present such information is to draw a 2-D histogram (or stereogram), which is formed as follows:

(a) the x-axis represents the forms of the categorical variable;
the y-axis represents the units of time;
the z-axis represents the values of the variable for the forms and the units of time;
(b) because it is a 2-D figure, the basis is drawn in perspective.

This graph gives the same information as a multiple bar chart, but it is not as easy to plot without computer software. Figure 3.17 shows a multiple bar chart based on the example of patents registration.

6. Statistical Maps or Cartograms

6.1. Introduction

Frequently, it is interesting to present data on maps, when the units of observation have a geographic relation among themselves. For example, such a map can be plotted (*cf.* Fig. 3.18) based on the crimes data set (*cf.* Appendix 2, §15).

6.2. Construction

The cartogram is formed as follows:

(a) consider *one* quantitative variable (here, one of the seven crime variables, denoted by murder);
(b) compute the range R of the variable;
(c) decide the number of hachure classes (in general, less than ten);
(d) assign a hachure to each unit of observation (here, a state of the USA);
(e) plot the hachures corresponding to each unit (here, a state of the USA).

This plot is done for one variable, so that there are as many statistical maps as there are variables involved in the data set. Figure 3.18 shows the cartogram of the variable "murder."

Chapter 4
2-D Statistical Data Analysis

1. Introduction

In practice, many users stop their statistical investigations after having studied the variables independently from each other. However, they have used only 1-D analysis, and usually cannot put forward any explanations of any causality for their data. For example, a questionnaire with two questions can be analyzed using two frequency distributions. However, studying each frequency distribution individually cannot provide any relation between the two questions. Another example is given by the study of two quantitative variables, for which as many statistical characteristics or graphics as required can be built (*cf.* Chapter 3). They cannot help, however, to explain the relation between the two variables. The only way to approach the explanation of how one variable is related to another is to build a relation between the two variables. That is the objective of 2-D statistical data analysis, where two variables are analyzed according to the following points of view:

1. To express and highlight the relationship between two variables, in order to show the statistical dependence between them.

2. When possible, to sum up the relations by a law of variation or a statistical dependence, and to characterize them by a numerical coefficient independent of the units of measure of the variables.

These studies vary according to the type of variables involved (quantitative, categorical, chronological, logical, etc.), and are presented in what follows.

2. 2-D Analysis of Two Categorical Variables

2.1. Contingency Data Sets

The way to express a relation between two categorical variables is to compute a *contingency data set* as follows: Let two categorical variables be denoted by V_1 and V_2:

V_1 has h forms denoted by $A_1, A_2, \ldots, A_h$;
V_2 has k forms denoted by $B_1, \ldots, B_k$.

For each couple of forms (A_i, B_j), we compute the number of observations, denoted by n_{ij}, that possesses the forms A_i and B_j *simultaneously*. The contingency data set (or data table) corresponding to the two categorical variables V_1 and V_2 is a rectangular table, where the rows correspond to the forms of V_1 and the columns correspond to those of V_2. The table contains the numbers n_{ij} corresponding to all of the couples of forms (A_i, B_j) from V_1 and V_2. Such a data set representing two categorical variables is called a *contingency data set* (or contingency data table). We give an example based on a questionnaire, whose questions are considered as categorical variables.

1. How much time do you spend each week using the Minitel service?
The replies (or forms) were the following:

1. less than 5 minutes weekly;
2. from 5 minutes to 10 minutes weekly;
3. from 11 minutes to 16 minutes weekly;
4. from 16 minutes to 30 minutes weekly;
5. from 31 minutes to 45 minutes weekly;
6. from 46 minutes to 60 minutes weekly;
7. from 1 hour to 2 hours weekly;
8. from 2 hours to 3 hours weekly;
9. more than 3 hours weekly.

2. Are you satisfied with the Minitel service?
The replies (or forms) were the following:

1. very satisfied;
2. rather satisfied;
3. rather unsatisfied;
4. very unsatisfied.

From these two questions, the contingency data set in Table 4.1 was computed.

2. 2-D Analysis of Two Categorical Variables

TABLE 4.1. Contingency data set from two questions of the Minitel survey.

	V_1	Very satisfied	Rather satisfied	Rather unsatisfied	Very unsatisfied	Total
	V_2					
Variable V_2	≤ 5'	110	449	132	13	704
	from 6' to 10'	83	262	24	1	370
	from 11' to 16'	70	170	13	3	256
	from 16' to 30'	76	146	9	3	234
	from 31 to 45'	31	49	0	0	80
	from 45' to 60'	11	39	1	0	51
	from 1h to 2h	15	13	0	0	28
	from 2h to 3h	12	5	0	0	17
	≥ 3h	4	2	0	0	6
	Total	412	1135	179	20	1746

For a contingency data set to be used to express the relationship that exists between two categorical variables, two characteristics need to be considered: statistical existence of the relation, and measure of the association. Some complementary notations are given first on contingency data sets before presenting these characteristics:

n_{ij} = crossed frequency of the two forms i and j;

$$n_i = \sum_{j \in J} n_{ij} = \text{frequency of form } i;$$

$$n_j = \sum_{i \in I} n_{ij} = \text{frequency of form } j;$$

$$n = \sum_{i \in I} \sum_{j \in J} n_{ij} = \text{total} = \text{population size};$$

$$f_{ij} = \frac{n_{ij}}{n} = \text{relative crossed frequency of the forms } i \text{ and } j;$$

$$f_i = \frac{n_i}{n} = \text{relative frequency of form } i;$$

$$f_j = \frac{n_j}{n} = \text{relative frequency of form } j.$$

2.2. Independence Testing

It is said that two variables are independent from each other if the following relation is satisfied:

$$\frac{n_{ij}}{n} = \frac{n_i}{n} \cdot \frac{n_j}{n} \quad \forall i \in I \text{ and } \forall j \in J,$$

or

$$f_{ij} = f_i f_j \quad \forall i \in I \text{ and } \forall j \in J.$$

It is not enough, however, to decide if the variables are independent from a statistical point of view. The following also has to be considered. Let n'_{ij} be the theoretical relative crossed frequency assuming independence of i and j, so that $n'_{ij} = f_i f_j = n_i n_j / n^2$. The deviations between the observed and theoretical values, $(n_{ij} - n'_{ij})$, characterize the disagreement between the observations and the hypothesis of independence. All of the deviations can be studied by computing the quantity denoted by χ^2 (chi-square) as follows:

$$\chi^2 = \sum_{i \in I, j \in J} \frac{(n_{ij} - n'_{ij})^2}{n'_{ij}}$$

$$= \left(\sum_{i \in I, j \in J} \left(\frac{n_{ij}^2}{n'_{ij}} \right) \right) - n.$$

Assuming that n'_{ij} is not too small, χ^2 is distributed according to Pearson's law with $(k-1)(h-1)$ degrees of freedom. This is the base of the *chi-square test*, which tests the *independence* of two categorical variables in their contingency data set.

The significance test proceeds as follows:

(a) compute χ^2, and $v = (k-1)(h-1)$ (the degree of freedom);
(b) read on the chi-square distribution table the value corresponding to the v (row) and the probability level p (column); if χ^2 is greater than this, then the hypothesis of independence is accepted (*cf.* Table of the χ^2 distribution in Appendix 2, §16).

2. 2-D Analysis of Two Categorical Variables

2.3. Global Measures of Association

Several different measures of association are commonly used to express the intensity of the relation between two categorical variables:

$$\chi^2 = \sum_{i \in I, j \in J} \frac{(n_{ij} - n'_{ij})^2}{n'_{ij}} \quad \text{(Pearson)};$$

$$\phi^2 = \frac{1}{n} \sum_{i \in I, j \in J} \frac{(n_{ij} - n'_{ij})^2}{n'_{ij}} = \frac{\chi^2}{n} \quad \text{(Pearson)};$$

$$C = \frac{\phi^2}{1 + \phi^2} = \frac{\chi^2}{n + \chi^2} \quad \text{(Cramer)};$$

$$T^2 = \frac{\phi^2}{(k-1)(h-1)} \quad \text{(Tschuprov)}.$$

2.4. Numerical Study of Contingency Data Sets

From the usual contingency data sets, other data sets can be computed and studied:

(a) row profiles data set: $\{n_{ij}/n_i ; i \in I; j \in J\}$;
(b) column profiles data set: $\{n_{ij}/n_j ; i \in I; j \in J\}$;
(c) relative frequency data set: $\{n_{ij}/n ; i \in I; j \in J\}$;
(d) independence ratios data set: $\{f_{ij}/f_i f_j ; i \in I; j \in J\}$;
(e) χ^2-contributions data set: $\{(f_{ij} - f_i f_j)^2/f_i f_j ; i \in I; j \in J\}$.

TABLE 4.2. Row profiles data set from two questions of the Minitel survey.

	V_1	Very satisfied	Rather satisfied	Rather unsatisfied	Very unsatisfied	Total
	$\leq 5'$	16	64	19	2	100
	from 6' to 10'	22	71	6	0	100
	from 11' to 16'	27	66	5	1	100
Variable V_2	from 16' to 30'	32	62	4	1	100
	from 31 to 45'	39	61	0	0	100
	from 45' to 60'	22	76	2	0	100
	from 1h to 2h	54	46	0	0	100
	from 2h to 3h	71	29	0	0	100
	$\geq 3h$	67	33	0	0	100

TABLE 4.3. Column profiles data set from two questions of the Minitel survey.

V_2 \ V_1	Very satisfied	Rather satisfied	Rather unsatisfied	Very unsatisfied
≤ 5'	27	40	74	65
from 6' to 10'	20	23	13	5
from 11' to 16'	17	15	7	15
from 16' to 30'	18	13	5	15
from 31' to 45'	8	4	0	0
from 45' to 60'	3	3	1	0
from 1h to 2h	4	1	0	0
from 2h to 3h	3	0	0	0
≥ 3h	1	0	0	0
Total	100	100	100	100

(columns labeled Variable V_1; rows labeled Variable V_2)

These different data sets highlight particular elements that are not considered by a global measure of association. They are represented in Tables 4.2–4.6, based on the Minitel telephone service survey.

2.5. Graphical study of contingency data sets

Each previous contingency data set can be associated with a graph-table that represents graphically the numerical information, and highlights *directly* the main features of the data set.

(a) Row profiles contingency data set. For each cell (i, j) of the data set, a hachure whose height is proportional to (n_{ij}/n_i) is plotted. An example of such a representation is given in Table 4.7, taken from the data set in Table 4.1.

(b) Column profiles contingency data set. For each cell (i, j) of the data set, a hachure whose height is proportional to n_{ij}/n_j is plotted. An example of such a representation is given in Table 4.8, taken from the data set in Table 4.1.

2. 2-D Analysis of Two Categorical Variables

TABLE 4.4. Relative frequency data set from two questions of the Minitel survey.

Variable V_1

V_2 \ V_1	Very satisfied	Rather satisfied	Rather unsatisfied	Very unsatisfied	Total
≤ 5'	6	26	8	1	40
from 6' to 10'	5	15	1	0	21
from 11' to 16'	4	10	1	0	15
from 16' to 30'	4	8	1	0	13
from 31' to 45'	2	3	0	0	5
from 45' to 60'	1	2	0	0	3
from 1h to 2h	1	1	0	0	2
from 2h to 3h	1	0	0	0	1
≥ 3h	0	0	0	0	0
Total	24	65	10	1	100

(Variable V_2 labels the rows.)

(c) Relative frequency data set. The hachures are determined as follows:

1. find the minimum, the maximum, and the range of values f_{ij};
2. decide the number of classes k, and divide the range into k classes;
3. assign a class and a hachure to each value f_{ij};
4. replace the value f_{ij} by its corresponding hachure.

TABLE 4.5. Independence ratios data set from two questions of the Minitel survey. (the values are multiplied by 1000)

Variable V_1

V_2 \ V_1	Very satisfied	Rather satisfied	Rather unsatisfied	Very unsatisfied
≤ 5'	625	1000	2000	25
from 6' to 10'	992	1000	476	0
from 11' to 16'	1100	10	666	0
from 16' to 30'	1300	946	769	0
from 31' to 45'	1600	923	0	0
from 45' to 60'	1400	1000	0	0
from 1h to 2h	1300	1000	0	0
from 2h to 3h	2000	769	0	0
≥ 3h	4100	0	0	0

TABLE 4.6. χ^2-contributions data set from two questions of the Minitel survey.

		Variable V_1			
	V_1 \ V_2	Very satisfied	Rather satisfied	Rather unsatisfied	Very unsatisfied
Variable V_2	≤ 5′	13.5	0	40	9
	from 6′ to 10′	0	1.4	5.7	2
	from 11′ to 16′	0.4	0.1	1.6	1
	from 16′ to 30′	2.6	1.9	6.9	1
	from 31′ to 45′	5.3	0.12	5	5
	from 45′ to 60′	1.3	4.6	3	3
	from 1h to 2h	5	13	2	0.1
	from 2h to 3h	2	6	1	0
	≥ 3h	0	0	0	0

An example of such a representation is given in Table 4.9, taken from the data set in Table 4.4.

(d) Independence ratios data set. The hachures are determined as in (c), with the independence ratio $(f_{ij}/f_i f_j)$ playing the role of (f_{ij}). An example of such a representation is given in Table 4.10, taken from the data set in Table 4.5.

(e) χ^2-contributions data set. The hachures are determined as in (c), with the values of contributions playing the role of f_{ij}. An example of

TABLE 4.7. Row profiles contingency data set. Hachures proportional in height to the values n_{ij}/n_i are plotted.

	V_1 \ V_2	very satisfied	rather satisfied	rather unsatisfied	very unsatisfied	total
Variable V_2	≤ 5′					100
	From 6′ to 10′					100
	From 11′ to 16′					100
	From 16′ to 30′					100
	From 31′ to 45′					100
	From 45′ to 60′					100
	From 1h to 2h					100
	From 2h to 3h					100
	≥ 3h					100

2. 2-D Analysis of Two Categorical Variables

TABLE 4.8. Column profiles contingency data set. Hachures, proportional in height to the values n_{ij}/n_j are plotted.

V_2 \ V_1	very satisfied	rather satisfied	rather unsatisfied	very unsatisfied
⩽ 5′				
From 6′ to 10′				
From 11′ to 16′				
From 16′ to 30′				
From 31′ to 45′				
From 45′ to 60′				
From 1 h to 2 h				
From 2 h to 3 h				
⩾ 3 h				

TABLE 4.9. Relative frequency data set. Hachures are determined by partioning the range of frequencies into five classes.

V_2 \ V_1	very satisfied	rather satisfied	rather unsatisfied	very unsatisfied
⩽ 5′				
From 6′ to 10′				
From 11′ to 16′				
From 16′ to 30′				
From 31′ to 45′				
From 45′ to 60′				
From 1 h to 2 h				
From 2 h to 3 h				
⩾ 3 h				

Keycode: from 0 to 5 ; from 6 to 10 ; from 11 to 15 ; from 16 to 20 ; from 21 to 26.

TABLE 4.10. Independence ratios data set. Hachures are determined by partitioning the range of ratios into five classes.

Variable V_1

V_2 \ V_1	very satisfied	rather satisfied	rather unsatisfied	very unsatisfied
≤ 5'			▓	
From 6' to 10'	▒	▒		
From 11' to 16'	▒			
From 16' to 30'	▒			
From 31' to 45'	▒			
From 45' to 60'	▒			
From 1 h to 2 h	▒			
From 2 h to 3 h	▓			
≥ 3 h	▓			

Keycode: ☐ from 0 to 820 ;   ▒ from 821 to 1640 ;
          ▒ from 1641 to 2460 ;   ☐ from 2461 to 3280 ;
          ▓ from 3281 to 4000.

TABLE 4.11. χ^2-contributions data set. Hachures are determined by partitioning the range of contributions into five classes.

Variable V_1

V_2 \ V_1	very satisfied	rather satisfied	rather unsatisfied	very unsatisfied
≤ 5'	▒		▒	▒
From 6' to 10'				
From 11' to 16'				
From 16' to 30'				
From 31' to 45'				
From 45' to 60'				
From 1 h to 2 h				
From 2 h to 3 h				
≥ 3 h				

Keycode: ☐ from 0 to 0.008 ;   ▒ from 0.009 to 0.0016 ;
          ☐ from 0.0017 to 0.0024 ;   ☐ from 0.025 to 0.032 ;
          ▓ from 0.032 to 0.040.

such a representation is given in Table 4.11, taken from the data set in Table 4.6.

(f) Dispersion stereogram. Instead of plotting graph tables as in the previous paragraphs, a 2-D histogram can be drawn to represent the frequency distribution of two categorical variables. It is plotted as follows:
  (a) on the plan formed by the x- and y-axes, a rectangle is drawn in perspective representing the contingency data set;
  (b) on each cell (i, j) of the rectangle is built a rectangular prism whose height is proportional to n_{ij}.

3. 2-D Analysis of Two Quantitative Variables

3.1. Dispersion Charts (or Scatter Diagrams)

To visualize satisfactorily two quantitative variables X and Y, a scatter diagram of the two variables can be plotted. A scatter diagram shows the points (X_i, Y_i) on a rectangular coordinate system, and is formed as follows:

  (a) draw two rectangular axes, x and y, with the same scale;
  (b) take every couple (X_i, Y_i) corresponding to X and Y values for the units of observation i;
  (c) plot all the points i on the scatter diagram and give them a symbol (.) to produce a *density scatter diagram*;
  (d) the previous symbol can be replaced by a label assigned to each unit of observation i in order to obtain a *labelled scatter diagram*.

A scatter diagram is also called a *dispersion chart* (because it highlights the dispersion of the cloud of points).

Figure 4.1 shows such a representation based on the car data set (*cf.* Appendix 2, §1). Two scatter diagrams are given: a density scatter diagram and a labelled scatter diagram.

3.2. Dispersion (or Scatter) Data Sets

The previous scatter diagram can be studied in two ways. The first is to study the shape of the set of plotted points, and if the scatter diagram seems to lie near a line (curved or straight), then an appropriate equation

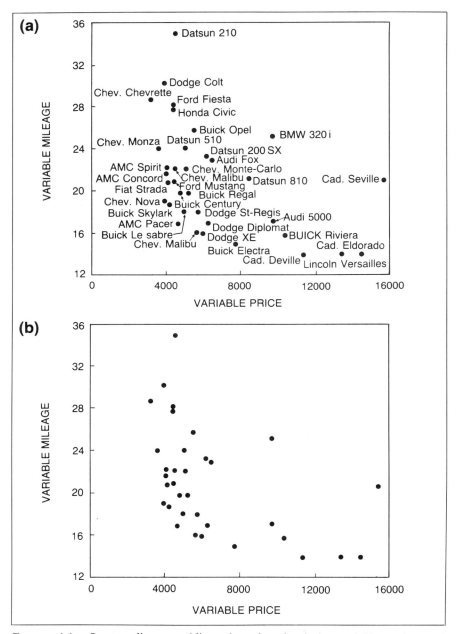

FIGURE 4.1. Scatter diagrams (dispersion charts) of the variables price and mileage, from the car models data set: (a) density scatter diagram; (b) labelled scatter diagram.

3. 2-D Analysis of Two Quantitative Variables

can be sought by *regression analysis* (*cf.* Section 4). The other way is to study the relation between two quantitative variables taken from a specific contingency data set. It is computed from the dummy categorical variables associated with the two quantitative ones. The specific contingency data set is called a scatter or a dispersion data set, and it is formed as follows:

(a) compute the range of each quantitative variable X and Y;
(b) divide the range of X into classes with equal range. Let $c_1, c_2, \ldots, c_i, \ldots, c_n$ be these classes;
(c) divide the range of Y into classes with equal range. Let $d_1, d_2, \ldots, d_j, \ldots, d_p$ be these classes;
(d) the forms of the dummy variable related to X are $c_1, c_2, \ldots, c_i, \ldots, c_n$;
(e) the forms of the dummy variable related to Y are $d_1, d_2, \ldots, d_j, \ldots, d_p$;
(f) compute the contingency data set corresponding to the two previous dummy variables. It has $c_1, c_2, \ldots, c_n$ as the rows and $d_1, d_2, \ldots, d_p$ as the columns. The element of the data set denoted by the pair (c_i, d_j) contains the number of units that have the form c_i and the form d_j simultaneously;
(g) the scatter diagram is ready to be studied.

3.3. Study of Dispersion Data Sets

Since the scatter data sets are contingency data sets, they can be studied as such completely, without any restriction. Thus, the following data sets and associated graphics can be computed:

(a) the row profiles data set and its associated hachured data set;
(b) the column profiles data set and its associated hachured data set;
(c) the relative frequency data set and its associated hachured data set;
(d) the independence ratios data set and its associated hachured data set;
(e) the χ^2-contributions data set and its associated hachured data set;
(f) the dispersion stereogram associated with the scatter data set (*cf.* Fig. 4.2).

3.4. Curve Fitting and Regression

3.4.1. Introduction

Sometimes, in practice, a relationship can exist between two (or more) variables. For example: weights of adults depend on their size; the

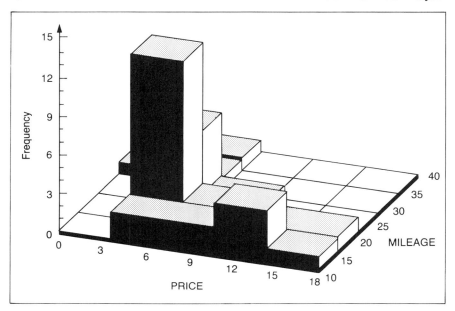

FIGURE 4.2. Dispersion stereogram for the variables price and mileage, from the car models data set.

pressure of a given gas depends on its temperature and volume. It is thus useful to express this relationship mathematically by determining an equation connecting the variables.

3.4.2. Curve Fitting—Principle
We wish to determine an equation connecting two (or more) variables. A scatter diagram can be useful (*cf.* Section 3.1) for visualizing a smooth curve that approximates the data. Such a curve is called an *approximating curve*. It can be linear if the relationship between variables is linear (*cf.* Fig. 4.3, left), or non-linear if the relationship between variables is nonlinear (*cf.* Fig. 4.3, right). The general problem of finding equations of approximating curves of given data sets is called *curve fitting*. Techniques in curve fitting include *the least squares and regression methods*.

3.4.3. The Method of Least Squares
To avoid the need for subjective judgement when constructing lines, parabolas, or other approximating curves to fit data sets, an unbiased definition of a *"best fitting line,"* or a *best fitting curve* is necessary.

3. 2-D Analysis of Two Quantitative Variables

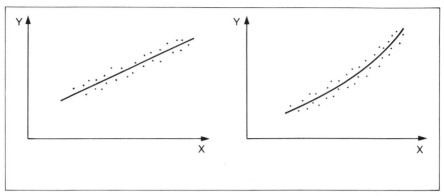

FIGURE 4.3. Two scatter diagrams and their approximating curves: (left) linear relation; (right) nonlinear relation.

As indicated in Fig. 4.4, the deviations of the points from the least square curve (C) are denoted by $D_1, D_2, \ldots, D_n$. A usual measure of "goodness of fit" of the curve (C) for a given data set is provided by the quantity $D = (D_1^2 + D_2^2 + \ldots D_n^2)$. If D is small, the fit is good but, if not, the fit is bad. The least square curve is the curved line whose D is minimum. Because the role of X and Y can be changed in the scatter diagram another least square curve can be obtained.

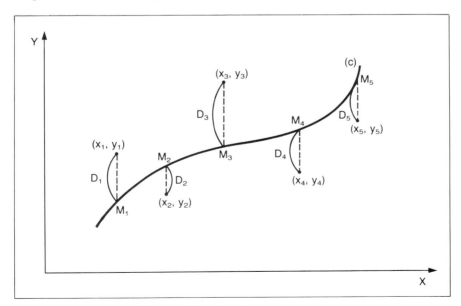

FIGURE 4.4. Least square curve.

3.4.4. Equations

To formalize the general equations associated with the least square methods, some supplementary notations need to be introduced.

3.4.4.1. Marginal Variables and Conditional Variables. Consider the model of scatter data set associated with two quantitative variables X and Y (*cf.* Table 4.12).

The *marginal variable* X (or variable X) is the variable that has the values $\{x_1, x_2, \ldots, x_n\}$, where x_i is the center of the class d_i (*cf.* Section 3.2), and with $\{n_1, n_2, \ldots, n_i, \ldots, n_n\}$ as the frequency distribution. The *marginal variable* Y (or variable Y) is the variable that has the values $\{y_1, y_2, \ldots, y_p\}$, where y_i is the center of the class c_i, and with $\{n_1, n_2, \ldots, n_j, \ldots, n_p\}$ as the frequency distribution. The following values are computed from these two variables: Let $\bar{X}$ be the mean of X, $\bar{Y}$ the mean of Y, $\bar{s}^2(X)$ the variance of X and $\bar{s}^2(Y)$ the variance of Y. Then:

$$\begin{cases} \bar{\bar{X}} = \frac{1}{n} \sum_{i \in I} n_i x_i, \\ \bar{s}^2(X) = \frac{1}{n} \sum_{i \in I} n_i (x_i - \bar{\bar{X}})^2; \end{cases}$$

TABLE 4.12. Scatter data set associated with two quantitative variables.

		Variable Y						
X $\diagdown$ Y		c_1, y_1	c_2, y_2	c_3, y_3	c_4, y_4	c_5, y_5	c_6, y_6	Total
Variable X	d_1, x_1							
	d_2, x_2							
	d_3, x_3			n_{ij}				n_i
	d_4, x_4							
	d_5, x_5							
	d_6, x_6							
	Total			n_j				n

3. 2-D Analysis of Two Quantitative Variables

$$\begin{cases} \bar{\bar{Y}} = \frac{1}{n} \sum_{j \in J} n_j y_j, \\ \bar{s}^2(Y) = \frac{1}{n} \sum_{j \in J} n_j (y_j - \bar{\bar{Y}})^2. \end{cases}$$

The *conditional variable* X knowing Y with value y_j, denoted by $X|_{Y=y_j}$, is the variable that has the values $\{x_1, x_2, \ldots, x_i, \ldots, x_n\}$ and has $\{n_{1j}, n_{2j}, \ldots, n_{ij}, \ldots, n_{nj}\}$ as its frequency distribution. The *conditional variable* Y knowing X having value x_i, denoted by $Y|_{X=x_i}$, is the variable that has the values $\{y_1, y_2, \ldots, y_p\}$ and has $\{n_{i1}, n_{i2}, \ldots, n_{ij}, \ldots, n_{ip}\}$ as its frequency distribution. Compute the following values from these two variables:

Let $\bar{X}_j$ be the mean of $X|_{Y=y_j}$ and s_j^2 be the variance of $X|_{Y=y_j}$.
Let $\bar{Y}_j$ be the mean of $Y|_{X=x_i}$ and s_i^2 be the variance of $Y|_{X=x_i}$.

Then:

$$\begin{cases} \bar{X}_j = \frac{1}{n_j} \sum_{i \in I} n_{ij} x_i = \sum_{i \in I} f_i^j x_i, \\ s_j^2 = \frac{1}{n_j} \sum_{i \in I} n_{ij} (x_i - \bar{X}_j)^2 = \sum_{i \in I} f_i^j (x_i - \bar{X}_j)^2; \end{cases}$$

$$\begin{cases} \bar{Y}_i = \frac{1}{n_i} \sum_{j \in J} n_{ij} y_j = \sum_{j \in J} f_j^i y_j, \\ s_i^2 = \frac{1}{n_i} \sum_{j \in J} n_{ij} (y_j - \bar{Y}_i)^2 = \sum_{j \in J} f_j^i (y_j - \bar{Y}_i)^2. \end{cases}$$

We see for the first time the notations f_i^j and f_j^i; f_i^j is the conditional frequency of i knowing j, and f_j^i is the conditional frequency of j knowing i. They are used in probability calculus and in 2-D correspondence analysis (*cf.* Chapters 8 and 9).

3.4.4.2. Relations between Marginal and Conditional Variables

The following relations then hold:

$$\bar{\bar{X}} = \sum_{j \in J} f_j \bar{X}_j \quad \text{with } f_j = n_j/n,$$

$$\bar{\bar{Y}} = \sum_{i \in I} f_i \bar{Y}_i \quad \text{with } f_i = n_i/n,$$

giving

$$\min_j(\bar{X}_j) \le \bar{\bar{X}} \le \max_j(\bar{X}_j)$$

and

$$\min_i(\bar{Y}_i) \le \bar{\bar{Y}} \le \max_j(\bar{Y}_i),$$

$$s^2(X) = \sum_{j\in J} f_j s_j^2 + \sum_{j\in J} f_j(\bar{X}_j - \bar{\bar{X}})^2,$$

and

$$s^2(Y) = \sum_{i\in I} f_i s_i^2 + \sum_{i\in I} f_i(\bar{Y}_i - \bar{\bar{Y}})^2.$$

3.4.4.3. Regression Curves

A regression curve of Y at x is the curve representing the conditional means $\bar{y}_i$ as a function of the values x_i of X. Similarly, a regression curve of X at y is the curve representing the conditional means $\bar{x}_i$ as a function of the values y_i of Y (cf. Fig. 4.5).

It can be proved that the center of gravity of the set of points, located on the regression curve of Y at x and weighted by the marginal frequencies f_i, is also the center of gravity of the pairs (x_i, y_i) weighted

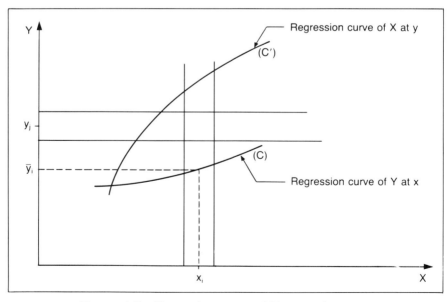

FIGURE 4.5. Regression curves of Y at x and X at y.

3. 2-D Analysis of Two Quantitative Variables

by the frequencies f_{ij}:

$$x_G = \sum_{i \in I} \sum_{j \in J} f_{ij} x_i = \sum_{i \in I} \sum_{j \in J} f_i f_j^i x_i$$

$$= \sum_{i \in I} \left(f_i x_i \left(\sum_{j \in J} f_j^i \right) \right) = \sum_{i \in I} f_i x_i = \bar{\bar{X}}.$$

Exchanging the roles of X and Y, we see that $Y_G = \bar{\bar{Y}}$, since

$$\bar{\bar{Y}} = \sum_{i \in I} f_i \bar{y}_i \quad \text{and} \quad \bar{\bar{X}} = \sum_{j \in J} f_j \bar{x}_j;$$

thus,

$$x_G = \bar{\bar{X}} = \sum_{j \in J} f_j \bar{X}_j,$$

$$y_G = \bar{\bar{Y}} = \sum_{i \in I} f_i \bar{Y}_i.$$

Thus, the center of gravity of the points $(x_i, \bar{y}_i)$, weighted by the marginal frequencies f_i, is the center of gravity, denoted by G, of the points (x_i, y_j) weighted by f_{ij}.

We obtain the following properties:

(a) if the regression curve of Y at x has a concavity with a constant sign, G is located inside the concavity.
(b) if the regression curve is a straight line, G belongs to that line.

Permuting the roles of X and Y shows that G is the center of gravity of the pairs of points $(\bar{x}_j, y_j)$, weighted by the marginal frequencies f_j.

(c) if the two regression curves are straight lines, they intersect at G, which then has the coordinates $(\bar{\bar{X}}, \bar{\bar{Y}})$;
(d) we have independence of the two variables X and Y when

$$\bar{X}_j = \bar{\bar{X}} \quad \forall j \in J,$$
$$\bar{Y}_i = \bar{\bar{Y}} \quad \forall i \in I.$$

Consequently, the regression curves are straight lines parallel to the coordinates axes and these lines intersect at G, the center of gravity (cf. Fig. 4.6).

(e) functional relations. Consider Y as a function of X (i.e., $Y = f(X)$). The mean of the conditional variable $Y|_{X=x_i}$ is equal to y_i,

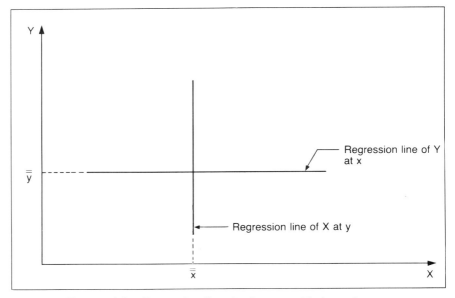

FIGURE 4.6. Regression lines in the case of independence.

so that the regression curve of the function Y is $f(X)$. This is called the *nonreciprocal functional relation*. When the relation between X and Y is reciprocal, the two regression curves are merged with the curve representing the function $Y = f(X)$. This is called the *reciprocal functional relation*.

3.4.4.4. Correlation Ratio

(a) Definition. The correlation ratio of Y at x, denoted by $\eta_{Y|x}$, is given by the following formula (expressed as its square):

$$\eta^2_{Y|x} = \frac{\sum_{i \in I} f_i(\bar{y}_i - \bar{\bar{Y}})}{s^2(Y)}$$

or

$$\eta^2_{Y|x} = 1 - \frac{\sum_{i \in I} f_i s_i^2}{s^2(Y)}.$$

3. 2-D Analysis of Two Quantitative Variables

The correlation ratio of X at y, denoted by $\eta_{X|y}$ is given by the following formula (expressed as its square):

$$\eta^2_{X|y} = \frac{\sum_{j \in J} f_j(\bar{x}_j - \bar{\bar{X}})}{s^2(X)}$$

$$= 1 - \frac{\sum_{j \in J} f_j s_j^2}{s^2(X)}.$$

(b) Properties. (cf. Table 4.13)

- In general, $\eta^2_{Y|x} \neq \eta^2_{X|y}$.
- $0 \leq \eta^2_{Y|x}, \eta^2_{X|y} \leq 1$.
- Geometrically, $\eta^2_{Y|x}$ is equal to the sine of the angle θ formed as described in Fig. 4.7.
- $\eta^2_{Y|x} = 0 \Rightarrow$ the variance of the conditional means is equal to zero
  $\Rightarrow \bar{y}_i = \bar{\bar{y}} \quad \forall i \in I$
  $\Rightarrow$ the regression curve of Y at x is parallel to the x-axis
  $\Rightarrow Y$ has no correlation with X
  $\Rightarrow$ conversely, Y having no correlation with X leads to
  $$\eta^2_{Y|x} = 0.$$
- $\eta^2_{Y|x} = 1 \Rightarrow$ the mean of the conditional variances s_i^2 is equal to zero
  $\Rightarrow s_i^2 = 0 \quad \forall i \in I$
  $\Rightarrow$ at x_i corresponds to one value of Y only
  $\Rightarrow Y$ is a function of X and conversely.

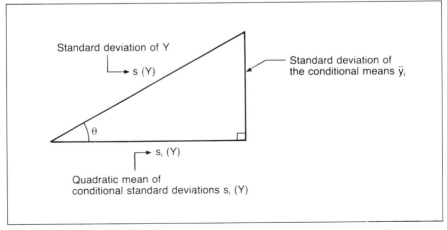

FIGURE 4.7. Geometric interpretation of a correlation ratio.

TABLE 4.13. Interpretation of the correlation ratios $\eta_{Y|x}$ and $\eta_{X|y}$.

Y \ X	$\eta^2_{X|y} = 0$	$0 < \eta^2_{X|y} < 1$	$\eta^2_{X|y} = 1$
$\eta^2_{Y|x} = 0$	reciprocal absence of correlation	general case of absence of correlation of Y with respect to X	functional relation $X = g(Y)$; no correlation of Y with respect to X
$0 < \eta^2_{Y|x} < 1$	general case of absence of correlation of X with respect Y	general case	general case of nonreciprocal functional relation $X = g(Y)$
$\eta^2_{Y|x} = 1$	functional relation $Y = f(X)$; no correlation of X with respect to Y	general case of nonreciprocal functional relation $Y = f(X)$	reciprocal functional relation

(Variable X across top, Variable Y down side)

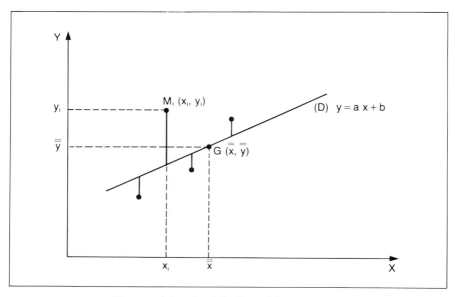

FIGURE 4.8. Straight line of least square.

3. 2-D Analysis of Two Quantitative Variables

3.4.4.5. Straight Lines of Least Squares

DEFINITION. The straight line of least square is the line nearest to the set of points (x_i, y_i); the goodness-of-fit is given by the mean of the squares of the distances of the points to the line, travelling parallel to the y-axis. A geometric representation of the definition is given in Fig. 4.8.

EQUATIONS OF THE LEAST SQUARE LINE. We need to compute the coefficients a and b in the equation $y = ax + b$ of the line of least square. We seek a and b such that the following quantity Q is minimal:

$$Q = \sum_{i \in I} \sum_{j \in J} f_{ij}(y_j - ax_i - b)^2.$$

For a given a,

$$\frac{\delta Q}{\delta b} = -2 \sum_{i \in I} \sum_{j \in J} f_{ij}(y_i - ax_i - b)$$
$$= -2(\bar{\bar{Y}} - a\bar{\bar{X}} - b).$$

Thus, the minimum is given for $b = \bar{\bar{Y}} - a\bar{\bar{X}}$, and the line (D) passes through the center of gravity of the set of points (x_i, y_i); its coordinates are $(\bar{\bar{X}}, \bar{\bar{Y}})$. Then, b is substituted in Q, yielding

$$Q = \sum_{i \in I} \sum_{j \in J} f_{ij}(y_j - ax_i - \bar{y} + a\bar{\bar{X}})$$
$$= \sum_{i \in I} \sum_{j \in J} f_{ij}(y_j - \bar{\bar{Y}} - a(x_i - \bar{\bar{X}}))^2.$$

To find the minimum, we compute $\delta Q / \delta a$:

$$\frac{\delta Q}{\delta a} = -2 \sum_{i \in I} \sum_{j \in J} f_{ij}(x_i - \bar{\bar{X}})(y_j - \bar{\bar{Y}}) - a(x_i - \bar{\bar{X}}) = 0.$$

Solving for a,

$$a = \frac{\sum_{i \in I} \sum_{j \in J} f_{ij}(x_i - \bar{\bar{X}})(y_j - \bar{\bar{Y}})}{\sum_{i \in I} \sum_{j \in J} f_{ij}(x_i - \bar{\bar{X}})^2}$$

$$= \frac{\sum_{i \in I} f_i(x_i - \bar{\bar{X}})(y_i - \bar{\bar{Y}})}{\sum_{i \in I} f_i(x_i - \bar{\bar{X}})^2}$$

$$= \frac{\sum_{i \in I} f_i(x_i - \bar{\bar{X}})(y_i - \bar{\bar{Y}})}{s(X)}.$$

We compute the following quantity r:

$$r = \frac{\sum_{i \in I} \sum_{j \in J} f_{ij}(x_i - \bar{\bar{X}})(y_i - \bar{\bar{Y}})}{s(X)s(Y)}$$

$$= \frac{\sum_{i \in I} f_i(x_i - \bar{\bar{X}})(y_i - \bar{\bar{Y}})}{s(X)s(Y)},$$

where $s(X)$ and $s(Y)$ are the standard deviations of X and Y, respectively. The quantity r is called the *linear coefficient of correlation* (*cf.* Section 3.4.4.6). Thus,

$$a = r\frac{s(Y)}{s(X)},$$

and the equation of the least square line D is

$$y = r\frac{s(Y)}{s(X)}(x - \bar{\bar{X}}) + \bar{\bar{Y}}.$$

The value of the minimum is given in substituting a and b in the expression for Q:

$$\text{Minimum of } Q = \sum_{i \in I} \sum_{j \in J} f_{ij}\left(y_j - \bar{\bar{Y}} - r\frac{s(Y)}{s(X)}(x_i - \bar{\bar{X}})^2\right).$$

Then,

$$\text{Minimum of } Q = s^2(Y) + r^2\frac{s^2(Y)}{s^2(X)}s^2(X) - 2r\frac{s(Y)}{s(X)}rs(Y)s(X)$$

$$= (1 - r^2)s^2(Y).$$

Thus, we have

$$0 \leq (1 - r^2)s^2(Y) \leq s^2(Y).$$

Consider now the regression curve of Y at x; the value of the optimum for the regression curve is

$$(1 - \eta^2_{Y|x})s^2(Y).$$

We then have

$$0 \leq (1 - \eta^2_{Y|x})s^2(Y) \leq (1 - r^2)s^2(Y) \leq s^2(Y),$$
$$0 \leq r^2 \leq \eta^2_{Y|x} \leq 1.$$

3. 2-D Analysis of Two Quantitative Variables

Switching the roles of X and Y, we obtain the least square line D', where we are now travelling parallel to the x-axis. We give the following results:

$$\text{line } (D): \begin{cases} y = r\dfrac{s(Y)}{s(X)}(x - \bar{\bar{X}}) + \bar{\bar{Y}}, \\ \text{value of the minimum} = (1 - r^2)s^2(Y); \end{cases}$$

$$\text{line } (D'): \begin{cases} x = r\dfrac{s(X)}{s(Y)}(y - \bar{\bar{Y}}) + \bar{\bar{X}}, \\ \text{value of the minimum} = (1 - r^2)s^2(X); \end{cases}$$

$$0 \le r^2 \le \eta^2_{Y|x};\ \eta^2_{X|y} \le 1.$$

3.4.4.6. The Linear Correlation Coefficient r. The linear correlation coefficient, denoted by r, between two quantitative variables X and Y is determined by the following formula:

$$r = r_{Y|x} = r_{X|y} = \frac{\sum\limits_{i \in I}\sum\limits_{j \in J} f_{ij}(x_i - \bar{\bar{X}})(y_j - \bar{\bar{Y}})}{\left(\left(\sum\limits_{i \in I} f_i(x_i - \bar{\bar{X}})^2\right)\left(\sum\limits_{j \in J} f_j(y_j - \bar{\bar{Y}})^2\right)\right)^{1/2}}.$$

One sees that r is a dimensionless number such that $-1 \le r \le 1$. By construction, r must be used *"only if"* the relation between X and Y is *linear*. Thus, the meaning of r depends on the linearity of the relation between X and Y:

$r \# 1 \Rightarrow$ strong linear correlation between X and Y;
$r \# 0 \Rightarrow$ it must not be concluded that there is no relation between X and Y; the shape of the set of points on the scatter diagram needs to be studied first. Figure 4.9 gives a parabolic relation with $r = 0$; Fig. 4.10. gives six scatter diagrams of points that have the same value of $r = 0.70$; they show that a linear correlation coefficient cannot be interpreted without studying the scatter diagrams.

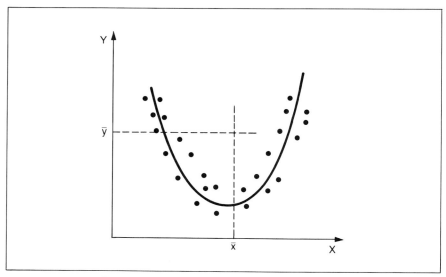

FIGURE 4.9. Parabolic relation where $r = 0$.

3.4.4.7. Relations between the Linear Correlation Coefficient r and the Correlation Ratios η. The following relations hold:
(a) $-1 \le r \le 1$;
(b) If $r^2 = \eta^2_{Y|x}$, then $(1 - r^2)s^2(Y) = (1 - \eta^2_{Y|x})s^2(Y)$;
the regression curve of Y at x is a straight line;
Y is said to be in linear correlation with X.
(c) If $r^2 = \eta^2_{Y|x} = \eta^2_{X|y}$, then the variables X and Y are in double linear correlation.
(d) If $r^2 = 1$, then the variables X and Y are functionally linked by a linear relation.
(e) If $r^2 = 0$, then there is no particular relation for the correlation ratios.

Figure 4.11 gives the relative positions of the least square lines according to different values of r, and Table 4.14 gives a summary of comparisons between the correlation ratios and the linear coefficient of correlation.

3. 2-D Analysis of Two Quantitative Variables

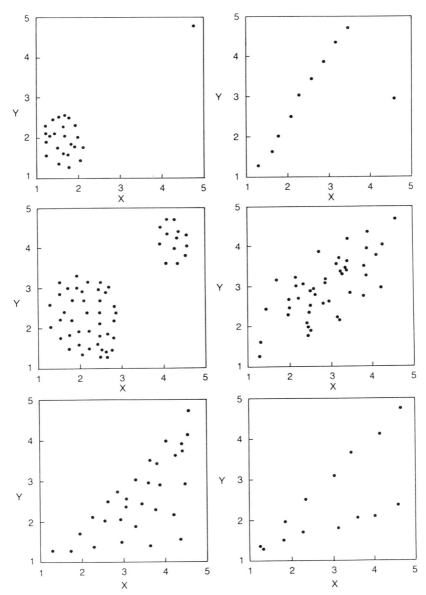

FIGURE 4.10. Scatter diagrams of points with the same value of $r = 0.70$. (From *Graphical Methods for Data Analysis*, by J.M. Chambers, W.S. Cleveland, B. Kleiner, and P.A. Tukey. Copyright © 1983 by Bell Telephone Laboratories Incorporated, Murray Hill, NJ. Reprinted by permission of Wadsworth & Brooks/Cole Advanced Books & Software, Pacific Grove, CA 93950.)

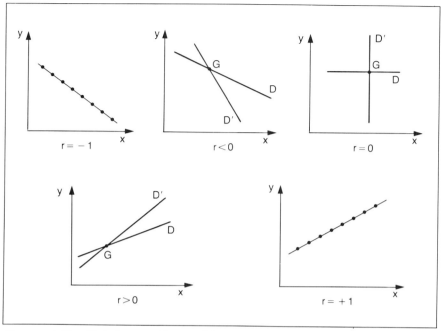

FIGURE 4.11. Relative positions of least square lines according to different values of r.

4. 2-D Analysis of a Quantitative Variable and a Categorical Variable

4.1. Introduction

The 2-D analysis of a quantitative and a categorical variable is similar to the previous 2-D analyses. These variables must be combined wisely in order to analyze them together.

4.2. The Quantitative Variable Is Taken as Such

On the one hand, a categorical variable is given with its forms (or modalities); on the other hand, the quantitative variable is given. The principle for analyzing them together is the following: The categorical variable plays the role of a *filter* for the quantitative variable. For each form of the categorical variable, a dummy quantitative variable is built with only the units of observation corresponding to that form. Thus, there are as many dummy quantitative variables as there are forms.

All the possibilities for a 1-D analysis of quantitative variable can now be applied to the dummy qualitative variable. If possible, however, all

4. 2-D Analysis of a Quantitative Variable and a Categorical Variable

TABLE 4.14. Summary of the comparisons of interpretation of r^2 and η^2 (Reprinted with permission from Dunod, Calot, G., *Cours de statistique descriptive* (1965).)

$r^2 \diagdown \eta^2$	No correlation of Y with X $\eta^2_{Y\|x}=0$, $\eta^2_{X\|y}\neq 0$	No reciprocal correlation between X and Y $\eta^2_{Y\|x}=0$, $\eta^2_{X\|y}=0$	Linear correlation of Y with X $\eta^2_{Y\|x}=r^2$, $\eta^2_{X\|y}\neq r^2$	Double linear correlation of X and Y $\eta^2_{Y\|x}=r^2$, $\eta^2_{X\|y}=r^2$	Nonreciprocal functional relation ($X \to Y$) $\eta^2_{Y\|x}=1$, $\eta^2_{X\|y}\neq 1$	Reciprocal functional relation $\eta^2_{Y\|x}=1$, $\eta^2_{X\|y}=1$	General case $0<r^2<1$, $\eta^2_{Y\|x}, \eta^2_{X\|y}<1$
$r^2=0$ — the least square lines are parallel to x-axis and the y-axis, respectively	Y is uncorrelated with X, but X is correlated with Y	X and Y are uncorrelated with each other (X and Y are not necessarily independent)			Y is a function of X but not conversely		
$r^2=1$ — X and Y are related by a linear function				X and Y are related by a linear function		X and Y are related by a linear function	
$0<r^2<1$ — the correlation of Y and X is linear but not conversely $\eta^2_{X\|y} > \eta^2_{X\|y}$			X and Y are in linear correlation with each other		Y is a function of X but not conversely		
general case							general case

TABLE 4.15. Presentation of statistics for two forms of a categorical variable. The example is the quality of service in the telephone network (cf. Appendix 2, §14).

Variable: IGQS	Form: north	Form: south	Total
Sample size	13	9	22
Average	92.84	86.46	90.23
Median	92.7	89.3	91.65
Mode	91.8	89.3	88
Geometric mean	92.78	85.85	89.88
Variance	11.87	101.35	55.70
Standard deviation	3.44	10.06	7.46
Minimum	87.4	61.4	61.4
Maximum	98.2	96	98.2
Range	10.8	34.6	36.8
Lower quartile	91.5	87.1	88
Upper quartile	95.7	90.2	93.8
Interquartile range	4.2	3.1	5.8

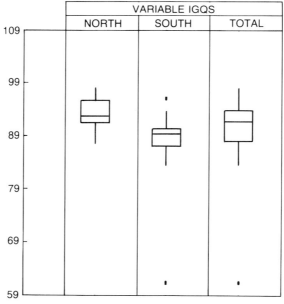

FIGURE 4.12. Presentation of dispersion box plots for two forms of a categorical variable. The example is the quality of service in the telephone network.

4. 2-D Analysis of a Quantitative Variable and a Categorical Variable

the computations and graphics should be represented together. The main computations and graphics commonly used are:

(a) statistical characteristics by forms (*cf.* Table 4.15);
(b) dispersion box plots by forms (*cf.* Fig. 4.12);
(c) frequency distributions and histograms by forms;

If the population size studied is large enough, frequency distributions and histograms are computed. To compare the distributions, the classes of the histograms need to be computed on the basis of the histogram associated with the original quantitative variable.

(d) sun ray plots of statistical characteristics by forms (*cf.* Fig. 4.13).

A circle is drawn, and as many radii are computed as there are forms. A given characteristic (such as mean, s, quartile, etc.), is computed for

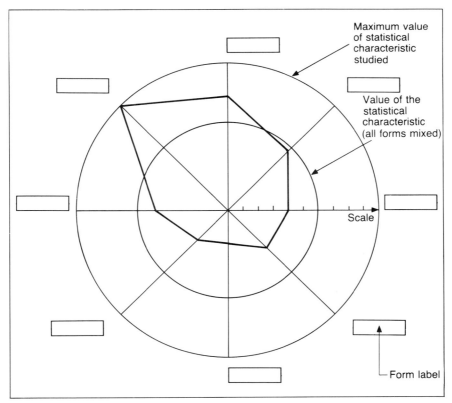

FIGURE 4.13. Presentation of a sun ray plot of statistical characteristics. One polygonal line for each characteristic.

each form, and put on each corresponding radius. A straight line joining all the values is then drawn as shown in Fig. 4.13. There are as many polygonal lines plotted as there are statistical characteristics studied.

4.3. Recoding the Quantitative Variable into a Dummy Categorical Variable

When a quantitative variable is recorded (i.e., transformed) into a dummy categorical one, it is clear that 2-D analysis of two categorical variables can be performed. In this case, the reader is referred to Chapter 4, Section 2 for such an analysis.

5. 2-D Analysis of a Quantitative Variable and a Categorical Variable with Multiple Forms

5.1. The Quantitative Variable is Taken as Such

The reader is referred to Section 4.2. The categorical variable with multiple forms is used as a *filter* for the quantitative variable. A 1-D analysis of a quantitative variable can be done for each form.

5.2. The Quantitative Variable is Recoded into a Dummy Categorical Variable

The reader is referred to Section 4.3. A 2-D analysis of two categorical variables can be done.

6. Conclusion

It has been shown that 2-D analyses are useful, easy to apply, easy to understand, and do not require a sophisticated training in statistics to be used fruitfully. Such analyses must be considered as preliminary analyses to further analyses involving N variables simultaneously. They provide the first step towards the understanding of a data set.

Chapter 5 N-D Statistical Data Analysis

1. Introduction

It is easy to understand why 2-D statistical data analysis does not provide a sufficient global vision of the interactions among variables and individuals; in order to study a set of N variables, we need $N(N-1)/2$ crossed tables in the case of qualitative variables, or $N(N-1)/2$ dispersion diagrams in the case of quantitative variables. But, how should the information be synthesized? There are two ways: The first is to represent graphically the relations among the N variables as done by Tukey (1977) and a group of associate researchers from Bell Laboratories. The second is also to represent graphically the relations among the N variables, but with a mathematical model for transforming the basic variables. The latter way was followed by Kruskal and Wish (1978), Shepard (1974), Schiffman *et al.* (1981), resulting in the technique of "multidimensional scaling," whose basic idea is to represent graphically distances between variables (or individuals) in a low dimensional space. Benzecri (1972, 1973) also adopted this way, and later on so did his students, Jambu (1983), Lebart (1984), and Greenacre (1984). They, however, used a specific geometric model so that any kind of data, from qualitative, quantitative, or mixed variables, could be analyzed and summarized. This method is known as correspondence analysis. It will be studied in detail in Chapters 8 and 9. The first steps of any N-D joint statistical data analysis are now studied, mainly using graphics.

2. Joint 3-D Statistical Data Analysis

2.1. Joint 3-D Analysis of a Quantitative Variable and Two Categorical Variables

2.1.1. Data Set of Statistical Characteristics

Consider three variables denoted by V_1, V_2, and V_3, where V_1 and V_2 are categorical variables with n and p forms, respectively, and V_3 is a

TABLE 5.1. Crossed table of statistical characteristics: V_1, V_2 = categorical variables; V_3 = quantitative variable; n_{ij} = crossed frequency of $V_1 \times V_2$; c_{ij} = statistical characteristic of V_3 related to n_{ij} (mean, s, IQR, range, Q_1, Q_3, etc.).

VARIABLE V_2

	V_2 V_1	j	V_1
VARIABLE V_1	i	c_{ij}	c_i
	V_2	c_j	c

quantitative variable. The data table given in Table 5.1 has as its rows the forms of V_1, and as its columns the forms of V_2. Each cell of the crossed table, denoted by (i, j) from the form i of V_1 and the form j of V_2, contains the subpopulation denoted by n_{ij} for the couple of forms (i, j). The usual statistical characteristics are computed for each subpopulation n_{ij}, if the subpopulation is large enough, and they are set in the crossed table as in Table 5.1. There are as many tables as there are statistical characteristics computed; sometimes, several of them can be presented in the same crossed table (means, standard deviations, etc.).

2.1.2. Stereograms and Hachured Data Sets

The information given in the previous crossed table can be represented graphically in two ways. First, the hachured data set can be used as presented in Chapter 4, Section 2.5. The maximum and minimum values of $\{c_{ij}, c_j, c_i, c\}$ are computed, and then a number of categories are chosen for the hachures. A hachure is assigned to each value c_{ij}, c_j, c_i, and then the crossed table and its dummy tables are hachured. The second way is to plot a stereogram for each characteristic based on the crossed table from V_1 and V_2 as given in Fig. 5.1.

2. Joint 3-D Statistical Data Analysis

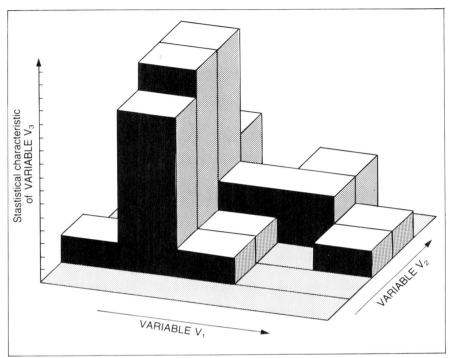

FIGURE 5.1. Stereogram of a statistical characteristic of a quantitative variable V_3 subordinate to the two categorical variables V_1 and V_2.

2.1.3. Dispersion Box Plots

As in the previous paragraph, the crossed table of V_1 and V_2 is used as a basis for the graph. Each cell (i, j) is filled in with a dispersion box plot computed from the subpopulation n_{ij} for the forms (i, j) of V_1 and V_2, as presented in Fig. 5.2.

2.1.4. Frequency Distributions

As before, the crossed table of V_1 and V_2 is used as a basis for the graph. Each cell (i, j) is filled in with the frequency histogram associated with the variable V_3 limited to the subpopulation n_{ij} for the forms i and j of V_1 and V_2, respectively. The graph is the same as the one given in Fig. 5.2, except that the dispersion box plot is replaced by the frequency histogram.

2.1.5. Sun Ray Plots of Statistical Characteristics

As before, the crossed table of V_1 and V_2 is used as a basis for the graph. Each cell (i, j) is filled in with a sun ray plot associated with the variable

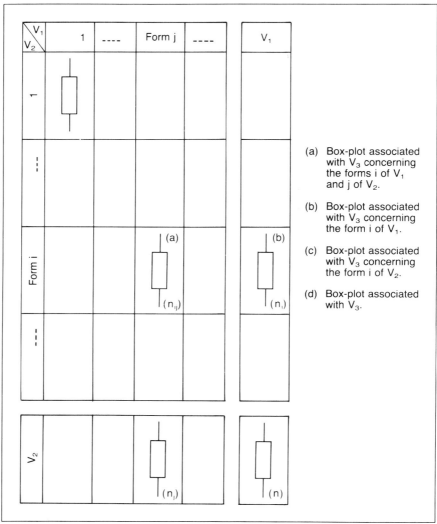

FIGURE 5.2. Crossed table of dispersion box plots for a quantitative variable V_3 subordinate to the two categorical variables V_1 and V_2.

V_3 limited to the subpopulation n_{ij} for the forms i and j of V_1 and V_2, respectively. The graph is the same as the one given in Fig. 5.2, except that the dispersion box plot is replaced by the sun ray plot. It is formed by drawing radii from a center. There are as many radii as there are statistical characteristics involved, and the scales can be different for each characteristic. Then, the values of each characteristic are transferred onto the rays, and a polygon is drawn joining all the extremes of the characteristics, as shown in Fig. 5.3.

2. Joint 3-D Statistical Data Analysis

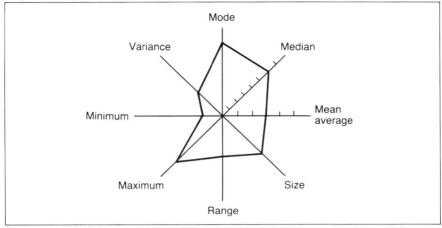

FIGURE 5.3. Sun ray plot of statistical characteristics.

2.2. Joint 3-D Analysis of a Categorical Variable and Two Quantitative Variables

2.2.1. Data Set of Statistical Characteristics

Consider three variables V_1, V_2, and V_3, where V_1 and V_2 are quantitative variables and V_3 is a categorical variable with p forms. These three variables are analyzed as follows: a 2-D joint analysis of the two quantitative variables is done for each form of the categorical variable. All the results obtained are summarized in a data set as in Table 5.2.

TABLE 5.2. Data set of statistical characteristics for the 2-D analysis of quantitative variables (V_1, V_2) subordinate to the forms of a categorical variable (V_3).

	Variable V_3	
coefficients \ forms	$1 \cdots\cdots j \cdots\cdots p$	(V_1, V_2)
$r = r_{(V_1, V_2)}$	$\cdots\cdots r_{j(V_1, V_2)} \cdots$	$r_{(V_1, V_2)}$
$\eta^2 = \eta^2_{(V_1, V_2)}$	$\cdots\cdots \eta^2_{j(V_2, V_1)} \cdots$	$\eta^2_{(V_2, V_1)}$
$\eta^2 = \eta^2_{(V_2, V_1)}$	$\cdots\cdots \eta^2_{j(V_2, V_1)} \cdots$	$\eta^2_{(V_2, V_1)}$
$r \dfrac{s(V_1)}{s(V_2)}$		$r \dfrac{s(V_1)}{s(V_2)}$
$\dfrac{1}{r} \dfrac{s(V_2)}{s(V_1)}$		$\dfrac{1}{r} \dfrac{s(V_2)}{s(V_1)}$

2.2.2. Scatter Diagrams (or Dispersion Charts)

Consider three variables V_1, V_2, and V_3, where V_1 and V_2 are quantitative variables and V_3 is a categorical variable with p forms. The scatter diagrams for V_1 and V_2 are plotted for each form of the categorical variable, giving as many scatter diagrams as there are forms in V_3. These are called *density scatter diagrams*, and each one represents the subclouds of points corresponding to a form. Another way is to plot a *labelled scatter diagram* of the variables V_1 and V_2, where the labels are given by the variable V_3 (i.e., to each point of the scatter diagram, a symbol associated with a form of V_3 is assigned). The density and labelled scatter diagrams for the car models data set is given in Fig. 5.4.

2.3. Joint 3-D Analysis of Three Quantitative Variables

2.3.1. Data Sets of Statistical Characteristics

We can compute a matrix with three rows and three columns that represents the statistical characteristics of V_1, V_2, and V_3 taken two by two. There are as many matrices as there are statistical characteristics computed. This can be presented as in Table 5.3.

2.3.2. Multiple Scatter Diagrams

The principle of the *density multiple scatter diagram* is to plot the scatter diagrams for variables taken two by two next to each other. Then, the three diagrams associated with the three variables can be visualized together as in Fig. 5.5. As in a previous paragraph, these scatter diagrams can be labelled using a supplementary variable to form a *labelled multiple scatter diagram*.

2.3.3. Triangular Diagrams

In some specific cases, triangular diagrams can be plotted, when the three quantitative variables have an equal sum. This occurs, for example, when each of the three variables represents a percentage, or the same quantity (ratios in mineralogy, rates in economics, etc.). A triangular diagram is formed as follows:

(a) an equilateral triangle is plotted;
(b) a scale is drawn on each side of the triangle (the same scale on each side). The scales are put in a specific order (*cf.* Fig. 5.6);
(c) each unit of observation is a point plotted inside the triangle. If the percentages are equal, the point is plotted at the centre of the equilateral triangle. If a point represents 100% of a variable, it is plotted at the vertex corresponding to that variable;

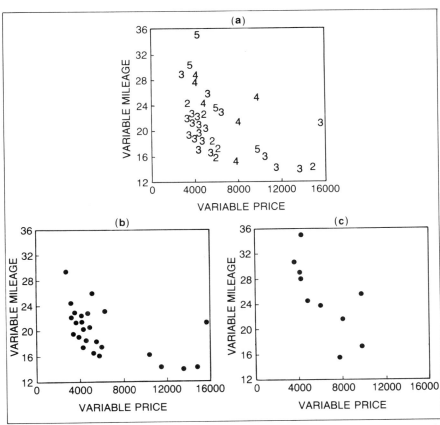

FIGURE 5.4. Density scatter diagram and labelled scatter diagrams associated with the 3-D analysis of V_1 (price), V_2 (mileage), and V_3 (number of repairs, 1978) from the car models data set: (a) labelled scatter diagram associated with the variable repairs in 1978 in four forms (2, 3, 4, 5); (b) density scatter diagram associated with the forms (2, 3) of the variable repairs in 1978; (c) density scatter diagram associated with the forms (4, 5) of the variable repairs in 1978.

TABLE 5.3. Data set of statistical characteristics for three quantitative variables (V_1, V_2, V_3).

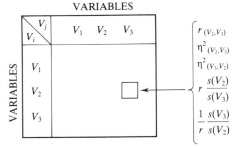

FIGURE 5.5. Multiple scatter diagram involving the following three variables: mileage (V_1); price (V_2); headroom (V_3); from the car models data set.

(d) a point is plotted as follows: it is the point formed by the three perpendiculars drawn from each side of the triangle. The sum of the lengths is necessarily equal to 100.

In Table 5.4, an example is given of an economic data set, and its associated triangular diagram is given in Fig. 5.6. The triangular diagram can be *labelled* (as for many dispersion charts), and when the points are set in a specific order (for example time), a polygonal line can be drawn, joining the points in order (*cf.* Fig. 5.6).

3. Joint N-D Statistical Data Analysis

3.1. Joint N-D Analysis of Quantitative Variables

3.1.1. Profile Diagrams

Consider N quantitative variables $V_1, V_2, \ldots, V_N$ that are not necessarily homogeneous since they do not have the same range nor the same units of measure. The principle of building a profile diagram is to plot the

3. Joint N-D Statistical Data Analysis

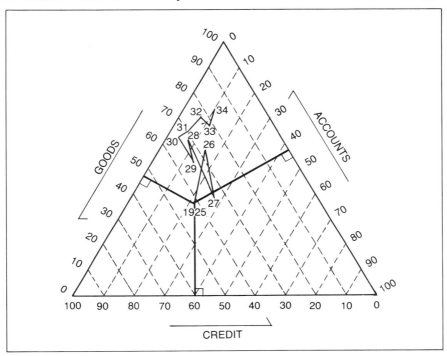

FIGURE 5.6. Triangular diagram for the financial data set presented in Table 5.4.

TABLE 5.4. Financial data set (cf. Appendix 2, §6).

		Branches		
I	J	Goods	Accounts	Credit
	1925	37	22	41
	1926	57	15	28
	1927	38	29	33
	1928	61	8	31
Years	1929	50	16	34
	1930	64	3	33
	1931	68	7	25
	1932	69	8	23
	1933	67	12	21
	1934	74	10	16

values of each variable on the same graph for *each* unit of observation and then set them next to each other.

The profile diagram is formed as follows:

(a) choose a scale for each variable (if possible, the same scale);
(b) for one unit of observation:
   (b_1) plot successively the straight lines corresponding to the values of each variable;
   (b_2) join the vertices of the lines;
(c) do the same for each unit of observation and place the diagrams next to each other.

An example of such a representation is given in Fig. 5.7, from the car models data set.

3.1.2. Sun Ray Plots

Consider N quantitative variables $V_1, V_2, \ldots, V_N$ that do not necessarily have the same scale nor units of measure. The principle for plotting a sun ray plot is the same as for a profile diagram. There is one sun ray plot for each unit of observation, in which the values of the variables are summarized for that unit of observation.

The sun ray plot is formed as follows:

(a) from a given center, draw as many rays as there are variables; these rays are scaled (if possible with the same scale);
(b) for one unit of observation:
   (b_1) plot successively the values of each variable on each scaled ray;
   (b_2) join the extremes of the rays to form a polygon;
(c) do the same for each unit of observation, and place the polygons next to each other.

Such a representation is given in Fig. 5.8, from the car models data set.

3.1.3. Quality Diagrams

When data involve percentage variables, indicators, or marks, as in many studies on quality or judgements, the two previous diagrams are of great interest. Plotting the maximum possible value (a line in a profile diagram or a circle in a sun ray plot) highlights the difference between this value and the real ones.

3.1.4. Multiple Scatter Diagrams

The principle of this graph is to plot the scatter diagrams for variables taken two by two, organized in a sort of triangular matrix (*cf.* Fig. 5.9).

3. Joint N-D Statistical Data Analysis

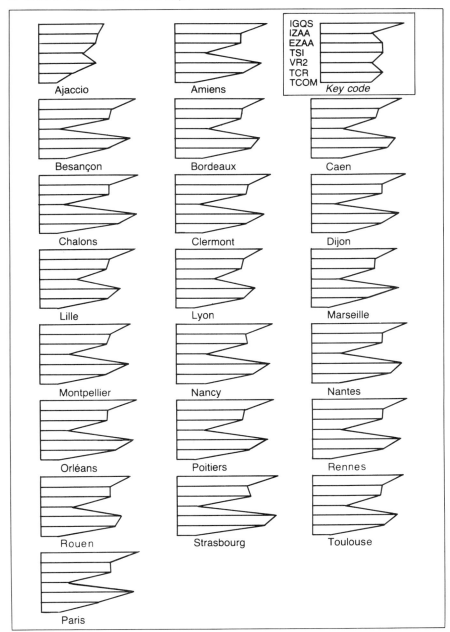

FIGURE 5.7. Profiles associated with the telephone quality of service data set. There is one profile diagram per statistical unit (here, a region of France identified by its representative town) (data in Appendix 2, §14).

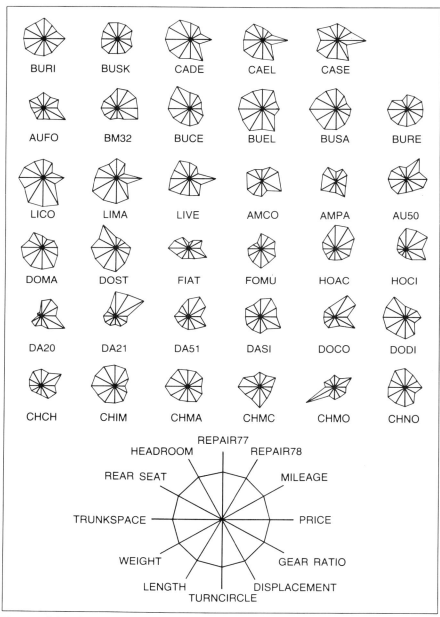

FIGURE 5.8. Sun ray plots associated with the car models data set. There is one sun ray plot by statistical unit (here, a model of car). The labels associated with sun ray plots are made with the first two letters of the makers, followed by the first two letters of the models (data in Appendix 2, §1).

3. Joint N-D Statistical Data Analysis 107

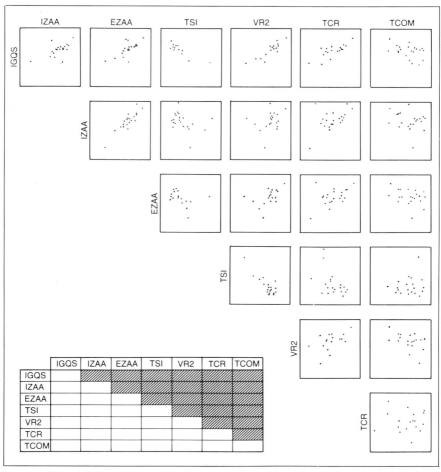

FIGURE 5.9. Multiple scatter diagram for the telephone quality of service data set (data in Appendix 2, §14).

Such a presentation has the advantage that all the relations between variables can be seen at a glance. The diagrams can be density scatter diagrams or labelled scatter diagrams. An example of such a graph is given in Fig. 5.9, from the telephone quality of service data set.

3.2. Joint N-D Analysis of Categorical Variables

3.2.1. Multiple Contingency Data Sets

Consider N categorical variables $V_1, V_2, \ldots, V_N$, and all the contingency data sets formed by all the variables, taken two by two. All of the N variables are crossed with themselves to give the crossed table containing

TABLE 5.5. Multiple contingency data set for four categorical variables. The contingency data sets on the diagonal are reduced to a diagonal table that contains the frequency distributions of the variables.

	V_1	V_2	V_3	V_4
V_1	0 / 0	(V_1,V_2)	(V_1,V_3)	(V_1,V_4)
V_2	(V_2,V_1)	0 / 0	(V_2,V_3)	(V_2,V_4)
V_3	(V_3,V_1)	(V_3,V_2)	0 / 0	(V_3,V_4)
V_4	(V_4,V_1)	(V_4,V_2)	(V_4,V_3)	0 / 0

all the contingency data sets of the variables taken two by two. This table is called a *multiple contingency data set* (*cf.* Table 5.5).

Consider the family timetables data set (*cf.* Appendix 2, § 10), where there are 10 quantitative variables; for each quantitative variable, a dummy categorical variable is created as follows:

the quantitative variable is divided into three intervals;

a category is associated with each interval, giving 10 categorical variables, each with three forms. From these variables, the multiple contingency data set is computed by crossing the variables with themselves (*cf.* Table 5.6).

3.2.2. Computations from Multiple Contingency Data Sets

Since a multiple contingency data set is an arrangement of contingency data sets, the same computations done for contingency data sets are available for a multiple contingency data set. The following may be computed:

(a) Multiple data set of profiles.
(b) Multiple data set of frequencies.
(c) Multiple data set of ratios to independence.
(d) Multiple data set of contributions to χ^2.

3. Joint N-D Statistical Data Analysis

(e) χ^2 data set; it contains the χ^2 associated with each simple contingency data set and its significance in terms of χ^2-test.

(f) ϕ^2 data set; it contains the ϕ^2 associated with each simple contingency data set.

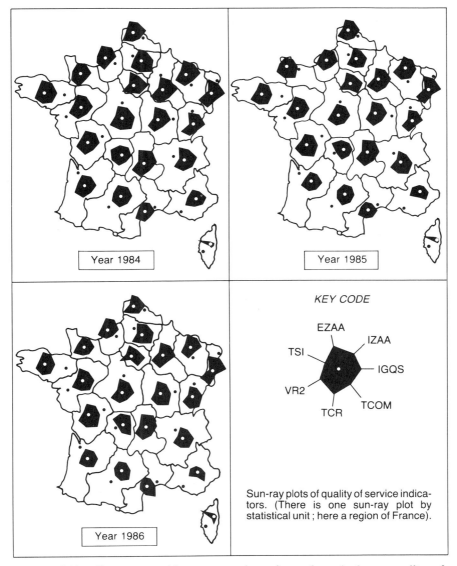

FIGURE 5.10. Cartogram with sun ray plots, from the telephone quality of service data set (data in Appendix 2, §14).

TABLE 5.6. Multiple contingency data set, from the family timetables data set. J = set of forms associated with the ten basic variables (data in Appendix 2, §10).

FORMS

J \ J	WORK₁	WORK₂	WORK₃	TRANSPORT₁	TRANSPORT₂	TRANSPORT₃	HOUSEHOLD₁	HOUSEHOLD₂	HOUSEHOLD₃	CHILDREN₁	CHILDREN₂	CHILDREN₃	SHOPPING₁	SHOPPING₂	SHOPPING₃	PERSONAL CARE₁	PERSONAL CARE₂	PERSONAL CARE₃	MEALS₁	MEALS₂	MEALS₃	SLEEP₁	SLEEP₂	SLEEP₃	TV₁	TV₂	TV₃	LEISURE₁	LEISURE₂	LEISURE₃
WORK₁	10	0	0	8	2	0	0	3	7	1	2	7	0	5	5	4	1	5	2	3	5	5	5	5	4	3	3	2	4	4
WORK₂	0	10	0	2	4	4	4	4	2	4	4	2	2	4	4	4	3	3	6	3	1	3	3	1	4	2	4	7	1	2
WORK₃	0	0	8	0	3	5	6	2	0	4	2	0	8	0	0	5	3	0	3	2	3	1	1	3	2	4	2	1	5	2
TRAN₁	8	4	0	10	0	0	0	2	8	1	1	8	1	5	4	5	2	3	0	4	6	4	4	6	3	4	3	4	3	3
TRAN₂	2	4	3	0	9	0	3	3	1	3	5	1	4	2	3	4	1	4	6	0	3	0	4	3	4	3	2	3	4	2
TRAN₃	0	4	5	0	0	9	7	2	0	7	2	0	5	2	2	4	4	1	5	4	0	2	1	0	3	2	4	3	3	3
HOUS₁	0	4	6	3	7	0	10	0	0	10	0	0	6	5	2	7	2	1	3	4	3	6	6	3	3	3	4	2	4	4
HOUS₂	3	4	2	2	5	2	0	9	0	1	8	0	3	3	3	0	4	5	7	0	2	3	4	2	3	4	2	4	3	2
HOUS₃	7	2	0	8	1	0	0	0	9	0	0	9	1	4	4	6	1	2	1	4	4	1	4	4	4	2	3	4	3	3
CHIL₁	1	4	6	1	3	7	10	1	0	11	0	0	4	5	2	7	2	2	3	4	4	4	1	4	3	4	4	2	4	5
CHIL₂	2	4	2	2	5	2	0	8	0	0	8	0	3	2	3	0	4	4	7	0	1	1	4	1	4	3	2	4	3	1
CHIL₃	7	2	9	8	1	0	0	0	9	0	0	9	1	4	4	6	1	2	1	4	4	3	4	4	4	2	3	4	3	2
SHOP₁	0	2	8	1	4	5	6	1	1	6	3	1	10	0	6	6	4	0	4	2	4	4	5	4	3	4	2	3	5	2
SHOP₂	5	4	0	5	2	2	2	3	0	3	2	4	9	9	0	6	1	3	3	3	3	5	3	1	5	3	2	5	1	3
SHOP₃	5	4	0	4	3	2	2	3	4	2	3	4	0	0	9	1	2	6	2	3	2	2	5	3	2	2	5	2	4	3
CARE₁	4	4	5	5	4	4	7	0	6	6	0	6	6	6	1	13	0	0	3	5	6	5	2	6	6	4	3	3	6	2

FORMS

CARE$_2$	1	3	3	2	1	4	2	4	1	2	4	1	4	1	2	0	7	0	4	2	1	4	1	3	2	3	2	4	1	2
CARE$_3$	5	3	0	3	4	1	5	2	2	2	2	2	0	2	6	0	8	5	1	2	6	3	1	2	4	3	1	3	3	4
MEAL$_1$	2	6	3	0	6	5	3	7	3	4	4	1	4	0	4	5	11	0	5	0	5	3	1	3	0	3	4	4	3	3
MEAL$_2$	3	3	2	4	0	4	4	0	4	4	3	2	3	3	5	2	1	0	8	0	4	3	1	2	3	3	3	4	3	1
MEAL$_3$	5	1	3	6	3	0	3	2	4	1	4	3	4	3	2	6	1	2	0	9	0	1	8	1	5	2	4	2	4	3
SLEE$_1$	0	6	4	0	3	8	6	3	1	6	3	1	5	0	2	5	4	1	6	4	0	10	0	0	4	2	4	5	3	2
SLEE$_2$	5	3	1	4	4	1	1	4	4	4	1	0	5	2	2	2	1	6	5	3	1	0	9	0	5	1	3	3	3	3
SLEE$_3$	5	1	3	6	3	3	2	4	1	4	4	3	2	6	2	2	1	0	2	1	8	0	0	9	1	6	2	2	4	3
TV$_1$	4	4	2	3	4	3	3	3	3	4	3	5	5	2	5	2	6	2	3	1	4	5	1	10	0	0	4	3	3	3
TV$_2$	3	2	4	4	3	2	3	4	4	2	2	2	2	4	2	4	3	2	2	2	5	2	1	6	0	0	3	3	3	3
TV$_3$	3	4	2	3	4	4	4	2	3	3	2	3	5	3	4	3	2	4	5	3	3	4	3	2	0	9	3	3	4	2
LEIS$_1$	2	7	1	4	3	3	2	4	4	3	4	5	2	5	2	5	4	1	4	2	5	3	3	3	2	4	5	10	0	0
LEIS$_2$	4	1	5	3	4	4	3	4	3	4	3	1	4	6	1	4	6	3	3	4	3	3	3	3	4	3	3	0	10	0
LEIS$_3$	4	2	2	3	2	3	2	4	5	2	2	3	2	3	2	2	4	1	3	3	3	2	3	3	4	2	3	0	0	8

(g) T^2 data set; it contains the T^2 associated with each simple contingency data set.

3.2.3. Graphics from Multiple Contingency Data Sets

Since a multiple contingency data set is an arrangement of contingency data sets, the graphics plotted for contingency data sets are available for multiple contingency data sets (*cf.* Chapter 4, Section 2.5).

3.3. Joint N-D Analysis of Mixed Variables

The variables can be quantitative or categorical, and need to be organized for analysis. One way is to transform all the variables into categorical ones as indicated in the previous paragraph (a dummy categorical variable is associated with the quantitative variable). Then, N-D analysis is performed on the resulting set of categorical variables.

4. Cartograms and N-D Analysis

When the units of observation are linked by a contiguity relation as a geographic relation, a map can be drawn with the results given by N-D analysis. If the variables are categorical, profiles can be transferred onto maps; when the variables are quantitative, sun ray plots or profiles diagrams can be transferred onto maps.

Such a representation is given in Fig. 5.10, for the telephone quality of service data set (*cf.* Appendix 2, §14).

Chapter 6 Factor Analysis of Individuals–Variables Data Sets

1. Introduction

The aim of factor analysis is to represent geometrically the information in a data set, in a low dimensional Euclidean space. This information may involve quantitative as well as qualitative variables, whatever the size or complexity of the data set. The fundamental goal of geometric representation is to highlight relations among elements, which are represented by points, as well as to highlight features displayed in factor maps. In the next section, the general principle of factor analysis is presented, followed by the basic technique of any factor analysis method, such as principal components analysis, 2-D and N-D correspondence analyses, and proximities data factor analysis.

2. From Linear Adjustment to Factor Analysis

The technique of linear adjustment is similar to factor analysis, but less complex. In the following example, the relationship between the mechanical resistance of steel and its carbon grade is examined; a series of measurements on steel samples was made (*cf.* Table 6.1). From this data set, a map-graphic is built where the points correspond to the steel samples (*cf.* Fig. 6.1). The x-coordinate is the carbon grade and the y-coordinate is the mechanical resistance. When the data set is plotted, the cloud of points (representing the steel samples) is elongated; the points are close to a straight line whose equation is given by $R = 0.873C + 30.08$. In practice, it is not easy to determine the coefficients a and b in the equation $R = aC + b$. The aim of linear fitting

TABLE 6.1. Steel samples data set.

J \ I	Steel Samples																
C in %	72	55	63	38	10	45	77	67	58	74	65	51	39	27	27	24	32
R kg/cm²	96	82	83	61	39	68	95	92	78	94	38	75	64	53	56	49	57

is to determine the coefficients a and b from the cloud of points only. The straight line corresponding to the equation $R = aC + b$ must be adjusted to fit as closely as possible to this cloud of points. To do this, we use the technique of least square, which gives an optimum solution. The basic principle is as follows: The line D is sought such that the sum of squares of the orthogonal projections of the points on that line is minimum; it has been proved that such a line passes through the centroid of the cloud of points (*cf.* Fig. 6.1).

To study a factor analysis model, this linear fitting technique needs to be modified in a few ways: the first is to study p variables instead of two; the second is to study how these p variables vary simultaneously.

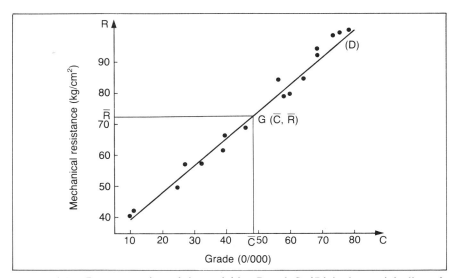

FIGURE 6.1. Representation of the variables R and C; (D) is the straight line of least squares (orthogonal).

2. From Linear Adjustment to Factor Analysis

Consider an example based on the marks given to a set of pupils in six subjects: mathematics, physics, foreign language, history, natural sciences, and English, denoted by $X_1, X_2, \ldots, X_6$, respectively (*cf.* Appendix 2, §2). Each pupil is assigned a mark in each subject. Analyzing the set of variables two by two does not provide a synthetic view of the relations between the subjects. Thus, for example, instead of showing if X_3 (foreign language) is dependent on the other variables X_1, X_2, X_4, X_5, X_6, we want to know how the variables vary simultaneously, and how these variables are dependent on one or several specific common variables, called *factors*, and described as follows:

$$\left.\begin{array}{l} X_1 = a_1 F_1, \\ X_2 = a_2 F_1, \\ X_3 = a_3 F_1, \\ X_4 = a_4 F_1, \\ X_5 = a_5 F_1, \\ X_6 = a_6 F_1, \end{array}\right\} \text{where } F_1 \text{ is the common factor for the whole set of variables.}$$

When the set of variables is dependent on two factors, then:

$$\left.\begin{array}{l} X_1 = a_1 F_1 + b_1 F_2, \\ X_2 = a_2 F_1 + b_2 F_2, \\ X_3 = a_3 F_1 + b_3 F_2, \\ X_4 = a_4 F_1 + b_4 F_2, \\ X_5 = a_5 F_1 + b_5 F_2, \\ X_6 = a_6 F_1 + b_6 F_2, \end{array}\right\} \text{where } F_1 \text{ and } F_2 \text{ are the common factors of the whole set of variables.}$$

In this case, the set of pupils can be summarized taking into account only the common factors that will replace the original variables. The pupils will have the following coordinates: pupil-point $= a_i F_1 + b_i F_2$ for two common factors. There exists a system of two factors representing at best the original cloud of points, which was set in a 6-D space.

To understand the significance of the factors, consider the usual 3-D space with only three of the original variables. Each pupil has three coordinates corresponding to his marks in each subject (*cf.* Fig. 6.2). The centroid is a point whose coordinates are the average of the marks in each subject. If the cloud is spherical around the center, there is no specific factor (each radius could be a factor). If the cloud is elongated along a line passing through the center, then that straight line is known as

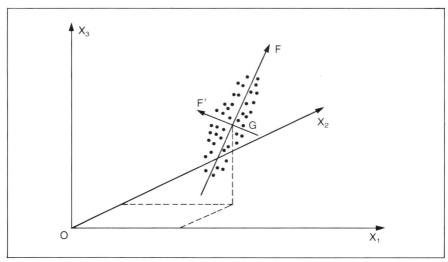

FIGURE 6.2. Representation of a cloud of points in a standard 3-D space.

a *factor axis*. The angles formed by the factor axes F_1, F_2, F_3 and the original axes $\overrightarrow{OX_1}$, $\overrightarrow{OX_2}$, $\overrightarrow{OX_3}$, can be studied; if F_1 is perpendicular to $\overrightarrow{OX_1}$, then X_1 varies little as F_1 varies, and X_1 has a weak relation with F_1; if the angle formed by F_1 and $\overrightarrow{OX_1}$ is small, then X_1 varies strongly with F_1, and X_1 has a strong relation with F_1. Some specific situations can arise:

- when the cloud is given only by a system of two factors F_1 and F_2, the cloud is flat and is elongated more according to F_1 than to F_2; F_1 is the principal direction of elongation and F_1 is the first factor axis; F_2 is the second direction of elongation, and F_2 is the second factor axis;
- when the cloud is reduced to one factor F_1, the cloud is elongated along a straight line; this is the case of linear adjustment.

Consider now the case where the number of variables is not three, but six as given in the initial example on pupils. The cloud of points associated with the data set is located in a 6-D Euclidean space (a point in the 6-D space is associated with each pupil). Several factor axes can be extracted from the cloud of points in such a way that one of them has a larger dispersion than the others. After computing the axes, the pupil-points are projected onto the coordinate space given by these axes.

3. From the Origin of Factor Analysis to Modern Factor Analysis Techniques

Originally, the problem of factor analysis was the following: consider p variables $X_1, X_2, \ldots, X_p$ that are correlated; the set of random variables X_i has to be expressed in terms of a smaller number of random variables, independent of each other two by two, which we call factors and denote by $F_1, F_2, \ldots, F_p$, such that the rests R_i are as small as possible and have little correlation among themselves:

$$X_i = \left(\sum_{j=1}^{p} a_{ij} F_j\right) + R_i.$$

Mathematically, the situation is as follows: With each variable X_i is associated a vector $\overrightarrow{OM}_i$ in the p-D Euclidean space. These vectors form angles two by two, which have as cosines the correlation coefficients of the variables. The first factor axis is the straight line $\overline{OF}_1$ such that the sum of the squares of the distances from the points M_i to that line is minimal. The second factor axis is the straight line, denoted by $\overline{OF}_2$ and perpendicular to $\overline{OF}_1$, such that the sum of the squares of the distances from the points M_i to the plane generated by $\overline{OF}_1$, $\overline{OF}_2$ is minimal. The coordinates X_i are expressed as a linear combination of the factor coordinates $F_1, F_2, \ldots, F_p$. Generally, the extraction of factors is stopped before. So q factor axes are obtained ($q \leq p =$ number of original variables). Mathematically, the computation of factors is related to the diagonalization of a symmetric matrix, which has a unique solution. The coordinates of the variables X_i are computed with respect to the orthonormal set of factor axes. Principal components analysis is basically similar to the general model of factor analysis. How should variables $X_1, X_2, \ldots, X_p$ be determined from a specific data set and which matrix must be diagonalized? For each type of factor analysis, a specific cloud of points exists associated with a metric, and also the variance matrix of the cloud of points which have different properties according to the metrics. The different types of factor analysis will be studied next (Chapters 7, 8, and 9).

4. Mathematical Description of Modern Factor Analysis

4.1. Presentation of the Problem

The basic principle is to give the best summary of a data set, represented by a matrix X, with n rows and p columns and whose the general term is

X_{ij}. To give the best summary of X_{ij} means to reconstruct the np initial values of X by a quantity of numerical values smaller than $n + p$, and such that the reconstruction is correct. In the particular case where X is reconstructed so that:

$$\underset{(n,p)}{X} = \underset{(n,1)}{u_1} \underset{(1,p)}{v_1},$$

where u_1 is the column vector of n components and v_1 is the column vector of p components (v_1' is the row vector of size $(1, p)$); the data set X has been reconstructed with only $(n + p)$ values.

In practice, an approximation of X with rank ≥ 1 is required such that:

$$\underset{(n,p)}{X} = u_1 v_1' + u_2 v_2' + \cdots + u_p v_p' + \underset{(n,p)}{\varepsilon},$$

where $\varepsilon_{(n,p)}$ is a matrix such that ε_{ij} is smaller than the other values $u_1 v_1'$, $u_2 v_2', \ldots, u_p v_p'$.

To solve this problem, we need to associate geometric representations with the matrix X; the n rows are considered as coordinates of vectors in a p-D Euclidean space and the p columns are considered as coordinates of vectors in a n-D Euclidean space.

4.2. Analysis of the Points of $N(I)$ in R^p

4.2.1. Adjustment by a 1-D Vector Subspace of R^p

Assume that a cloud of points $i \in I$ denoted by $N(I)$, associated with X is included in a 1-D vector subspace. The coordinates of the points $i \in I$ will be reconstructed via only one coordinate. Therefore, the vector subspace of R^p giving the best approximation of the cloud of points i of I is required. Assuming that R^p is provided with the usual Euclidean metric, the straight line F_1 passing through the origin that fits the cloud the best is sought. Let u be the unit vector of F_1. u is represented by the column matrix $u_{(1,p)}$. Then, $u'u = 1 (\sum_{j=1}^{p}; u_j^2 = 1)$. The product $X_{(n,p)} u_{(p,1)}$ is a vector whose components are the scalar products of the vectors $\overrightarrow{OM}_i$ with u. Therefore, it gives the coordinates of the points $i \in I$ on the straight line F_1 (cf. Fig. 6.3). Xu is a vector whose components are the lengths of the projections of the n points on the straight line F_1.

We require F_1 such that the sum of the squares of the orthogonal projections is maximal. This is equivalent to seeking F_1 so that $(Xu)'(Xu)$ is the maximum possible with $u'u = 1$ (since $(Xu)'(Xu)$ represents the sum of the lengths of the projections along the line F_1). Thus, to find a vector subspace of dimension 1 giving the best linear fit to the cloud of

4. Mathematical Description of Modern Factor Analysis

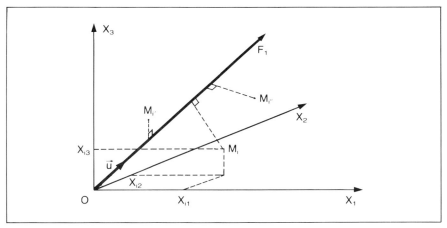

FIGURE 6.3. Representation of the points i of $N(J)$ in R^p.

points according to the least square criterion is equivalent to having $u'X'Xu$ at the maximum possible with $u'u = 1$. If u_1 is the vector that satisfies this maximum condition, then u_1 generates the desired vector subspace F_1 of dimension 1.

4.2.2. Adjustment by a 2-D Vector Subspace of R^p

The vector subspace of dimension 2 is required that best fits the cloud of points $i \in I$ in R^p. Now:

- It is proved that such a vector subspace contains u_1.
- The vector u_2, orthogonal to u_1, is required such that $u_2'X'Xu_2$ is the maximum possible with $u_1'u_1 = 1$, $u_2'u_2 = 1$, and $u_1'u_2 = 0$.

The two vectors u_1 and u_2 obtained generate the desired 2-D vector subspace sought.

4.2.3. Adjustment by a q-D Vector Subspace of R^p

The vector subspace of dimension q is required that best fits the cloud of points $i \in I$ in R^p. It is found inductively:

- It is proved that such a vector subspace contains $u_1, u_2, \ldots, u_{q-1}$.
- The vector u_q is required so that $u_q'X'Xu_q$ is the maximum possible with $u_q'u_q = 1$ and $u_l u_k = 0$ for $l \neq k = 1, 2, \ldots, q$.

The q vectors $u_1, u_2, \ldots, u_{q-1}, u_q$ generate the q-D vector subspace sought.

Only a numerical aspect remains to be solved: To find the maximum of the quadratic form $u'X'Xu'$ under the condition $u'u = 1$. Since $X'X$ is a real symmetric matrix, the eigenvalues $\lambda_1, \lambda_2, \ldots, \lambda_q$ can be extracted from $X'X$, and the corresponding orthonormal basis of eigenvectors $u_1, u_2, \ldots, u_q$ can be formed to generate the eigenspace of dimension q (if the rank r of the matrix $X'X$ is less than q, then the eigenspace will have dimension r). Diagonalization of $X'X$ yields

$$X'Xu_\alpha = \lambda_\alpha u_\alpha \quad \text{for } \alpha = 1, 2, \ldots, q.$$

The eigenvalues are labelled in decreasing order: $\lambda_1, \lambda_2, \ldots, \lambda_q$. λ_1 is associated with u_1, λ_2 with $u_2, \ldots, \lambda_q$ with u_q, and the following relations hold:

$$u_1'X'Xu_1 = \lambda_1 u_1,$$
$$u_2'X'Xu_2 = \lambda_2 u_2,$$
$$\vdots$$
$$u_q'X'Xu_q = \lambda_q u_q.$$

$\lambda_1, \lambda_2, \ldots, \lambda_q$ represent the dispersion along each factor axis.

4.3. Analysis of the Points of $N(J)$ in R^n

In R^n, the data set X also gives a cloud of points, denoted by $N(J)$. Initially, the data set X is transposed, giving X'. We seek the straight line G_1 that best fits the cloud of points $j \in J$ in R^n according to the least square criterion. Let v be the unit vector associated with G_1. The solution is given by the line G_1 such that the sum of the squares of the orthogonal projections on G_1 is maximum with $v'v = 1$. The p points j of J are the row vectors of X' (i.e., the column vectors of X). $X'_{(n,p)}v_{(p,1)}$ is a vector whose components are the scalar products of the vectors $\overrightarrow{OP_j}$ with v, and therefore contain the coordinates of the points $j \in J$ on the straight line G_1 (cf. Fig. 6.4). As in the previous subsection, computing the v_1 associated with G_1 amounts to computing the maximum of $(X'v)'(Xv)$ with $v'v = 1$. Finally, the maximum of the quadratic form $v'XX'v$ is sought under the constraint $v'v = 1$. The solutions are found by diagonalizing XX', giving the eigenvectors $v_1, v_2, \ldots, v_q$ and the eigenvalues $\mu_1, \mu_2, \ldots, \mu_q$ ($q \leq r$, the rank of XX').

4.4. Relations between the Points of $N(I)$ and $N(J)$

Let r be the rank of X; it is known that r is also the rank of $X'X$ and XX'. Consider only the r eigenvectors and eigenvalues with $r \leq$

4. Mathematical Description of Modern Factor Analysis

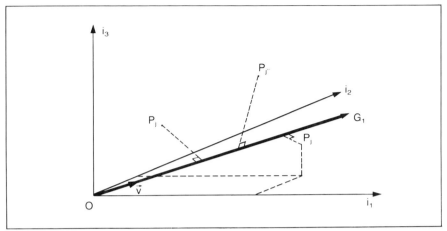

FIGURE 6.4. Representation of the points j of $N(J)$ in R^n.

minimum(n, p). From the two previous subsections, it follows that
$$v'_\alpha XX' v_\alpha = \mu_\alpha \quad \text{for } \alpha = 1, 2, \ldots, r;$$
thus,
$$XX' v_\alpha = \mu_\alpha v_\alpha;$$
so
$$X'(XX') v_\alpha = X'(\mu_\alpha v_\alpha);$$
and
$$(X'X)(X' v_\alpha) = \mu_\alpha v_\alpha.$$

Thus, $X' v_\alpha$ is an eigenvector of $X'X$; since u_α is an eigenvector of $X'X$ (*cf.* Section 4.2), it follows that $X' v_\alpha$ and u_α are collinear and the set of μ_α is included in the set of λ_α.
Thus,
$$u_\alpha = k_\alpha (X' v_\alpha)$$
and
$$\{\mu_\alpha\} \subset \{\lambda_\alpha\}.$$

By analogy, the same can be done for $u'_\alpha X' X u_\alpha$:
$$u'_\alpha X' X u_\alpha = \lambda_\alpha \quad \text{(by definition)},$$
$$X' X u_\alpha = \lambda_\alpha u_\alpha,$$
$$(XX')(X u_\alpha) = X(\lambda_\alpha u_\alpha),$$
$$(XX')(X u_\alpha) = \lambda_\alpha (X u_\alpha).$$

Thus, Xu_α is an eigenvector of XX'; since u_α is an eigenvector of XX', it follows that Xu_α and v_α are collinear. Thus,

$$\begin{cases} v_\alpha = k'(Xu_\alpha), \\ \{\lambda_\alpha\} \subset \{\mu_\alpha\}; \end{cases}$$

$$\Rightarrow \lambda_\alpha = \mu_\alpha \quad \text{for } \alpha = 1, 2, \ldots, r.$$

From the constraints associated with the analyses in R^p and R^n, it follows that

$$u'_\alpha u_\alpha = v'_\alpha v_\alpha = 1 \Rightarrow k_\alpha^2 (X'v_\alpha)'(X'v_\alpha) = 1$$
$$\Rightarrow k_\alpha^2 v'_\alpha X'Xv_\alpha = 1$$
$$\Rightarrow k_\alpha^2 v'_\alpha (\mu_\alpha v_\alpha) = 1 \quad (\text{since } X'Xv_\alpha = \mu_\alpha v_\alpha)$$
$$\Rightarrow \mu_\alpha k_\alpha^2 v'_\alpha v_\alpha = 1$$
$$\Rightarrow k_\alpha^2 = 1/\mu_\alpha \quad (\text{since } \lambda_\alpha = \mu_\alpha)$$
$$\Rightarrow k_\alpha = 1/\sqrt{\mu_\alpha}.$$

This implies the following relations:

$$\begin{cases} u_\alpha = \lambda_\alpha^{-1/2} X' v_\alpha, \\ v_\alpha = \lambda_\alpha^{-1/2} X u_\alpha. \end{cases}$$

They can be described as follows:

$$u_\alpha(j) = \lambda_\alpha^{-1/2} \sum_{i=1,\ldots,n} X_{ij} v_\alpha(i),$$

$$v_\alpha(i) = \lambda_\alpha^{-1/2} \sum_{j=1,\ldots,p} X_{ij} u_\alpha(j).$$

- The straight line F_α generated by the unit vector u_α is called the αth factor axis of R^p; the straight line G_α generated by the unit vector v_α is called the αth factor axis of R^n.
- The coordinates of the points i of $N(I)$ on the α-axis of R^p are the components of the vector Xu_α (by definition).
  The coordinates of the points j of $N(J)$ on the α-axis of R^n are the components of the vector $X'v_\alpha$ (by definition).
- The αth component of the unit vector associated with one space is proportional to the unit vector associated with the other space.

4.5. Reconstruction and Summary of the Data Set

To reposition the data set optimally means to reconstruct the points of the cloud in R^p or in R^n by the new set of coordinates, given by the

4. Mathematical Description of Modern Factor Analysis

factors F_α or G_α, respectively. When u_1 is the only axis of reconstruction, then for a correct repositionning the λ_1 associated with u_1 needs to be distinctly larger than the other eigenvalues $\lambda_2, \lambda_3, \ldots, \lambda_r$. In other words, it means that the cloud is elongated along a straight line. In projection, the variance of the cloud on the F_1-axis, associated with u_1, is $\lambda_1 = u_1' X' X u_1$. Not only the points in R^p or in R^n are repositionned, but the total variance of the cloud in R^p or in R^n is affected by the repositionning. When the first q factors are used to reposition of i in R^p, the first q eigenvectors are associated with $\lambda_1, \lambda_2, \ldots, \lambda_q$; $\lambda_1 + \lambda_2 + \cdots + \lambda_q$ represents a part of variance of the cloud in R^p since $\sum_{\alpha=1}^{r} \lambda_\alpha = M^2(N(I))$. The greater $\sum_{\alpha=1}^{q} \lambda_\alpha$ is, the better the reconstruction of the variance is. The reconstruction will be exact only when r eigenvalues are considered since r is the rank of the matrix X.

For reconstructing the original cloud of points set in R^p or R^n, the coordinates of points on factor axes are needed as well as cosine directions of points or factor axes. Consider again the following formula:

$$v_\alpha = \lambda_\alpha^{-1/2} X u_\alpha.$$

Thus,

$$X u_\alpha = \lambda_\alpha^{1/2} v_\alpha,$$

$$X u_\alpha u_\alpha' = \lambda_\alpha^{1/2} v_\alpha u_\alpha',$$

$$\sum_{\alpha=1}^{p} X u_\alpha u_\alpha' = \sum_{\alpha=1}^{p} \lambda_\alpha^{1/2} v_\alpha u_\alpha',$$

$$X \underbrace{\left(\sum_{\alpha=1}^{p} u_\alpha u_\alpha'\right)}_{=I} = \sum_{\alpha=1}^{p} \lambda_\alpha^{1/2} v_\alpha u_\alpha'.$$

Thus,

$$\boxed{X_{(n,p)} = \sum_{\alpha=1}^{p} \lambda_\alpha^{1/2} v_\alpha u_\alpha',}$$

$$\boxed{X_{ij} = \sum_{\alpha=1}^{p} \lambda_\alpha^{1/2} v_\alpha(i) u_\alpha'(j).}$$

This formula gives the p-reconstruction of the data set X; if only $q \leq p$ eigenvectors are used, then a q-reconstitution is described as follows:

$$\boxed{X_{ij} = \sum_{\alpha=1}^{q} \lambda_\alpha^{1/2} v_\alpha(i) u_\alpha'(j).}$$

The reconstruction is better when q is greater. The quality of the reconstruction in q factors, denoted by τ_q, is measured as follows:

$$\tau_q = \left(\sum_{\alpha=1}^{q} \lambda_\alpha\right) \Big/ \left(\sum_{\alpha=1}^{p} \lambda_\alpha\right)$$

Remark. The differences between the techniques of factor analysis depend on the values of X and the associated metric. These will be studied in the next three chapters (Chapters 7, 8, and 9).

5. Factor Analysis Formulas

(a) Matrices to be diagonalized

X reference individuals–variables data set;

X' transposed data set of X;

$X'X$ and XX' matrices that give the eigenvectors and eigenvalues associated with factor analysis.

$\{\lambda_1, \lambda_2, \ldots, \lambda_q\}$: eigenvalues of $X'X$ and XX';

$\{u_1, u_2, \ldots, u_q\}$: eigenvectors of $X'X$;

$\{v_1, v_2, \ldots, v_q\}$: eigenvectors of XX'.

(b) Coordinates of the points of $N(I)$. $F_\alpha(i) = i^{\text{th}}$ coordinate of i of $N(I)$ on the α-factor axis. The α-coordinates of the points i of I are the components of the vector Xu_α.

(c) Coordinates of the points of $N(J)$. $G_\alpha(j) = j^{\text{th}}$ coordinate of j of $N(J)$ on the α-factor axis. The α-coordinates of the points j of J are the components of the vector $X'v_\alpha$.

(d) Relations between $N(I)$ and $N(J)$

$$v_\alpha(i) = \lambda_\alpha^{-1/2} \sum_{j=1}^{p} X_{ij} u_\alpha(j),$$

$$u_\alpha(j) = \lambda_\alpha^{-1/2} \sum_{i=1}^{n} X_{ij} v_\alpha(i).$$

(e) Data reconstruction

$$X_{ij} = \sum_{\alpha=1}^{p} \lambda_\alpha^{1/2} v_\alpha(i) u'_\alpha(j),$$

$$M^2(N(I)) = \text{total variance of } I = \sum_{\alpha=1}^{p} \lambda_\alpha.$$

Chapter 7 Principal Components Analysis

1. Basic Data Sets

The types of data that can be analyzed by principal components analysis are quantitative variables, which may be continuous, homogeneous or not, and a priori correlated. The corresponding data sets are similar to the one in Table 7.1.

In Appendix 2, some data sets that can be analyzed by principal components analysis are presented.

2. Different Patterns of Principal Components Analysis

The different patterns of principal components analyses are distinguished by the different possible values given to X for the diagonalization of $X'X$. The main patterns are presented in the following sections.

2.1. Centered Principal Components Analysis

Suppose that the variables are heterogeneous with regard to their means, because they are measured in different units. If such a data set was analyzed by the usual factor analysis model, only the effects resulting from the row measures would be seen, due to the variation in the units of measure. To reduce this "size" effect, the matrix X is replaced by $\{(X_{ij} - \bar{X}_j)/\sqrt{n}\}$, where $\bar{X}_j$ is the arithmetic mean of X_j, and $\sqrt{n}$ is a normalization coefficient used for graphical representation. This transformation gives the centered principal components analysis.

TABLE 7.1. Data set $X_{IJ} = \{X_{ij}; i \in I; j \in J\}$; p = number of variables; n = number of units.

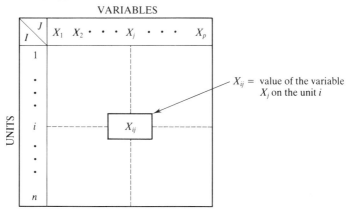

X_{ij} = value of the variable X_j on the unit i

2.2. Standardized Principal Components Analysis

Suppose that the variables are heterogeneous with regard to their means and also their dispersions. These variables are given in noncomparable units of measures, so that a framework of comparison is needed in order to apply the usual factor analysis model. This framework of comparison is given by the following transformation, in which each of the values of the variables X_{ij} is replaced by $(X_{ij} - \bar{X}_j)/\sigma_j\sqrt{n}$, where $\bar{X}_j$ is the arithmetic mean of X_j, σ_j the standard deviation of X_j, and $\sqrt{n}$ is a normalization coefficient used for graphical representations. Thus, the matrix to be diagonalized, $X'X$, is the matrix of the linear coefficients of correlation for the set of variables X_j. This transformation gives the standardized principal components analysis, usually meant by "principal components analysis," and this method is the most widely used.

2.3. Ranks Factor Analysis

Suppose that the values X_{ij} of each variable X_j are unanalyzable as their are, while their associated ranks are analyzable. In this case, X_{ij} is replaced by r_{ij}, where r_{ij} is the rank value of i associated with the variable X_j, giving the following relations: $\bar{X}_j = (n+1)/2$; $\sigma_j^2 = (n^2-1)/12$.

In computer programs it is sufficient to replace X_{ij} by r_{ij} to obtain the result.

3. Standardized Principal Components Analysis

2.4. Concluding Remarks

The use of each of the patterns depends on statistical considerations made by the analyst. Standardized principal components analysis is, however, probably the most often used. So it is studied in detail in what follows.

3. Standardized Principal Components Analysis

3.1. Analysis of the Points i of N(I)

The data set analyzed is $X_{IJ} = \{(X_{ij} - \bar{X}_j)/\sigma_j\sqrt{n}\,; i \in I; j \in J\}$, where X_{ij} is the row value of X_j on the unit i of I.

(a) The cloud of points $N(I)$ is presented in Fig. 7.1.

$$N(I) = \{i \in I; j^{\text{th}} \text{ coordinate of } i = (X_{ij} - \bar{X}_j)/\sigma_j\sqrt{n}\,\}.$$

(b) The distance between each pair of points i and i' of $N(I)$ is

$$d^2(i, i') = \sum_{j=1}^{p} (X_{ij} - X_{i'j})^2/\sigma_j^2 n.$$

Thus, each of the variables X_j gives an equal contribution to the total dispersion of the cloud of points $N(I)$.

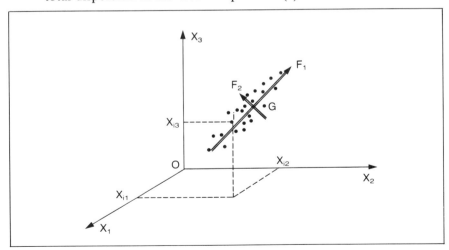

FIGURE 7.1. Geometric representation of $N(I)$ in R^p according to the factor axes F_1 and F_2.

(c) The matrix to be diagonalized is

$$X'X = \left\{ \sigma_{jk} = \sum_{i=1}^{n} (X_{ij} - \bar{X}_j)(X_{ik} - \bar{X}_K)/\sigma_j\sigma_j n; j \in J; k \in K \right\}.$$

XX' is the matrix of the linear coefficients of correlation between variables.

(d) The coordinates of the points i of $N(I)$ on the α-axis are given by the components of the vector Xu_α, where X is the data set $\{(X_{ij} - \bar{X}_j)/\sigma_j\sqrt{n}; i \in I; j \in J\}$ (cf. Chapter 6).

3.2. Analysis of the Points j of $N(J)$

(a) The cloud of points $N(J)$ is presented in Fig. 7.2.

$$N(J) = \{j \in J; i^{\text{th}} \text{ coordinate of } j = (X_{ij} - \bar{X}_j)/\sigma_j\sqrt{n}\}.$$

(b) The distances between any $X_j \equiv j$ and the center O are given by the following:

$$d^2(O, X_j) = d^2(O, j) = \sum_{i=1}^{n} (X_{ij} - \bar{X}_j)^2/\sigma_j^2 n$$

$$= 1.$$

Therefore, the variables X_j (or the points j of $N(J)$) are set on a sphere with radius 1 centred at O.

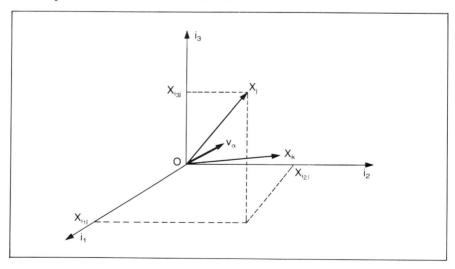

FIGURE 7.2. Geometric representation of $N(J)$ in R^n.

3. Standardized Principal Components Analysis

(c) The distance between X_k and X_j is

$$d^2(X_j, X_k) = d^2(j, k) = \left(\frac{1}{n}\right) \sum_{i=1}^{n} \left(\frac{(X_{ij} - \bar{X}_j)}{\sigma_j} \cdots \cdots \frac{(X_{ij} - \bar{X}_k)}{\sigma_k}\right)$$
$$= \sigma_j^2/\sigma_j^2 + \sigma_k^2/\sigma_k^2 - 2r_{jk}$$
$$= 2(1 - r_{jk}),$$

where r_{jk} is the coefficient of linear correlation between the variables X_j and X_k.

Therefore, the proximity between two variables is given by its linear coefficient of correlation:

$r_{jk} = 1 \Leftrightarrow$ the points j and k are close;

$r_{jk} = -1 \Leftrightarrow$ the points j and k are opposite.

(d) The coordinates of the points j of $N(J)$ on the α-axis result from the construction of the components of the vector $X'v_\alpha (= u_\alpha \lambda_\alpha^{1/2})$. Consider

$$X_j v_\alpha = r_{\alpha j} = (1/n) \sum_{i=1}^{n} v_{\alpha i}((X_{ij} - \bar{X}_j)/\sigma_j)$$

where v_α is a unit vector. In other words, the α-coordinates of the points j of $N(J)$ are the coefficients of correlation of the variable X_j on the α-axis: $r_{\alpha j}$.

3.3. Reconstruction of the Data Set X and the Matrix X'X

The matrix X is reconstruct by the following formula:

$$X_{(n,p)} = \sum_{\alpha=1}^{p} \lambda_\alpha^{-1/2} v_\alpha u_\alpha'.$$

The values of $X'X$ are derived from the values of X.

3.4. Relations between N(I) and N(J)

These relations are based on those in Chapter 6:

$$v_\alpha(i) = \lambda_\alpha^{-1/2} \sum_{j=1}^{p} X_{ij} u_\alpha(j),$$

$$u_\alpha(j) = \lambda_\alpha^{-1/2} \sum_{i=1}^{n} X_{ij} v_\alpha(i).$$

Remark: The values of X_{ij} could be negative because $X_{ij} = (X_{ij} - \bar{X})/\sigma_j\sqrt{n}$.

TABLE 7.2. Activities timetables data set involving 28 population groups related to 10 major activities.

Groups of Population	J	Work	Transport	Household	Children	Shopping	Personal Care	Meals	Sleep	Television	Leisure
Employed men in U.S.A.	EM. USA	610	140	60	10	120	95	115	760	175	315
Employed women in U.S.A.	EW. USA	475	90	250	30	140	120	100	775	115	305
Unemployed women in U.S.A.	UW. USA	10		495	110	170	110	130	785	160	430
Married men in U.S.A.	MM. USA	615	140	65	10	115	90	115	765	180	305
Married women in U.S.A.	MW. USA	179	29	421	87	161	112	119	776	143	373
Single men in U.S.A.	SM. USA	585	115	50		150	105	100	760	150	385
Single women in U.S.A.	SW. USA	482	94	196	18	141	130	96	775	132	336
Employed men in Western countries	EM. WEST	653	100	95	7	57	85	150	808	115	330
Employed women in Western countries	EW. WEST	511	70	307	30	80	95	142	816	87	262
Unemployed women in Western countries	UW. WEST	20	7	568	87	112	90	180	843	125	368
Married men in Western countries	MM. WEST	655	97	97	10	52	85	152	807	122	320
Married women in Western countries	MW. WEST	168	22	528	69	102	83	174	824	119	311
Single men in Western countries	SM. WEST	643	105	72		62	77	140	813	100	388
Single women in Western countries	SW. WEST	429	34	262	14	92	97	147	849	84	392
Employed men in Eastern countries	EM. EAST	650	142	122	22	76	94	100	764	96	334
Employed women in Eastern countries	EW. EAST	578	106	338	42	106	94	92	752	64	228
Unemployed women in Eastern countries	UW. EAST	24	8	594	72	158	92	128	840	86	398
Married men in Eastern countries	MM. EAST	652	133	134	22	68	94	102	763	122	310
Married women in Eastern countries	MW. EAST	436	79	433	60	119	90	107	772	73	231
Single men in Eastern countries	SM. EAST	627	148	68		88	92	86	720	58	463
Single women in Eastern countries	SW. EAST	434	86	297	21	129	102	94	799	58	380
Employed men in Yugoslavia	EM. YUG	630	140	120	15	85	90	105	745	70	365
Employed women in Yugoslavia	EW. YUG	560	105	375	45	90	90	95	815	60	235
Unemployed women in Yugoslavia	UW. YUG	10	10	710	55	145	85	130	760	80	380
Married men in Yugoslavia	MM. YUG	650	145	112	15	85	90	105	775	65	358
Married women in Yugoslavia	MW. YUG	260	52	576	59	116	85	117	760	40	295
Single men in Yugoslavia	SM. YUG	615	125	95		115	90	85	775	45	475
Single women in Yugoslavia	SW. YUG	433	89	318	23	112	96	102	774		408

4. Interpretation of Principal Components Analysis

3.5. Reconstruction of Distances

$$d^2(i, i') = \sum_{\alpha=1}^{p} \lambda_\alpha (v_\alpha(i) - v_\alpha(i'))^2,$$

$$d^2(j, j') = \sum_{\alpha=1}^{p} (u_\alpha(j) - u_\alpha(j'))^2.$$

4. Interpretation of Principal Components Analysis

4.1. Introduction

The principal components u_α can be considered as new variables since they are a linear combination of pairwise, uncorrelated row variables. It has been shown that the points i can be projected onto a space generated by the set of principal components, whose origin is the center of gravity of $N(I)$. It also has been shown that the points j (identifying the row variables) can be projected onto the set of principal components whose origin is O (the origin of the row variables), and that the coordinates of the variables must be interpreted in terms of correlations. The terms principal components, factors, or factor axes are used indiscriminately. Since the clouds $N(I)$ and $N(J)$ have different meanings, their interpretation is distinct, although the results for I and J are visualized on the same graph. The example used in this section is based on activities timetables data set (cf. Appendix 2, §10). The results are presented in the order given by the principal components analysis program used (cf. Chapter 12).

4.2. Basic Data Set Analyzed (cf. Table 7.2)

4.3. Matrix of Correlations (cf. Table 7.3)

TABLE 7.3. Matrix of linear coefficients of correlation (all the coefficients are multiplied by 1000).

	WORK	TRANS	HOLS	CHILD	SHOP	CARE	MEALS	SLEEP	TV	LEISURE
WORK	1000									
TRANSPORT	933	1000								
HOUSEHOLD	−908	−869	1000							
CHILDREN	−870	−809	861	1000						
SHOPPING	−658	−503	501	543	1000					
PERSONAL CARE	−112	−79	−35	124	593	1000				
MEALS	−455	−613	361	367	−184	−360	1000			
SLEEP	−538	−702	433	277	−30	−217	817	1000		
TV	−59	−44	−206	122	216	322	316	11	1000	
LEISURE	−190	−105	−113	−109	235	73	−40	201	−95	1000

4.4. Eigenvalues of Correlation Matrix

Since the sum of eigenvalues is equal to p = number of row variables, the average of the eigenvalues is 1. Therefore, the four principal components whose eigenvalues are greater than 1 are studied first. (cf. Table 7.4)

TABLE 7.4. Table of eigenvalues of $X'X$ and associated parameters:
Column 1 = Num: rank-order of eigenvalues = α.
Column 2 = iter: number of iterations to obtain the numerical result.
Column 3 = eigenvalue: value of λ_α, eigenvalue of $X'X$.
Column 4 = percent: percentage of variance of the principal component α.
Column 5 = cumul: cumulative proportion of variance for 2, then 3, ..., then p principal components.

Num	iter	eigenvalue	percent	cumul
1	0	4.58869171	45.886	45.884
2	1	2.11981201	21.198	67.084
3	1	1.32097912	13.210	80.294
4	2	1.19526100	11.052	92.246
5	1	0.46850914	4.685	96.931
6	1	0.19903100	1.990	98.921
7	4	0.04684873	0.468	99.390
8	2	0.03706244	0.371	99.761
9	1	0.02393039	0.239	100.000
10	2	0.00001172	0.000	100.000

4.5. Factor Coordinates of the Variable-points of N(J)

The coordinates of the points j of J, denoted by $G_\alpha(j)$, are given by the following formula:

$$G_\alpha(j) = r_{\alpha j}, \text{ where } r_{\alpha j} \text{ is the correlation coefficient between the } \alpha\text{-axis and the variable } j.$$

The values of the coordinates are given in Table 7.5, which also contains the contributions of the variables to the variance and to the eccentricity, from which the factors axes can be interpreted.

4. Interpretation of Principal Components Analysis

TABLE 7.5. Factor coordinates and contributions associated with the set J for five first factors axes (all the values are multiplied by 1000).

Label[a]	qlt$_s$[b]	weig[c]	inr[d]	1st factor			2nd factor			3rd factor			4th factor			5th factor		
				G_1[e]	cor$_1$[f]	ctr$_1$[g]	G_2	cor$_2$	ctr$_2$	G_3	cor$_3$	ctr$_3$	G_4	cor$_4$	ctr$_4$	G_5	cor$_5$	ctr$_5$
Work	990	1	100	−977	955	208	121	15	7	−85	7	5	67	4	4	−96	9	20
Transport	978	1	100	−980	960	209	−56	3	2	−8	0	0	46	2	2	111	12	26
Household	988	1	100	900	810	176	−23	1	0	362	131	99	214	46	38	−4	0	0
Children	922	1	100	872	761	166	−179	32	15	84	7	5	294	87	73	190	36	77
Shopping	918	1	100	564	318	69	−761	579	273	−5	0	0	−121	15	12	85	7	16
Personal care	972	1	100	80	6	1	−818	669	316	−302	91	69	−64	4	3	−448	201	429
Meals	976	1	100	588	346	75	669	448	211	−426	182	138	14	0	0	−2	0	0
Sleep	971	1	100	644	415	90	569	324	153	−191	36	28	−312	98	82	−313	98	209
Television	993	1	100	99	10	2	−195	37	18	−930	845	655	152	23	19	241	58	124
Leisure	985	1	100	92	8	2	−110	12	6	30	1	1	−957	917	767	215	46	99
		1000				1000			1000			1000			1000			1000

[a]Label(j) = label associated with the variable j;
[b]qlt$_s(j)$ = quality of representation of the variable j in s factor axes (here, $s = 5$);
[c]weig(j) = weight associated with the variable j;
[d]inr(j) = variance of the variable j with respect to the total variance of the cloud $N(J) = 1$;
[e]$G_\alpha(j)$ = α-coordinate of the variable j = coefficient of correlation between the variable j and the α-axis;
[f]cor$_\alpha(j) = G_\alpha^2(j)$ = relative contribution of the α-axis to the eccentricity of the variable j = cosine squared of the angle formed by the vector radius of j and the α-axis;
[g]ctr$_\alpha(j) = G_\alpha^2(j)/\lambda_\alpha$ = relative contribution of the variable j to the variance of the α-axis;

4.6. Rules for Interpreting Factor Axes by the Variable-points j of $N(J)$

By construction, the principal component u_α (or the factor axis) depends on the variables j that are the most correlated with it. The set of the outermost points correlated with each axis are divided into two parts; one contains the points having positive coordinates and the other the points having negative coordinates. The outermost points j of J are called explicative points of the axes. The explicative points are selected according to the values of ctr_α (cf. Table 7.5). The factors are interpreted as follows:

(1) Interpretation of the first factor axis ($\lambda_1 = 4.58$; $\tau_1 = 45.8\%$).

explicative points with negative coordinates	explicative points with positive coordinates
Household (176) Children (166)	Professional work (208) Transport (209)

The first axis characterizes an opposition between different types of time: on one hand, those including activities outside the home (work and associated transportation); on the other hand, those including activities related to the home (household and children). The numbers given in parentheses are the contributions to the variance denoted by $\text{ctr}_1(j)$.

(2) Interpretation of the second factor axis ($\lambda_2 = 2.11$; $\tau_2 = 21.1\%$).

explicative points with negative coordinates	explicative points with positive coordinates
Shopping (273) Personal care (316)	Meals (211) Sleep (153)

The second factor highlights an opposition between two pairs of activities: one hand, shopping and personal care; on the other hand, meals and sleep. The first is concerned with "modern organized life," and the second with "traditional organized life." The numbers given in parentheses are the contributions to the variance denoted by $\text{ctr}_2(j)$.

4. Interpretation of Principal Components Analysis

(3) Interpretation of the third factor axis ($\lambda_3 = 1.32$; $\tau_3 = 13.2\%$).

explicative points with negative coordinates	explicative points with positive coordinates
Meals (138) Television (655)	Household (99)

The third factor is mainly concerned with television. In 1965, the date of the survey, television was widely used in West European countries and the USA, but not in Eastern European countries. The numbers given in parentheses are the contributions to the variance denoted by $ctr_3(j)$.

(4) Interpretation of the fourth factor axis ($\lambda_4 = 1.19$; $\tau_4 = 11.9\%$).

explicative points with negative coordinates	explicative points with positive coordinates
Leisure (767) Sleep (82)	

It is interesting to note the association of leisure and sleep explaining the fourth factor; this relates to single people who have more leisure time and sleep more than others. The numbers given in parentheses are the contributions to the variance denoted by $ctr_4(j)$.

4.7. Factor Coordinates of the Individual-points of N(I)

The coordinates of the points i of $N(I)$, denoted by $F_\alpha(i)$, are given by the following formula: $F_\alpha(i) = \lambda^{1/2} v_{\alpha i}$. The values of the coordinates are given in Table 7.6, which also contains the contributions to the variance and to the eccentricity, helping with the interpretation of the factor axes.

4.8. Rules for Interpreting Factor Axes by the Individual-points of N(I)

To interpret the dispersion of the axes λ_α, the points i of $N(I)$ that can explain the λ_α are sought. The points i of I, explicative of λ_α, are those that have the greatest values of $ctr_\alpha(i)$, for each λ_α. A subset of points i can then be selected whose contributions in terms of $ctr_\alpha(i)$ are greater than the average of the contributions for the whole set. The subset of explicative points is then divided into two columns: the first has the

Principal Components Analysis

TABLE 7.6. Factor coordinates and contributions associated with the set I for five first factor axes (all the values are multiplied by 1000).

Label[a]	qlt$_x$[b]	weig[c]	inr[d]	1st factor			2nd factor			3rd factor			4th factor			5th factor		
				F_1[e]	cor$_1$[f]	ctr$_1$[g]	F_2	cor$_1$	ctr$_2$	F_3	cor$_3$	ctr$_3$	F_4	cor$_4$	ctr$_4$	F_5	cor$_5$	ctr$_5$
EM. USA	946	1	31	−1773	364	24	−684	54	8	−1871	405	95	576	38	10	854	84	56
EW. USA	996	1	26	−172	4	0	−2215	686	83	−661	61	12	438	27	6	−1252	219	119
UW. USA	964	1	89	4053	658	128	−2278	208	87	−1061	45	30	−520	11	8	1037	43	82
MM. USA	920	1	33	−1779	346	25	−293	9	1	−1885	388	96	734	59	16	1038	118	82
MW. USA	970	1	47	2514	316	53	−2283	395	88	−797	48	17	108	1	0	371	10	10
SM. USA	917	1	33	−1503	246	18	−1892	389	60	−1363	202	50	−782	67	18	355	14	10
SW. USA	990	1	46	−465	17	2	−2844	634	136	−1297	132	45	−147	2	1	−1617	205	199
EM. WEST	994	1	30	−1176	166	11	2368	672	94	−1117	149	34	−46	0	0	−232	6	4
EW. WEST	996	1	17	312	20	1	1495	459	38	−272	15	2	943	183	27	−1248	320	119
UW. WEST	984	1	81	4323	827	145	1633	118	45	−890	35	21	−144	1	1	228	2	4
MM. WEST	989	1	33	−1125	138	10	2464	663	102	−1286	180	45	151	2	1	−218	3	4
MW. WEST	994	1	53	3131	660	76	1989	266	67	−388	23	9	735	36	16	346	8	9
SM. WEST	993	1	36	−1370	188	15	2572	664	111	−327	28	7	−1023	105	31	287	8	6
SW. WEST	988	1	31	1099	140	9	1655	317	46	−344	34	8	−1496	259	67	−1430	237	156
EM. EAST	963	1	20	−2163	834	36	241	10	1	709	90	14	−240	10	2	324	19	8
EW. EAST	989	1	30	−1005	122	8	−180	4	1	1616	315	71	2132	548	136	44	0	0
UW. EAST	947	1	39	3537	753	97	377	9	2	1635	161	72	−530	17	8	276	3	6

4. Interpretation of Principal Components Analysis

	qlt	weig	F_1	cor_1	ctr_1	F_2	cor_2	ctr_2	F_3	cor_3	ctr_3	F_4	cor_4	ctr_4	F_5	cor_5	ctr_5
MM. EAST	971	1	20	-2221	890	38	212	8	1	483	42	6	-110	2	0	397	28	12
MW. EAST	978	1	24	1540	351	18	216	7	1	1622	390	71	1165	201	41	439	29	15
SM. EAST	989	1	45	-2135	360	35	-581	27	6	1609	204	70	-2178	374	142	554	24	23
SW. EAST	958	1	13	-336	30	1	-418	47	3	1490	596	60	-1025	282	31	-120	4	1
EM. YUG	921	1	18	-2147	893	36	69	1	0	131	3	0	322	20	3	148	4	2
EW. YUG	992	1	27	-968	130	8	-585	46	6	1366	249	50	2040	556	124	-268	10	5
UW. YUG	955	1	63	3919	875	120	-49	0	0	672	26	12	-976	54	28	2	0	0
MM. YUG	895	1	21	-2077	743	34	170	5	0	-425	31	5	792	108	19	216	8	4
MW. YUG	959	1	21	490	40	2	-284	7	1	1184	235	38	2012	677	121	-43	0	0
SM. YUG	978	1	42	-2529	543	50	-147	2	0	1062	96	30	-1964	327	115	352	11	9
SW. YUG	933	1	13	-55	1	0	-804	183	11	1002	284	27	-968	265	28	-842	201	54
		1000			1000			1000			1000			1000			1000

[a] Label(i) = label of the point i.
[b] qlt$_s(i)$ = quality of representation of the point i in s factor axes (here, $s = 5$);
[c] weig(i) = weight assigned to the point i;
[d] inr(i) = variance of the point i with respect to the total variance $M^2(N(I))$;
[e] $F_\alpha(i) = \lambda_\alpha v_{\alpha i}$ = α-coordinate of the point i;
[f] $\text{cor}_\alpha(i) = F_\alpha^2(i)/\rho^2(i)$ = relative contribution of the α-axis to the eccentricity of the point i = cosine squared of the angle formed by the vector radius of i and the α-axis;
[g] $\text{ctr}_\alpha(i) = F_\alpha^2(i)/\lambda_\alpha$ = relative contribution of the point i to the variance of the α-axis;

negative coordinates and the second has the positive coordinates. The numbers given in parentheses in the following tables are the contributions to the variance denoted by $\text{ctr}_\alpha(i)$. The factors are interpreted as follows:

(1) Interpretation of the first factor axis ($\lambda_1 = 4.58$; $\tau_1 = 45.8\%$).

explicative points with negative coordinates	explicative points with positive coordinates
EM. YUG (36)	UW. USA (128)
MM. YUG (38)	MW. USA (53)
EM. YUG (36)	UW. WEST (145)
SM. EAST (50)	MW. WEST (76)
MM. YUG (38)	UW. YUG (97)
	UW. EAST (120)

The first factor is explained by the four groups of unemployed women, whatever their country of origin. The contributions of the groups of men are four times less than those of the unemployed women. This feature, associated with the contributions of the points j of J to the first factor, confirms that the first factor represents the type of activity, involving different timetables.

(2) Interpretation of the second factor ($\lambda_2 = 2.11$; $\tau_2 = 21.1\%$).

explicative points with negative coordinates	explicative points with positive coordinates
EW. USA (83)	
UW. USA (87)	
MW. USA (88)	
SW. USA (80)	
SM. USA (136)	
EM. WEST (94)	
EW. WEST (38)	
UW. WEST (45)	
MW. WEST (102)	
SM. WEST (111)	
SW. WEST (46)	
SW. WEST (67)	

The second factor axis is easy to interpret; it separates the Western countries from the Eastern countries; but the highest contributions are given by Western countries' points.

4. Interpretation of Principal Components Analysis

(3) Interpretation of the third factor ($\lambda_3 = 1.32$; $\tau_3 = 13.2\%$).

explicative points with negative coordinates	explicative points with positive coordinates
EM. USA (95)	EW. YUG (71)
MM. USA (96)	UW. YUG (72)
SM. USA (50)	MW. YUG (71)
SW. USA (45)	SM. YUG (70)
MM. WEST (45)	SW. YUG (70)
	EW. EAST (50)
	MW. EAST (38)

The third factor is related to the level of equipment (television) and, to a lesser degree, the level of life (meals) for the two groups of population. The points associated with Yugoslavia are of special interest, and are due to the fact that this country had very few TV sets at the time of the survey.

(4) Interpretation of the fourth factor ($\lambda_4 = 1.19$; $\tau_4 = 11.9\%$).

explicative points with negative coordinates	explicative points with positive coordinates
SW. WEST (67)	EW. YUG (136)
SM. YUG (142)	MW. YUG (41)
SM. EAST (115)	EW. EAST (142)
	MW. EAST (121)

This factor represents the status of single people, although one exception is given by the single population in the USA. At the period of the survey (1965), the single people in USA appreciated TV as well as "out-of-home" leisure. Since the factors are ordered according to their importance in terms of variance, the fourth factor is four times as less important as the first.

4.9. Joint Interpretation of Factor Axes by the Points of N(I) and N(J)

The information selected in Tables 7.5 and 7.6 on the contributions to the factor axes is presented together in Table 7.7, so that the most significant points of the analysis can be seen clearly.

TABLE 7.7. Joint interpretation of the factor axes by the points i of $N(I)$ and j of $N(J)$ that are the most explicative of the variance of the factor axes. The values of the contributions are in the parentheses.

1ˢᵗ factor axis ($\lambda_1 = 4.58$; $\tau_1 = 45.8\%$)	
Household (176), Children (166)	Professional work (208), Transport (209)
EM. YUG (36), MM. YUG (38),	UW. USA (128), MW. USA (53),
EM. YUG (36), SM. EAST (50),	UW. WEST (145) MW. WEST (76),
MM. YUG (38)	UW. YUG (97), UW. EAST (120)

2ⁿᵈ factor axis ($\lambda_2 = 2.11$; $\tau_2 = 21.1\%$)	
Shopping (273), Personal care (316)	Meals (211), Sleep (153)
EW. USA (83), UW. USA (87),	
MW. USA (88), SW. USA (80),	
SM. USA (136), EM. WEST (84)	
EW. WEST (38), UW. WEST (45),	
MW. WEST (102), SM. WEST (111)	
SW. WEST (46), SW. WEST (67)	

3ʳᵈ factor axis ($\lambda_3 = 1.32$: $\tau_3 = 13.2\%$)	
Meals (138), Television (655)	Household (99)
EM. USA (95), MM. USA (86),	EM. YUG (71), UW. YUG (72),
SM. USA (50), SW. USA (45)	MW. YUG (71), SM. YUG (70),
MM. WEST (45)	SM. YUG (70), EW. EAST (50),
	MW. EAST (38)

4ᵗʰ factor axis ($\lambda_4 = 1.19$: $\tau_4 = 11.9\%$)	
Leisure (767), Sleep (82)	
SM. WEST (67), SM. YUG (142),	EW. YUG (136), MW. YUG (41),
SM. EAST (115)	EW. EAST (142), MW. EAST (121)

4.10. Quality of Representation

From a numerical point of view, the last feature to study is the quality of representation of the factor space (generated by the factor axes), and the reconstructing of the positions of i or j in the factor space.

4. Interpretation of Principal Components Analysis

4.10.1. Quality of Representation of the Factor Space

The quality of representation of the factor space, denoted by qge$_s$, is given by

$$\text{qge}_s = \left(\sum_{\alpha=1}^{s} \lambda_\alpha\right) \Big/ \left(\sum_{\alpha=1}^{p} \lambda_\alpha\right) = \sum_{\alpha=1}^{s} \tau_\alpha,$$

where s is the number of factor axes in the factor space.

In the example on the activities timetables, the quality of representation is:

for a 1-D factor space: 45.8%;

for a 2-D factor space: $(45.8 + 21.1)\% = 66.9\%$;

for a 3-D factor space: $(4.8 + 21.1 + 13.2)\% = 80.1\%$;

for a 4-D factor space: $(45.8 + 21.1 + 13.2 + 11.9)\% = 92\%$.

4.10.2. Quality of Representation of the Points in the s-D Factor Space

For each of the points (i or j), we want to measure the quality of representation in the factor space generated by the principal components analysis. It is given as follows:

$$\text{qlt}_s(i) = \sum_{\alpha=1}^{s} \text{cor}_\alpha(i) = \sum_{\alpha=1}^{s} F_\alpha^2(i)/\rho^2(i),$$

$$\text{qlt}_s(j) = \sum_{\alpha=1}^{s} \text{cor}_\alpha(j) = \sum_{\alpha=1}^{s} G_\alpha^2(j)/\rho^2(j).$$

In Tables 7.5 and 7.6, the quality of representation over five factors for the points of $N(I)$ and $N(J)$ is given. Note that the whole set of points is well represented over four or five factor axes.

4.11. Factor Graphics

Recall that the objective of principal components analysis is to represent points in a small dimensional space with respect to the original space, and to highlight relations unseen by computations. To do this, graphics are required. Therefore, we need to plot the points of $N(I)$ and $N(J)$ in a s-D factor space generated by the first s factor axes. Since relations cannot be easily seen in a s-D space, projections of the points are studied in a 2-D space. The factor axes for the 2-D space are chosen from the s factor axes, and related to the quality of representation qge$_s$.

4.11.1. Graphics Associated with the Variable-points of $N(J)$

Since the α-coordinates of the variable points are similar to a correlation coefficient (similar to a cosine square), these points can be projected onto a circle (called a correlation circle) whose axes are pairs of factor axes. The variables are represented in a correlation circle generated by the two first factor axes in Fig. 7.3, and the first and the third axis in Fig. 7.4.

4.11.2. Graphics Associated with the Individual-points of $N(I)$

The points i of $N(I)$ are projected onto a 2-D factor space, and the axes for the factor planes are selected from the s factors. The graphics associated with the factors are seen in Figs. 7.5, 7.6, and 7.7.

4.11.3. Joint Graphics Associated with the Variables of $N(J)$ and the Individuals of $N(I)$

The points of $N(I)$ and $N(J)$ can be projected onto the same graphics, though the meaning of each set is not the same. The proximity between one point i of $N(I)$ and one point j of $N(J)$ cannot be interpreted, however, because the points of $N(I)$ are centered at G (the center of gravity of $N(I)$), whereas the points of $N(J)$ are centered at O (the origin

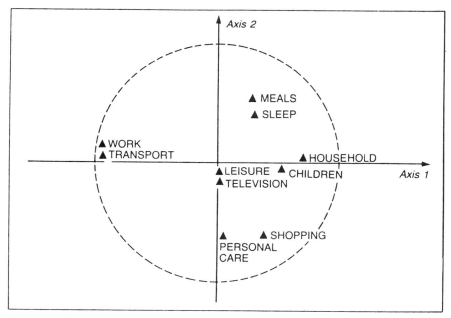

FIGURE 7.3. Activities timetables. Principal components analysis. Representation of the variables in the factor space $(1, 2)$ within the correlation circle.

5. Classifying Supplementary Points into Graphics

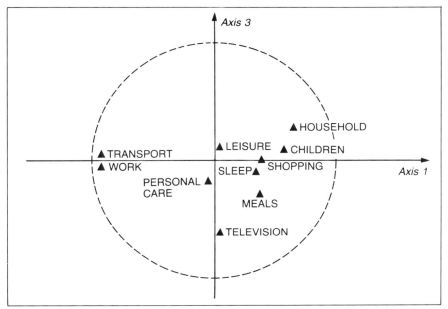

FIGURE 7.4. Activities timetables. Principal components analysis. Representation of the variables in the factor space (1, 3) within the correlation circle.

of the row axes). The joint graphics associated with the pairs of factors (1, 2), (1, 3), and (1, 4), are presented in Figs. 7.8, 7.9, 7.10.

5. Classifying Supplementary Points into Graphics

5.1. Introduction

The most fruitful part of factor analysis techniques is the possibility of introducing supplementary elements (variables or individuals) into factor graphics. Before explaining this feature, its interest is shown using two case models. Consider the first case model in which a set of variables is composed of *explicative variables and variables to be explained*. It is better to analyze first the variables to be explained without including the explicative variables, because the factor axes must be computed without taking the latter into account. If included, they would have an influence on the factor axis, and this is not desired by the analyst who wants to explain a posteriori the relations between variables. The explicative variables are included after as supplementary points. Consider the second case model for an individuals–variables data set; for groups of in-

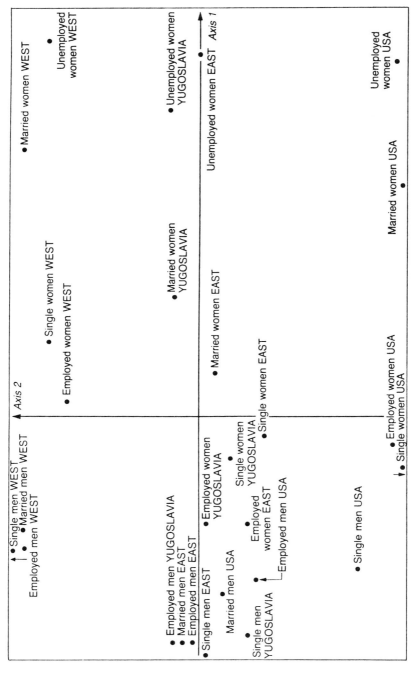

FIGURE 7.5. Activities timetables. Principal components analysis. Representation of the individuals (●) in the factor space (1, 2).

5. Classifying Supplementary Points into Graphics 145

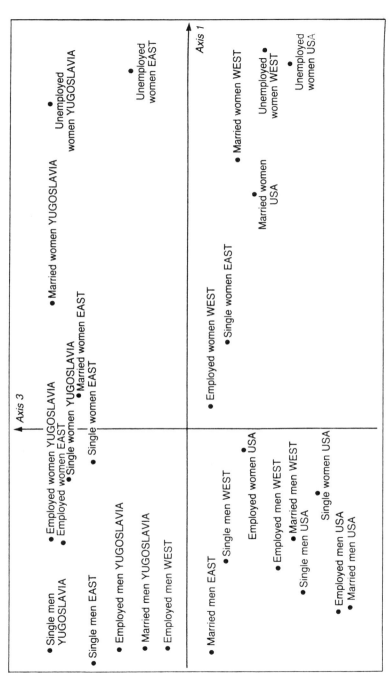

FIGURE 7.6. Activities timetables. Principal components analysis. Representation of the individuals (●) in the factor space (1, 3).

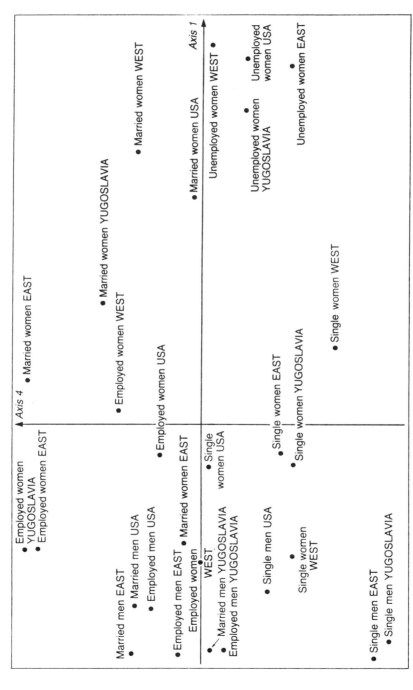

FIGURE 7.7. Activities timetables. Principal components analysis. Representation of the individuals (●) in the factor space (1, 4).

5. Classifying Supplementary Points into Graphics

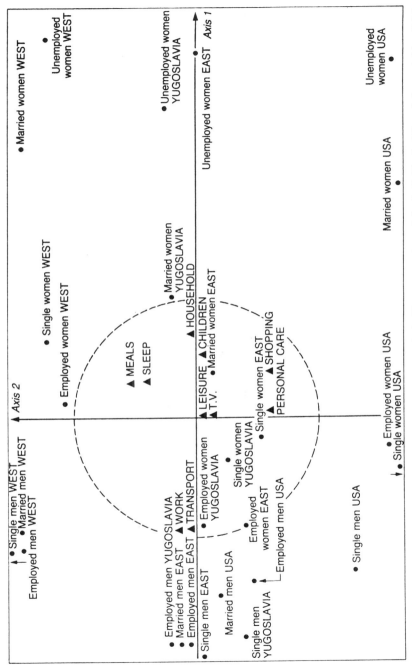

FIGURE 7.8. Activities timetables. Principal components analysis. Representation of the individuals (●) and the variables (▲) in the factor space (1, 2).

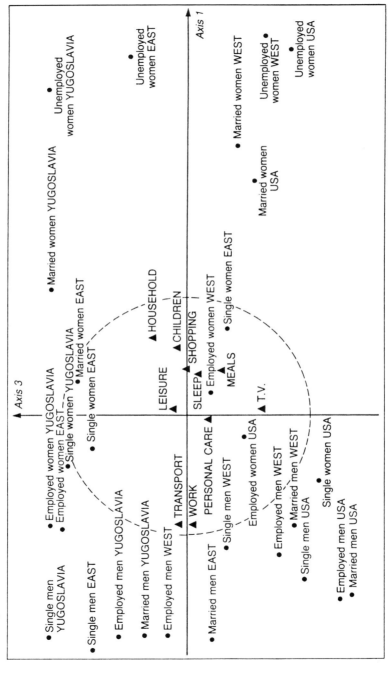

FIGURE 7.9. Activities timetables. Principal components analysis. Representation of the individuals (●) and the variables (▲) in the factor space (1, 3).

5. Classifying Supplementary Points into Graphics 149

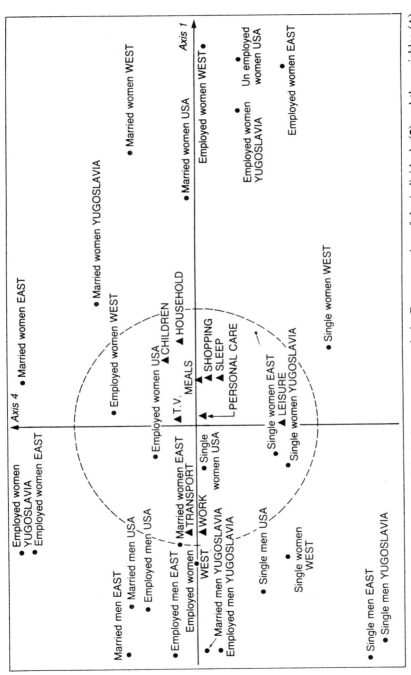

FIGURE 7.10. Activities timetables. Principal components analysis. Representation of the individuals (●) and the variables (▲) in the factor space (1, 4).

dividuals, the positions of the groups needs to be identified by their centers of gravity. After doing factor analysis of the individuals-variables data set (*cf.* Section 5.4), the points "centers of gravity" are introduced in the factors graphics as supplementary individual-points.

5.2. Coordinates of Supplementary Elements

Consider the data set X_{IJ} and two dummy data sets, denoted by X_{SJ} and X_{IT} as represented in Table 7.8.

5.2.1. Factor Coordinates of Supplementary Variables

From the original coordinates of principal variables, the coordinates are normalized as follows:

$$X_{ij} = (X_{ij} - \bar{X})/\sigma_j\sqrt{n}.$$

From the original coordinates of the supplementary variables, the coordinates are normalized as follows:

$$X_{it} = (X_{it} - \bar{X}_t)\sigma_t\sqrt{n}.$$

TABLE 7.8. Representation of principal and supplementary data sets for classifying supplementary individuals or variables.

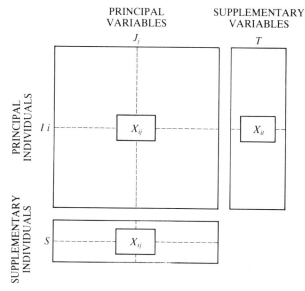

5. Classifying Supplementary Points into Graphics 151

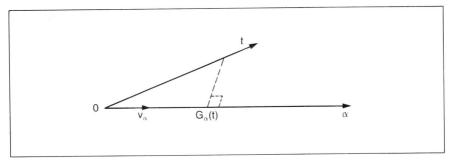

FIGURE 7.11. Coordinate of the supplementary point t with respect to the axis α.

The factor coordinates of the supplementary variables are given by the scalar product $\langle t \cdot v_\alpha \rangle$, with v_α a unit vector. They are also the components of $X'_{it} \cdot v_\alpha$ (cf. Fig. 7.11).

5.2.2. Factor Coordinates of Supplementary Individuals

As in the previous subsection, the coordinates of supplementary individuals are transformed as follows:

$$X_{sj} = (X_{sj} - \bar{X}_j)/\sigma_j \sqrt{n}.$$

Then the factor coordinates of principal individuals are given by the scalar product $\langle s \cdot u_\alpha \rangle$, with u_α a unit vector. They are also the components of $X_{sj} u'_\alpha$ (cf. Fig. 7.12).

5.3. Graphics of Supplementary Elements

The supplementary elements (individuals or variables) are plotted in the factor space generated by explicative factor axes. To avoid confusion,

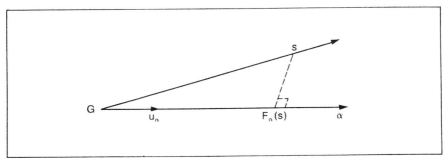

FIGURE 7.12. Coordinate of the supplementary individual s with respect to the axis α.

principal points and supplementary points must be plotted with different symbols.

5.4. Applications

(a) From the activities timetables data set, classes of individuals are formed as follows:

The class of men, denoted by Men;

The class of women, denoted by Women;

The class of married (women or men), denoted by Married;

The class of single people (women or men), denoted by Single;

The class of people from Yugoslavia, denoted by Yugo;

The class of people from USA, denoted by USA;

The class of people from Western countries, denoted by West;

The class of people from Eastern countries, denoted by East;

TABLE 7.9. Principal and supplementary data sets associated with the activities timetables data set.

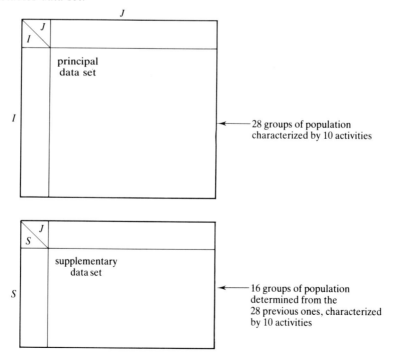

6. Rules for Selecting Significant Axes and Elements

The class of married men, denoted by Married.Men;

The class of married women, denoted by Married.Women;

The class of employed men, denoted by Em.Men;

The class of employed women, denoted by Em.Women;

The class of unemployed women, denoted by Ue.Women;

The class of single men, denoted by Single Men;

The class of single women, denoted by Single Women.

The different classes formed are considered as supplementary individuals; their α-coordinates are computed, and they are plotted with the principal points on the factor graphics (*cf.* Table 7.9).

(b) Coordinates of supplementary individual-points (*cf.* Table 7.10).

(c) Graphics.

The applications of classifying supplementary elements into graphics for the activities timetables example are presented in Figs. 7.13, 7.14, and 7.15.

6. Rules for Selecting Significant Axes and Elements

6.1. Introduction

In practice, it often has happened that data are so numerous that is impossible to process the results of factor analysis, coordinate by coordinate, contribution by contribution, or factor by factor. To avoid hard and tedious work, rules are proposed that can be included in any computer program to obtain automatic help in interpretation. It does not mean an automatic interpretation, but only a selection of significant elements that will help in interpreting data. Here, significant must be understood as meaning relevant or to be studied.

6.2. Rules for Selecting the Number of Significant Factor Axes

6.2.1. Rules for Selecting First Order Significant Factor Axes
Let N be the number of axes to be retained.

Rule 1. N is the number of factor axes such that $\lambda_N \geq 1$.

Rule 2. N is the number of factor axes such that $\sum_{\alpha=1}^{N} \tau_\alpha \geq p$ (where p is the percentage of cumulative variance).

154   **Principal Components Analysis**

TABLE 7.10. α-coordinates of supplementary variables and associated parameters. The values ctr are not relevant for supplementary elements. (The keycodes of columns are given in Table 7.4)

Label	qlt$_s$	weig	inr	1st factor			2nd factor			3rd factor			4th factor			5th factor		
				F_1	cor$_1$	ctr$_1$	F_2	cor$_2$	ctr$_2$	F_3	cor$_3$	ctr$_3$	F_4	cor$_4$	ctr$_4$	F_5	cor$_5$	ctr$_5$
Men	1000	1	14	−1844	878	0	360	32	0	−580	38	0	−308	24	0	315	26	0
Women	1000	1	8	1370	873	0	−261	32	0	200	39	0	234	25	0	−257	31	0
Married	994	1	3	75	8	0	289	112	0	−207	57	0	708	669	0	335	149	0
Single	997	1	9	−907	328	0	−322	41	0	181	13	0	−1198	569	0	−337	45	0
USA	998	1	17	143	4	0	−1780	650	0	−1289	541	0	58	1	0	105	2	0
WEST	998	1	19	746	104	0	2041	776	0	−726	98	0	−128	3	0	−301	17	0
YUGO	961	1	7	−395	78	0	11	0	0	1311	863	0	−109	4	0	291	42	0
EAST	904	1	3	−450	260	0	−223	54	0	709	550	0	179	35	0	−61	4	0
Em Men	994	1	14	−1894	830	0	497	60	0	−544	75	0	156	6	0	282	20	0
Em Women	998	1	11	−442	71	0	392	51	0	301	80	0	1384	634	0	−685	154	0
Married men	999	1	58	3961	970	0	−60	0	0	89	0	0	−542	15	0	397	10	0
Married women	995	1	14	−1890	711	0	635	89	0	−792	138	0	382	32	0	345	26	0
Single men	997	1	18	1949	760	0	−107	2	0	347	24	0	994	196	0	254	13	0
Single women	994	1	21	−1876	589	0	−8	0	0	190	6	0	−1494	374	0	384	25	0
	996	1	8	61	2	0	−598	164	0	161	12	0	−906	377	0	−979	441	0

6. Rules for Selecting Significant Axes and Elements

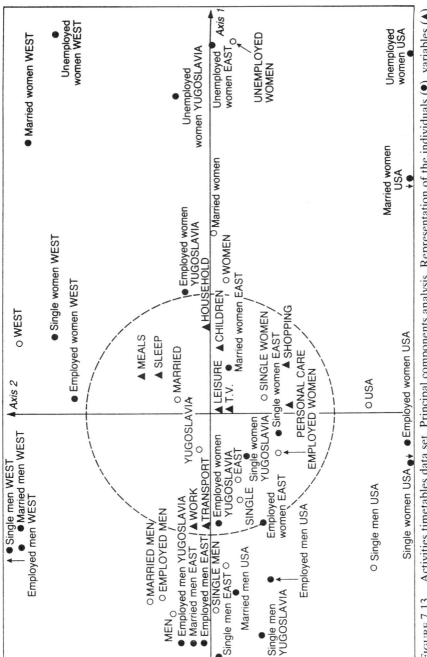

FIGURE 7.13. Activities timetables data set. Principal components analysis. Representation of the individuals (●), variables (▲), and supplementary individuals (○) in the factor space (1, 2).

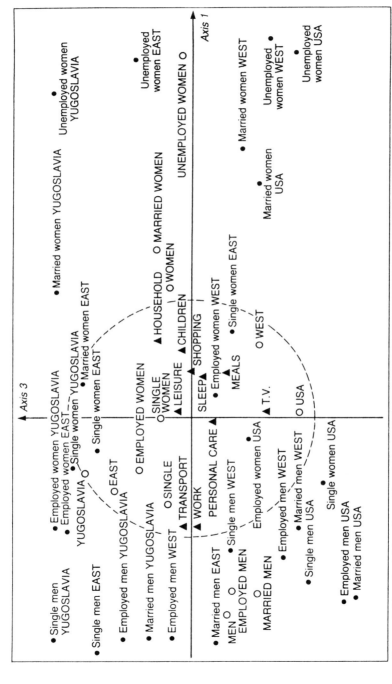

FIGURE 7.14. Activities timetables data set. Principal components analysis. Representation of the individuals (●), variables (▲), and supplementary individuals (○) in the factor space (1, 3).

6. Rules for Selecting Significant Axes and Elements

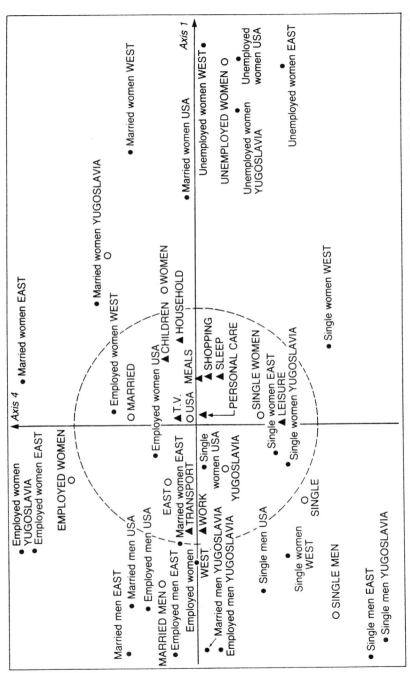

FIGURE 7.15. Activities timetables data set. Principal components analysis. Representation of the individuals (●), variables (▲), and supplementary individuals (○) in the factor space (1, 4).

Rule 3. Consider the values $F_\alpha^2(i)$ and $G_\alpha^2(j)$, ordered in decreasing value, denoted by $O(F_\alpha^2(i))$ and $O(G_\alpha^2(j))$. Consider the following double sums, and determine α such that

$$\sum_i \sum_\alpha O(F_\alpha^2(i)) \geq p$$

and

$$\sum_j \sum_\alpha O(G_\alpha^2(j)) \geq p,$$

where $p = \%$ of variance specified by the analyst. N is taken as equal to α.

The third rule is less applicable than the two first ones. Generally, only one of the two first rules is applied.

6.2.2. Rules for Selecting Second Order Significant Factor Axes

Rule 4. Let N be the rank of the α-axis to be retained. N is chosen so that at least one variable j or one individual i exists such that $\text{cor}_N(j)$ or $\text{cor}_N(i)$ is greater than a given k (k is similar to a cosine squared). This procedure allows us to retain as significant those factor axes that highlight local effects.

At the end of this selection, the user has only a small number of factor axes to study, corresponding to the first order and second order factor axes.

6.3. Rules for Selecting Explicative Elements of Factor Axes

6.3.1. Rules for Selecting Explicative Variables of Factor Axes
The α-coordinates of the variables are expressed as coefficients of correlation. Therefore, the rule for selecting significant variables is based on the variables that are the most correlated with the factor axes.

Rule 5. The variable j is selected as significant when $\text{cor}_\alpha(j)$ is greater than a given value k (k is similar to a cosine squared).

6.3.2. Rules for Selecting Explicative Individuals of Factor Axes
The α-coordinates of the individuals involve the following relations:

$$\sum_{i \in I} F_\alpha^2(i) = \lambda_\alpha.$$

7. Standardized Principal Components Analysis Formulas

The α-axis is dependent on the individuals i with the greatest values $F_\alpha^2(i)$. This gives the following rules:

Rule 6. The individuals i are selected as significant when their values are greater than the mean contribution.

Rule 7. The individuals i are selected as significant when:

$$\sum_{i \in I} F_\alpha^2(i) \geq p,$$

when $F_\alpha^2(i)$ is ordered by decreasing importance, and p is a given value fixed by the analyst (p is similar to a percentage of variance).

6.4. Concluding Remarks

- The selection procedure for interpretation is based on two given values fixed by the user, according to his knowledge of the data processed. Without any experience, p is often taken equal to 80% and k is taken equal to 0.5.
- The rules can be modified through experience.
- To test the robustness of the rules, it is sufficient to study how the interpretation procedures are modified when the values of p or k are changed step by step.

7. Standardized Principal Components Analysis Formulas

(a) Data sets.
$$X_{IJ} = \{(X_{ij} - \bar{X}_j)/\sigma_j \sqrt{n}; i \in I; j \in J\};$$

X_{ij} is the current value from the row data set.

(b) Matrix to be diagonalized.
$$X'X = \left\{ \sigma_{jk} = \sum_{i \in I} (X_{ij} - - \bar{X}_j)(X_{ik} - \bar{X}_k)/\sigma_j \sigma_k n; j \in J; k \in J \right\}.$$

(c) α-coordinates of variables j of J. Components of the vector $X'v_\alpha = u_\alpha \lambda_\alpha^{1/2}$.

(d) α-coordinates of individuals i of I. Components of the vector $Xu_\alpha = v_\alpha \lambda_\alpha^{1/2}$.

(e) Reconstruction of data set.

$$X_{IJ} = \left\{ (X_{ij} - \bar{X}_j)/\sigma_j\sqrt{n} = \sum_{\alpha=1,\ldots,p} \lambda_\alpha^{1/2} u_\alpha(j) v_\alpha(i); i \in I; j \in J \right\},$$

where X_{ij} is the current value of the row data set.

8. Applications and Case Studies

8.1. Study of a Marks Data Set

Consider the marks data set associated with a classroom of pupils; each pupil has a mark in the following subjects: Mathematics, Physics, English, History, Natural Sciences, and Foreign language (*cf.* Appendix 2, §2). The principal components analysis was applied, giving the results shown in Fig. 7.16.

8.2. Study of a Preference Data Set

Based on the previous data set the pupils were requested to order the subjects according to their preferences (1 indicates the favorite subject; 6

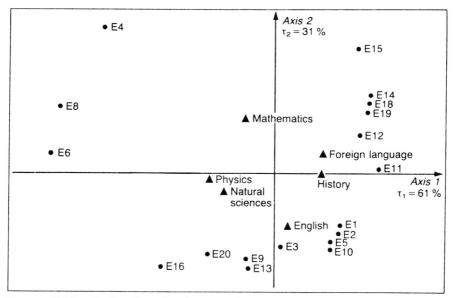

FIGURE 7.16. Study of marks data set. Representation of the pupils and the subjects in a 2-D factor space generated by the factor axes 1 and 2.

8. Applications and Case Studies

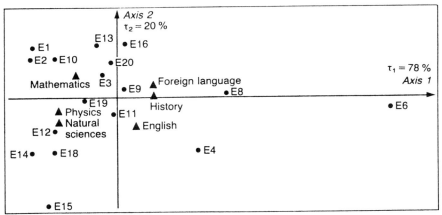

FIGURE 7.17. Study of a preference data set. Representation of the pupils and the subjects in a 2-D factor space generated by the factor axes 1 and 2.

the least favorite; *cf.* Appendix 2, §4). Principal components analysis was applied, giving the results shown in Fig. 7.17.

8.3. Study of a Measurements Data Set

The aim of the study was to classify an unknown skull within a population of dog and wolf skulls. To do this, principal components analysis was applied. The unknown skull was classified into the graphics as a supplementary element (*cf.* Fig. 7.18). The data set is given in Appendix 2, §7, along with a description of the skull variables.

8.4. Applications for Plotting a Statistical Map

It has been seen how a statistical map can be plotted for each variable. The same principle is used here, with a statistical map drawn for each significant factor axis. Such maps are interesting because instead of having as many statistical maps as there are variables, only a few statistical maps are required. They provide a greater synthesis than variable-by-variable maps, because the factor axes are a linear combination of variables. The global effect of the whole set of variables can thus be highlighted. An example is given based on a study of crimes in the USA covering seven crimes variables: Murder, Rape, Robbery, Assault, Burglary, Larceny, and Auto theft (*cf.* Appendix 2, §15). Principal components analysis was applied (*cf.* Fig. 7.19), and a statistical map based on the first factor was plotted (*cf.* Fig. 7.20).

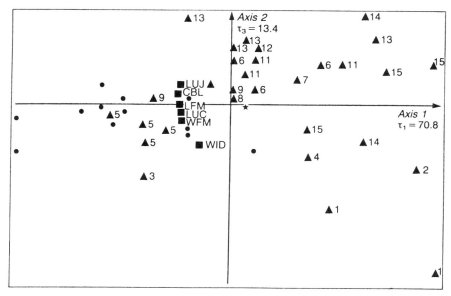

FIGURE 7.18. Study of skulls. Classifying an unknown fossil skull (★). Representation of the dog skulls (●) and wolf skulls (▲) and variables (■) in the 2-D factor space generated by the factor axes 1 and 2.

8.5. Applications to Cartography

The principle of cartography is similar to the plotting of statistical maps though the control unit is the pixel, whose value varies between 0 and 255. A multispectral image given by satellites can be analyzed by principal components analysis and, from the factor axes, a number of classes can be determined; each class is then assigned a color to represent a pixel. A map is then plotted by replacing each basic pixel by its assigned color.

8.6. Applications to Taxonomy and Pattern Recognition

Certain taxonomic studies also use principal components analysis, such as for plant or animal descriptions based on quantitative variables (descriptors or attributes). On these individuals–variables data sets, principal components analysis can be performed to describe the relations between basic variables as well as the relations between individuals. The factor maps allow detection of classes or groups of individuals associated with descriptors. By using a classifying procedure, unknown individuals can be easily introduced into factor graphics and then identified. This is known

8. Applications and Case Studies

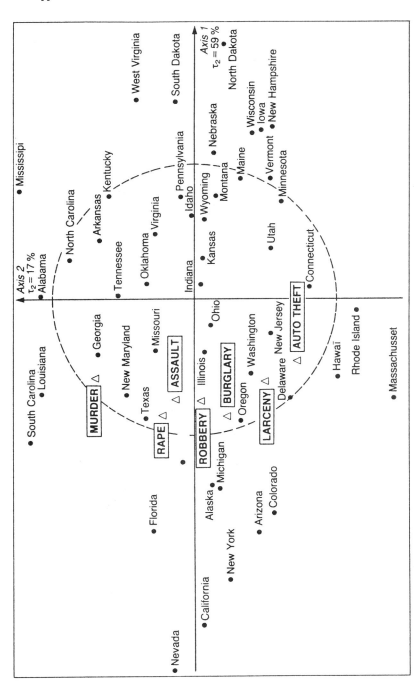

FIGURE 7.19. Principal components analysis of the crimes data set. Representation in the 2-D factor space generated by the factor axes 1 and 2.

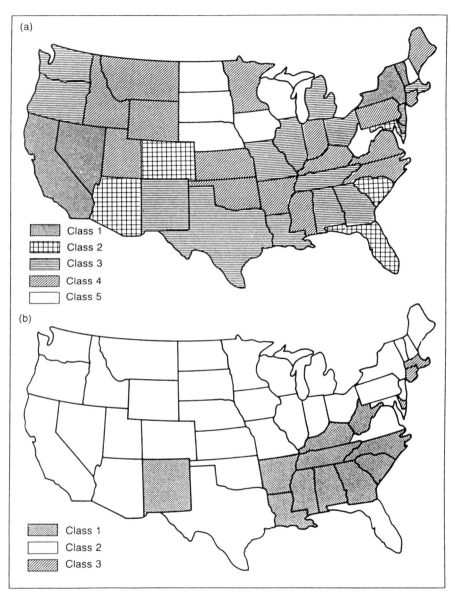

FIGURE 7.20. Statistical maps derived from the principal components analysis of the crimes data set. The maps are plotted according to the first factor (a) and the second factor (b).

8. Applications and Case Studies

as pattern recognition, and it can be described in a few steps; the first step is the analysis of a sample individuals–variables data set; the second step is the reduction step, giving a small number of factors (the factor axes) and the factor maps of individuals; the third step is the recognition step: transforming the original coordinates into factor coordinates, and then computing the exact location of the unknown individual in the factor space.

8.7. Applications to Chronological Data Sets

8.7.1. Introduction

Numerical statistical data sets involve variables that must be studied according to one specific variable: time. This variable can be studied in two ways. It can be considered as a continuous variable, which requires a specific type of analysis known as spectral analysis used, for example, in a time series data set for signal analysis.

It is mainly, however, not considered as a quantitative variable but a qualitative one, because typically, information is known only for specific intervals of time (day, month, year etc.). For certain data sets, the choice of the time intervals obviously depends on the specific problem. These latter data sets are called chronological data sets instead of time series data sets in order to avoid confusion between the two types of analyses.

8.7.2. How to Use Principal Components Analysis

Consider the following data set, denoted by X_{IJT}, where I is a set of regions in France, J is a set a variables observed on these regions, and T

TABLE 7.11. Model of a chronological data set.

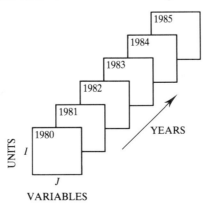

is a set of years:

$$X_{IJT} = \{X_{ijt} : i \in I; j \in J; t \in T\}.$$

The data set X_{IJT} is a 3-entry data set, for which a model is given in Table 7.11.

This chronological data set consists of a series of standard individuals–variables data sets. There are two ways to handle them; the first one is used for describing the positions of each unit for every year. One year is considered as a reference year, for example, the first year t_1. Principal components analysis of the data set X_{IJt_1} is performed, and then all of the other data sets, $X_{IJt_2}, X_{IJt_3}, \ldots, X_{IJt_n}$, are classified into the factor maps as supplementary elements of the reference data analysis (of the t_1 data set). Thus, there are as many points for each region as there are years, so that each region point is labelled by its reference year. A path can therefore be plotted starting with the first region point at year t_1, and finishing at the last region point at year t_n (cf. Fig. 7.21). The second way is used for describing the year-by-year evolution of the variables. As before, one year is taken as a reference year, for example, t_1; principal components analysis is performed, and then the other data sets, X_{IJt_2},

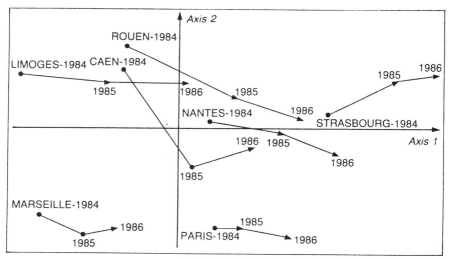

FIGURE 7.21. Example of principal components analysis of chronological data sets. The analysis was performed on the 1984 data set, while the 1985 and 1986 data sets were classified as supplementary elements. A path joining the points 1984, 1985, and 1986 shows the evolution of each region of France related to the two first factor axes. The variables are the quality of service indicators at France Telecom.

8. Applications and Case Studies

TABLE 7.12. Arrangement of chronological data sets in the case of analysis a chronological data set.

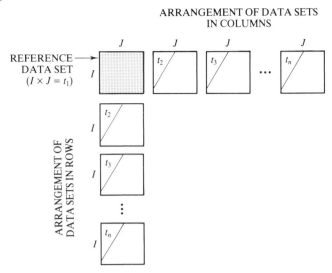

$X_{IJt_3}, \ldots, X_{IJt_n}$, are classified into the factor maps as supplementary elements of the reference data set (the t_1 data set). Thus, there are as many points for each variable as there are years; each variable point is labelled by its reference year. Thus, a path joining each year point of the same variable can be plotted. The difference in these two methods is the way that the data are arranged in order to use supplementary elements; the first needs to have the data sets arranged "in rows" whereas the second "in columns," as shown in Table 7.12. The results of the two methods can be presented on the same graph. These procedures show how principal components analysis can be used profitably in the case of chronological data sets.

Chapter 8

2-D Correspondence Analysis

1. Introduction

The fame of correspondence analysis is associated with Benzecri (1972, 1973, 1980a, 1980b, 1981, 1982, 1984) as well as the statistics department at the University of Paris. The basic model, presented for the first time in 1962 at the Collège de France, concerned the mathematical analysis of contingency data sets in linguistics. After visiting Princeton University and Bell Laboratories, where Shepard and Carroll were working on multidimensional scaling, Benzecri came to the University of Rennes. One of his students presented the mathematical relation known as "distributional equivalence." From a mathematical point of view, this was the beginning of correspondence analysis. At the University of Paris, in 1965, Benzecri then founded the statistics department, where he continued to teach data analysis. From 1965 onwards, correspondence analysis has been applied to various types of data sets related to many fields, e.g., marks data sets, measurements data sets, logical data sets, and questionnaire data sets. The basic method has been improved continually by adding contributions computations, classifying supplementary elements, and recoding data in dummy data sets. The study of two qualitative variables is known as 2-D correspondence analysis (e.g., the analysis of contingency data sets), and the study of N qualitative variables is known as N-D correspondence analysis. It will be shown that quantitative variables need to be recoded into dummy qualitative variables before they are used in correspondence analysis. Since the study of data sets by correspondence analysis is much more involved than that required by any kind of mathematical factor analysis, two chapters are devoted to it. The first describes correspondence analysis of contingency data sets, giving the original basic mathematical model with the original notations. Many examples of applications are presented, and a user's guide to correspondence analysis is given. The second chapter is devoted to N-D correspondence analysis, whose main application is the analysis of questionnaires or similar data sets. A detailed study is presented as an example.

2. Basic Correspondence Data Sets

A correspondence data set is a data set on which correspondence analysis can be applied. We distinguish contingency data sets and specific data sets considered similar to contingency data sets.

Contingency data sets are usually presented as follows: Consider two qualitative variables, denoted here by I and J instead of X and Y as in previous chapters;

The forms $1, 2, \ldots, i, \ldots, n$ are associated with I;
The forms $1, 2, \ldots, j, \ldots, p$ are associated with J.

Thus, there are p forms for J and n forms for I.

Consider the crossed table formed as follows: $A_1, A_2, \ldots, A_n$ are the labels of the rows; $B_1, \ldots, B_p$ are the labels of the columns; then consider the crossed table denoted by $k_{IJ} = \{k_{ij}; i \in I; j \in J\}$, where k_{ij} is

TABLE 8.1. Model of a contingency data set.

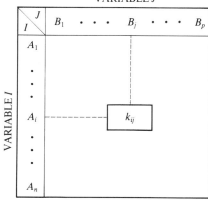

3. Mathematical Description of Correspondence Analysis

the number of units that have simultaneously the values i of A_i and j of B_j. This table k_{IJ} is the frequency data set associated with the two qualitative variables data set, or a correspondence data set as presented in Table 8.1.

Many other data sets that are not contingency data sets can be analyzed by correspondence analysis: marks data sets, logical values data sets, measurements data sets, ratios data sets, etc. In Sections 8.10 and 8.11, examples of such situations are analyzed. However, in the next section, which is devoted to the mathematical presentation of correspondence analysis, only contingency data sets are described so as to avoid any confusion.

3. Mathematical Description of Correspondence Analysis

3.1. Correspondence Data Sets

A correspondence data set is a data table that represents the relationship (called a correspondence) between two sets, denoted by I and J; it is a data table on which correspondence analysis can be performed. An example based on the data set given in Table 8.2 will serve as an illustration of the concerns of the chapter. Consider a set I of countries: {USA, Japan, West Germany, France, Great Britain, Italy, Holland,

TABLE 8.2. Data set of statistics of patents registration by countries for the period 1980–1986 in the economic area of telecommunications.

		YEARS						
	J	1986	1985	1984	1983	1982	1981	1980
I								
	USA	986	774	711	591	467	404	258
	Japan	653	552	361	307	195	129	43
COUNTRIES	FRG	405	357	347	254	295	313	208
	France	189	158	200	171	153	184	147
	Great Britain	204	182	158	137	92	86	67
	Italy	31	28	21	21	22	29	15
	Holland	64	59	61	64	33	30	11
	Sweden	25	19	31	25	15	12	13
	Switzerland	23	34	19	30	17	30	25

Sweden, Switzerland}; and a set of reference years: {1980, 1981, ..., 1986}. The number of patents registered by the country i at the year j is denoted by k_{ij}. The contingency data set k_{IJ} contains the registration of patents during the period 1980 to 1986, showing the relationships between the two sets.

Mathematical analysis consists in describing the evolution of patent registration based on the data set given in Table 8.2.

3.2. Marginal Data Sets

Two 1-D data sets are associated with a correspondence data set: the first, denoted by k_I, is called the row-marginal data set; and the second, denoted by k_J, is called the column-marginal data set:

$$k_I = \{k_i; i \in I\}, \quad k_i = \text{number of units of the form } i,$$

$$k_J = \{k_j; j \in J\}, \quad k_j = \text{number of units of the form } j,$$

$$k_i = \sum_{j \in J} k_{ij}, \quad \text{for } i \in I,$$

$$k_j = \sum_{i \in I} k_{ij}, \quad \text{for } j \in J,$$

$$k = \sum_{i \in I} k_i = \sum_{j \in J} k_j = \sum_{i \in I} \sum_{j \in J} k_{ij}.$$

TABLE 8.3 The correspondence data set k_{IJ} and its associated marginal data sets k_I and k_J.

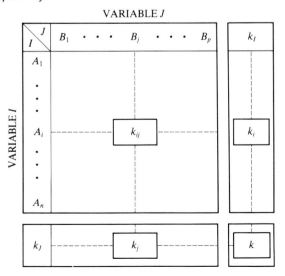

3. Mathematical Description of Correspondence Analysis 173

TABLE 8.4. Correspondence data set. Statistics of patents registration in telecommunications.

	J				YEARS				
I		1986	1985	1984	1983	1982	1981	1980	k_J
COUNTRIES	USA	986	774	711	591	467	404	258	4191
	Japan	653	552	361	307	195	129	43	2239
	FRG	405	357	347	254	295	313	208	2180
	France	189	158	200	171	153	184	147	1202
	Great Britain	204	182	158	137	92	86	67	926
	Italy	31	28	28	21	22	29	15	174
	Holland	64	59	61	64	33	30	11	322
	Sweden	25	19	31	25	15	12	13	140
	Switzerland	23	34	19	30	17	30	25	178
	k_I	2580	2163	1916	1597	1287	1222	787	11552

Usually, k_{IJ}, k_I, and k_J are presented as in Table 8.3; an example is given in Table 8.4.

3.3. Profiles Data Sets

Definition. *The profile of an element i of I on J is the* 1-D *data set, denoted by f^i_J, formed as follows*:

$$f^i_J = \{f^i_j; j \in J\}, \quad \text{with } f^i_j = \frac{f_{ij}}{f_i} \text{ if } f_i \neq 0,$$

$$f^i_j = \frac{k_{ij}}{k_i} \text{ if } k_i \neq 0.$$

Definition. *The profile of an element j of J on I is the* 1-D *data set, denoted by f^j_I, formed as follows*:

$$f^j_I = \{f^j_i; i \in I\}, \quad \text{with } f^j_i = \frac{f_{ij}}{f_j} \text{ if } f_j \neq 0,$$

$$f^j_i = \frac{k_{ij}}{k_j} \text{ if } k_j \neq 0.$$

Definition. *The marginal profile of the set I is the 1-D data set, denoted by f_I, formed as follows:*

$$f_I = \{f_i; i \in I\}, \quad \text{with } f_i = \frac{k_i}{k}.$$

Definition. *The marginal profile of the set J is the 1-D data set, denoted by f_J, formed as follows:*

$$f_J = \{f_j; j \in J\}, \quad \text{with } f_j = \frac{k_j}{k}.$$

Definition. *The transition of I to J is the 2-D data set, denoted by f_J^I, formed by the profiles of i as follows:*

$$f_J^I = \{f_J^i; i \in I\}$$
$$= \{f_j^i; i \in I; j \in J\}.$$

Definition. *The transition of J to I is the 2-D data set, denoted by f_I^J, formed by the profiles of j as follows:*

$$f_I^J = \{f_I^j; j \in J\}$$
$$= \{f_i^j; j \in J; i \in I\}.$$

Definition. *The relative frequency data set is the 2-D data set, denoted by f_{IJ}, formed as follows:*

$$f_{IJ} = \{f_{ij}; i \in I; j \in J\} \quad \text{with } f_{ij} = \frac{k_{ij}}{k}.$$

The previous data sets f_I^i, f_J^i, f_I, f_J, f_J^I, f_I^J, and f_{IJ} are the basic data sets from which the correspondence analysis is performed. From the probability calculus point of view, f_J^i is the conditional frequency of I at $j \in J$, f_J^i is the conditional frequency of J at $i \in I$. Tables 8.5 and 8.6 give the f_I^J and f_J^I data sets for the patent registration study.

3.4. Importance of Profiles

Any element i of I is represented geometrically by a vector of R^p. The j-components of this vector have the values f_j^i, which are the values of the profile of i on J. Conversely, any element j of J is represented

3. Mathematical Description of Correspondence Analysis

TABLE 8.5. The f_I^J, f_I, and f_J data sets associated with the statistics of patents registration in telecommunications.

I \ J	1986	1985	1984	1983	1982	1981	1980	f_J
USA	0.234	0.184	0.169	0.111	0.111	0.096	0.061	0.362
Japan	0.291	0.246	0.161	0.137	0.087	0.057	0.019	0.193
FRG	0.185	0.163	0.159	0.115	0.135	0.146	0.095	0.188
France	0.157	0.131	0.166	0.142	0.127	0.153	0.122	0.140
Great Britain	0.220	0.196	0.170	0.147	0.999	0.928	0.072	0.080
Italy	0.178	0.160	0.160	0.120	0.126	0.166	0.086	0.015
Holland	0.198	0.183	0.189	0.198	0.102	0.093	0.034	0.027
Sweden	0.178	0.135	0.321	0.178	0.107	0.085	0.092	0.012
Switzerland	0.129	0.191	0.106	0.168	0.095	0.168	0.140	0.015
f_I	0.223	0.187	0.165	0.138	0.111	0.105	0.068	

COUNTRIES (row label for I)

TABLE 8.6. f_J^I, f_I, and f_J data sets for the statistics of patents registration in telecommunications.

I \ J	1986	1985	1984	1983	1982	1981	1980	f_J
USA	0.382	0.357	0.370	0.370	0.362	0.330	0.327	0.362
Japan	0.253	0.255	0.188	0.192	0.151	0.104	0.054	0.193
FRG	0.156	0.165	0.181	0.157	0.229	0.261	0.264	0.188
France	0.073	0.073	0.104	0.107	0.118	0.150	0.186	0.140
Great Britain	0.079	0.084	0.082	0.085	0.071	0.070	0.085	0.080
Italy	0.012	0.012	0.012	0.013	0.017	0.023	0.019	0.015
Holland	0.024	0.013	0.031	0.040	0.025	0.024	0.014	0.027
Sweden	0.001	0.001	0.016	0.013	0.001	0.001	0.016	0.012
Switzerland	0.001	0.015	0.001	0.018	0.001	0.003	0.031	0.015
f_I	0.223	0.187	0.165	0.138	0.111	0.105	0.068	

geometrically by a vector of R^n. The i-components of this vector have the values f_i^j, the values of the profile of j on I. The basic model of correspondence analysis involves computations based on the profiles on I and on J. Thus, two elements i and i' of I are considered to be similar if and only if their associated profiles f_J^i and $f_J^{i'}$ are similar, i.e., if the ith row and the i'^{th} row from the contingency data set are proportional. Conversely, two elements j and j' of J are considered as similar if and only if their associated profiles f_I^j and $f_I^{j'}$ are similar, i.e., if the jth column and the j'^{th} column from the contingency data set are proportional.

4. Geometric Representation of the Sets I and J

4.1. Space of Profiles

The set I is represented in the space of profiles on J. Conversely, the set J is represented in the space of profiles on I:

$$\{f_J^i; i \in I\} = \text{set of profiles of } i \in I;$$

$$\{f_I^j; j \in J\} = \text{set of profiles of } j \in J.$$

In what follows, specific spaces of profiles are presented.

Consider the simplex of probability laws on the set J, denoted by $\mathcal{P}(J)$. An element of the simplex is a profile on J. Thus,

$$\pi_J = \{\pi_1, \pi_2, \ldots, \pi_i, \ldots, \pi_n\}$$

such that $\pi_1 + \pi_2 + \cdots + \pi_n = 1$, with $\pi_i \geq 0 \ \forall i \in I$. When the dimension of J is equal to 2, then:

$$\pi_J = \{\pi_1, \pi_2\}, \quad \text{with } \pi_1 + \pi_2 = 1 \text{ and } \pi_1, \pi_2 \geq 0.$$

Thus, the simplex of the probability laws is a line segment, as is shown in Fig. 8.1.

When the dimension of J is equal to 3, then:

$$\pi_J = \{\pi_1, \pi_2, \pi_3\}, \quad \text{with } \pi_1 + \pi_2 + \pi_3 = 1 \text{ and } \pi_1, \pi_2, \pi_3 \geq 0.$$

Thus, the simplex of the probability laws is an equilateral triangle, as is shown in Fig. 8.2.

In the general case, the space of profiles on J is the simplex of probability laws on J because $f_J^i = \{f_1^i, f_2^i, \ldots, f_p^i\}$, with $\sum_{j \in J} f_j^i = 1$ and $f_j^i \geq 0 \ \forall j \in J$.

4. Geometric Representation of the Sets *I* and *J*

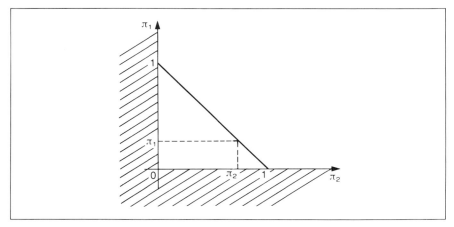

FIGURE 8.1. Simplex of probability laws; dimension of $J = 2$.

4.2. The Clouds of Points $N(I)$ and $N(J)$

Definition. *The cloud of points $N(I)$ is the set of elements of $i \in I$, whose coordinates are the components of the profile $f_J^i = \{f_j^i; j \in J\}$ and whose mass is f_i. $N(I)$ is a subset of $\mathcal{P}(J)$, i.e., the simplex of probability laws on J, itself a subset of R^p.*

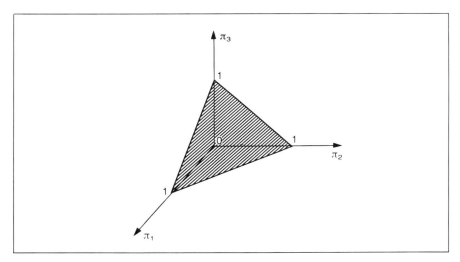

FIGURE 8.2. Simplex of probability laws; dimension of $J = 3$.

Definition. *The cloud of points $N(J)$ is the set of elements of $j \in J$, whose coordinates are the components of the profile $f_i^j = \{f_i^j; i \in I\}$ and whose mass is f_i. $N(J)$ is a subset of $\mathcal{P}(I)$, i.e., the simplex of probability laws on I, itself a subset of R^p.*

Thus,
$$N(I) = \{f_J^i, f_i\} \subset \mathcal{P}(J) \subset R^p,$$

$$N(J) = \{f_I^j, f_j\} \subset \mathcal{P}(I) \subset R^n.$$

A geometric representation of $N(I)$ is given in Fig. 8.3. A similar representation of $N(J)$ can be given in R^n.

4.3. Distance in 2-D Correspondence Analysis

The distance in the space of profiles on J for the set $N(I)$ is defined by

$$d^2(i, i') = d^2(f_J^i, f_J^{i'}) = \sum_{j \in J} \alpha_j (f_j^i - f_j^{i'})^2.$$

The distance in the space of profiles on I for the set $N(J)$ is defined by

$$d^2(j, j') = d^2(f_I^j, f_I^{j'}) = \sum_{i \in I} \alpha_i (f_i^j - f_{i'}^j)^2.$$

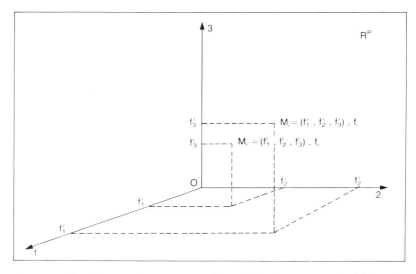

FIGURE 8.3. Geometric representation of the elements i of $N(I)$ in R^p.

4. Geometric Representation of the Sets I and J

The coefficients α_i and α_j must be computed. To determine them the following *principle*, often called *distributional equivalence principle*, is used.

- *If two rows (i and i') of I of the data table k_{IJ} are proportional and if they are replaced by only one, which is the sum, column-by-column, then the distances between columns ($j \in J$) are not changed in $N(J)$.*
- *If two columns (j and j') of J of the data table k_{IJ} are proportional and if they are replaced by only one, which is the sum, row-by-row, then the distances between rows ($i \in I$) are not changed in $N(I)$.*

From this principle, the coefficients α_i and α_j are determined as follows, based on the data table k_{IJ} with two proportional columns j and j'. The columns j and j' are replaced by a column, denoted by j_0, to give:

$$J_0 = J - \{j, j'\} \cup \{j_0\}.$$

The distances are computed in $N(I)$ from J_0 and J:

$$d_J^2(i, i') = \sum_{j \in J} \alpha_j (f_j^i - f_j^{i'})^2;$$

$$d_{J_0}^2(i, i') = \sum_{j \in J} \alpha_j (f_j^i - f_j^{i'})^2 - \alpha_j (f_j^i - f_j^{i'})^2$$

$$- \alpha_{j'} (f_{j'}^i - f_{j'}^{i'})^2$$

$$+ \alpha_{j_0} (f_{j_0}^i - f_{j_0}^{i'})^2,$$

with $f_{j_0} = f_j + f_{j'}$ and $f_{ij_0} = f_{ij} + f_{ij'}$. Then

$$d_J^2(i, i') = d^2(i, i') \Leftrightarrow \left(\alpha_j = \frac{1}{f_j}\right).$$

A parallel relation exists when the roles of I and J are interchanged:

$$d_I^2(j, j') = d_{I_0}^2(j, j') \Leftrightarrow \left(\alpha_i = \frac{1}{f_i}\right).$$

So, in 2-D correspondence analysis, the distances are the following:

$$\boxed{\begin{aligned} d^2(i, i') &= d^2(f_J^i - f_J^{i'})^2 = \sum_{j \in J} (f_j^i - f_j^{i'})^2 / f_j; \\ d^2(j, j') &= d^2(f_I^j - f_I^{j'})^2 = \sum_{i \in I} (f_i^j - f_i^{j'})^2 / f_i. \end{aligned}}$$

From the probability calculus point of view, the distance $d^2(i, i')$ is the χ^2 distance between conditional frequency distributions with respect to the frequency distribution f_J. The analogous statement holds for $d^2(j, j')$.

4.4. Analysis of the Points i of N(I)

4.4.1. Introduction

The general analysis given in Chapter 6 may only be applied after a mathematical transformation. Since the profiles are Euclidean coordinates, what we seek is a new system of Euclidean coordinates for which the general analysis can be applied.

Consider the following transformation:

$$\forall i \in I, \quad \forall j \in J \quad f_j^i \text{ is transformed into } f_j^i f_j^{-1/2}.$$

Thus,

$$d^2(i, i') = \sum_{j \in J} (f_j^i f_j^{-1/2} - f_j^{i'} f_j^{-1/2})^2 = \sum_{j \in J} (f_j^i - f_j^{i'})^2 / f_j;$$

therefore, $\{f_j^i f_j^{-1/2}; j \in J\}$ is a set of Euclidean coordinates. A new cloud of points i, denoted by $N'(I)$, is thus determined:

$$N'(I) = \{i \in I; f_j^i f_j^{-1/2}, \text{ mass} = f_i\} \subset \mathcal{P}(J) \subset R^p.$$

The centers of gravity of $N(I)$ and $N'(I)$, denoted by G and G', respectively, are computed:

G has $\{g_1, g_2, \ldots, g_j, \ldots, g_p\}$ as coordinates in R^p;

G' has $\{g'_1, g'_2, \ldots, g'_j, \ldots, g'_p\}$ as coordinates in R^p.

It is known that

G has $\{\{f_1, f_2, \ldots, f_j, \ldots, f_p\}$ as values,

because f_j is the marginal distribution associated with the profile $\{f_j^i; i \in I\}$.

For G', the following holds:

$$g'_j = \frac{\sum_{i \in I} f_i (f_j^i f_j^{-1/2})}{\sum_{i \in I} f_i}$$

$$= \sum_{i \in I} f_i \left(\frac{f_{ij}}{f_i}\right) f_j^{-1/2}$$

$$= \sum_{i \in I} f_{ij} f_j^{-1/2}$$

$$= f_j f_j^{-1/2}$$

$$= f_j^{1/2} = \sqrt{f_j}.$$

4. Geometric Representation of the Sets I and J

Thus,

$$G' \text{ has } \{\sqrt{f_1}, \sqrt{f_2}, \ldots, \sqrt{f_j}, \ldots, \sqrt{f_p}\} \text{ as values.}$$

The coordinates of the points i of $N(I)$ are computed with respect to G'.

Consider $\overrightarrow{G'M_i} = \{(f_j^i f_j^{-1/2} - f_j^{1/2}); \ j = 1, 2, \ldots, p\}$; the mass associated with the point i is f_i.

Now, the general analysis can be applied to the data set X, where X_{ij} is defined as follows:

$$\boxed{X_{ij} = (f_j^i f_j^{-1/2} - f_j^{1/2}) \quad \text{for } i = 1, 2, \ldots, n \text{ and } j = 1, 2, \ldots, p}$$

According to the general analysis, we require the vector u such that $\sum_{i=1}^n f_i \psi_i^2$ is the maximum possible $\psi_i = \overrightarrow{G'M_i} \cdot u$ and $\|u\| = 1$;

$$\overrightarrow{G'M_i} \cdot u = \sum_{j=1}^p f_i^{1/2}(f_j^i f_j^{-1/2} - f_j^{1/2}) u_j.$$

4.4.2. The Matrix to be Diagonalized

The matrix $X'X$, denoted by T (with general term $t_{jj'}$) is diagonalized:

$$t_{jj'} = \sum_{i=1}^n f_i (f_j^i f_j^{-1/2} - f_j^{1/2})(f_{j'}^i f_{j'}^{-1/2} - f_{j'}^{1/2}),$$

giving the eigenvalues λ_α and the eigenvectors u_α. Using certain mathematical properties, however, the eigenvectors and eigenvalues can be computed in a simpler way.

Notice that $u_0 = \{f_1^{1/2}, f_2^{1/2}, \ldots, f_j^{1/2}, \ldots, f_p^{1/2}\}$ is the eigenvector of T associated with the eigenvalue $\lambda_\alpha = 0$. Consider the equation $Tu_0 = 0$: j being given,

$$\sum_{j' \in J} t_{jj'} f_{j'}^{1/2} = 0.$$

Now, $u_0 = \{f_1^{1/2}, f_2^{1/2}, \ldots, f_j^{1/2}, \ldots, f_p^{1/2}\}$ is orthogonal to any eigenvector u' of T:

$$u_0 \cdot u' = \sum_{j=1}^p u_0 u'_j = \sum_{j=1}^p u'_j f_j = 0.$$

Consider now the following matrix, denoted by T^*, with general term $t_{jj'}^*$:

$$t_{jj'}^* = \sum_{i=1}^n (f_{ij} f_i^{-1/2} f_j^{-1/2})(f_{ij'} f_i^{-1/2} f_{j'}^{-1/2}).$$

Now, $T^* = Y'Y$, where Y is a matrix with general term y_{ij}:

$$y_{ij} = f_{ij} f_i^{-1/2} f_j^{-1/2} = \frac{f_{ij}}{\sqrt{f_i f_j}}.$$

It can be proved that $u_0 = \{f_1^{1/2}, f_2^{1/2}, \ldots, f_j^{1/2}, \ldots, f_p^{1/2}\}$ is the eigenvector of T^* associated with the eigenvalue $\lambda_0 = 1$. It follows that the coordinates of i of $N(I)$ on the α-axis, with respect to the eigenvector u_α and the eigenvalue λ_α, have as values

$$\boxed{\psi_{\alpha i} = \sum_{j=1}^{p} (f_j^i f_j^{-1/2}) u_{\alpha j},}$$

where $u_{\alpha j}$ is the jth component of the eigenvector u_α. These values replace the previous ones:

$$\psi_{\alpha i} = \sum_{j=1}^{p} (f_j^i f_j^{-1/2} - f_j^{1/2}) u_{\alpha j}.$$

The coordinates of the points i of $N(I)$ are now computed with respect to the new system of axes formed by the eigenvectors u_α of the matrix T^*, and so the analysis of the points i of $N(I)$ is finished.

4.5. Analysis of the Points i of N(I). Summary

Analysis in R^p	
diagonalized matrix $T = X'DX$ with $D = \text{diag}(f_i)$ $X_{ij} = f_j^i f_j^{-1/2} - f_j^{1/2}$	diagonalized matrix $T^* = Y'Y$ $Y_{ij} = f_{ij} f_i^{-1/2} f_j^{-1/2}$
u_0 eigenvector of T $\lambda_0 = 0$ eigenvalue of T $u_\alpha\ (\alpha \neq 0)$ eigenvector of T $\sum_{j=1}^{p} u_{\alpha j} f_j^{1/2} = 0$	u_0 eigenvector of T^* $\mu_0 = 1$ eigenvalue of T^* $u_\alpha\ (\alpha \neq 0)$ eigenvector of T^*
λ_α eigenvalue of T (for $\alpha \neq 0$)	$\mu_\alpha = \lambda_\alpha$ eigenvalue of T^* (for $\alpha \neq 0$)
coordinates of i on the α-axis $\psi_{\alpha i} = \sum_{j=1}^{p} (f_j^i f_j^{-1/2} - f^{1/2}) u_{\alpha j}$	coordinates of i on the α-axis $\psi_{\alpha i} = \sum_{j=1}^{p} (f_j^i f_j^{-1/2}) u_{\alpha j}$

4. Geometric Representation of the Sets I and J

4.6. Analysis of the Points j of $N(J)$

Since the sets I and J play symmetrical roles, the analysis of the points j of $N(J)$ is easily deduced from that of the points i of $N(I)$. The different steps of the analysis are given with no details.

4.6.1. Preliminary Transformation

$\forall i \in I$, $\forall j \in J$, f_i^j is transformed into $f_i^j f_i^{-1/2}$.

$\{f_i^j f_i^{-1/2}; i \in I\}$ is a set of Euclidean coordinates.

$N'(J) = \{j \in J; f_i^j f_i^{-1/2}, \text{mass} = f_j\} \subset \mathcal{P}(I) \subset R^n$.

The center of gravity H' of $N'(J)$ has as coordinates:

$$H' = \{\sqrt{f_1}, \sqrt{f_2}, \ldots, \sqrt{f_i}, \ldots, \sqrt{f_n}\}.$$

The coordinates of the points i of $N'(J)$ with respect to H' are

$$\overrightarrow{H'P_j} = \{f_i^j f_i^{-1/2} - f_i^{1/2}; i = 1, 2, \ldots, n\}.$$

The masses associated with the points j of J are f_j. Thus,

$$\boxed{X_{ij} = f_i^j f_i^{-1/2} - f_i^{1/2} \quad \text{for } i = 1, 2, \ldots, n \text{ and } j = 1, 2, \ldots, p}$$

4.6.2. The Matrix to be Diagonalized

According to the general analysis, we require the vectors v so that $\sum_{j=1}^{p} f_j \phi_j^2$ is the maximum possible with $\phi_j = \overrightarrow{H'P_j} \cdot v$ and $\|v\| = 1$;

$$\phi_j = \overrightarrow{H'P_j} \cdot v = \sum_{i=1}^{n} (f_i^j f_i^{-1/2} - f_i^{1/2}) u_i.$$

Analogously to the previous subsection, the matrix $S = XDX'$ is computed and diagonalized:

$$S_{ii'} = \sum_{j=1}^{p} f_j \underbrace{(f_i^j f_i^{-1/2} - f_i^{1/2})}_{(H'P_i)_j} \underbrace{(f_{i'}^j f_{i'}^{-1/2} - f_{i'}^{1/2})}_{(H'P_{i'})_j}.$$

From S, the eigenvalues μ_α and the eigenvectors v_α are obtained. Notice that $v_0 = \{f_1^{1/2}, f_2^{1/2}, \ldots, f_i^{1/2}, \ldots, f_n^{1/2}\}$ is the eigenvector of S associated with the eigenvalue μ_0, and that v_0 is orthogonal to any other eigenvector v' of S.

Consider the following matrix, denoted by S^*, with general term $s_{ii'}^*$:

$$s_{ii'}^* = \sum_{j=1}^{p} (f_{ij} f_i^{-1/2} f_j^{-1/2})(f_{i'j} f_{i'}^{-1/2} f_j^{-1/2}).$$

Now, $S^* = YY'$ where Y is a matrix of general term y_{ij}:

$$y_{ij} = f_{ij} f_i^{-1/2} f_j^{-1/2} = \frac{f_{ij}}{\sqrt{f_i f_j}}.$$

It can be proved that $v_0 = \{f_1^{1/2}, f_2^{1/2}, \ldots, f_i^{1/2}, \ldots, f_n^{1/2}\}$ is the eigenvector of S^* associated with the eigenvalue $\mu_0 = 1$. It follows that the coordinates of j of $N(J)$ on the α-axis, with respect to the eigenvector v_α and the eigenvalue μ_α, have as values

$$\boxed{\phi_{\alpha j} = \sum_{i=1}^{n} (f_i^j f_i^{-1/2}) v_{\alpha i},}$$

where $v_{\alpha i}$ is the ith component of the eigenvector v_α. These values replace the previous ones:

$$\boxed{\phi_{\alpha j} = \sum_{i=1}^{n} (f_i^j f_i^{-1/2} - f_i^{1/2}) v_{\alpha i}.}$$

Since the coordinates of the points j of $N(J)$ are now computed with respect to the new system of axes formed by the eigenvectors v_α of the matrix S^*, the analysis of the points j of $N(J)$ is finished.

4.7. Analysis of the Points j of $N(J)$. Summary

Analysis in R^n	
diagonalized matrix	diagonalized matrix
$S = XDX'$	$S^* = Y'Y$
with $D = \text{diag}(f_I)$	with
$X_{ij} = f_i^j f_i^{-1/2} - f_i^{1/2}$	$Y_{ij} = f_{ij} f_i^{-1/2} f_j^{-1/2}$
v_0 eigenvector of S	v_0 eigenvector of S^*
$\lambda_0 = 0$ eigenvalue of S	$\mu_0 = 1$ eigenvalue of S^*
v_α ($\alpha \neq 0$) eigenvector of S	v_α ($\alpha \neq 0$) eigenvalue of S^*
$\sum_{i=1}^{n} v_{\alpha i} f_i^{1/2} = 0$	
λ_α ($\alpha \neq 0$) eigenvalue of S	$\mu_\alpha = \lambda_\alpha$ ($\alpha \neq 0$) eigenvalue of S^*
coordinates of j on the α-axis	coordinates of j on the α-axis
$\phi_{\alpha j} = \sum_{i=1}^{n} (f_i^j f_i^{-1/2} - f_i^{1/2}) v_{\alpha i}$	$\phi_{\alpha j} = \sum_{i=1}^{n} (f_i^j f_i^{-1/2}) v_{\alpha i}$

4. Geometric Representation of the Sets I and J

4.8. Relations Between the Points i of $N(I)$ and j of $N(J)$

From the general analysis (*cf.* Chapter 6), the following results are recalled: Let T^* and S^* be such that $T^* = Y'Y$ and $S^* = YY'$. Then T^* and S^* have the same nonzero eigenvalues; consider $\{v_\alpha\}$, the eigenvectors of S^* and $\{u_\alpha\}$, the eigenvectors of T^*. Then,

$$\begin{cases} v_\alpha = \lambda^{-1/2} Y u_\alpha, \\ u_\alpha = \lambda^{-1/2} Y' v_\alpha. \end{cases}$$

In developing these relations, the following are obtained:

$$\begin{cases} v_{\alpha i} = \lambda_\alpha^{-1/2} \sum_{j=1}^{p} (f_{ij} f_i^{-1/2} f_j^{-1/2}) u_{\alpha j} & \text{for } i = 1, 2, \ldots, n, \\ u_{\alpha j} = \lambda_\alpha^{-1/2} \sum_{i=1}^{n} (f_{ij} f_i^{-1/2} f_j^{-1/2}) v_{\alpha i} & \text{for } j = 1, 2, \ldots, p \end{cases}$$

Since

$$\psi_{\alpha i} = \sum_{j=1}^{p} (f_j^i f_j^{-1/2}) u_{\alpha j} \qquad \text{for } i = 1, 2, \ldots, n,$$

it follows that

$$\psi_{\alpha i} = \lambda_\alpha^{1/2} f_i^{-1/2} v_{\alpha i} \qquad \text{for } i = 1, 2, \ldots, n.$$

Also, since

$$\phi_{\alpha j} = \sum_{i=1}^{n} (f_i^j f_i^{-1/2}) v_{\alpha i} \qquad \text{for } j = 1, 2, \ldots, p,$$

we have

$$\phi_{\alpha j} = \lambda_\alpha^{1/2} f^{-1/2} u_{\alpha j} \qquad \text{for } j = 1, 2, \ldots, p.$$

$$\phi_{\alpha j} = \lambda_\alpha^{-1/2} \sum_{i=1}^{n} f_i^j \psi_{\alpha i} \qquad \text{for } j = 1, 2, \ldots, p.$$

From $\phi_{\alpha j}$ and $v_{\alpha i}$, the following relation is deduced:

$$\boxed{\phi_{\alpha j} = \lambda_\alpha^{-1/2} \sum_{i=1}^{n} f_i^j \psi_{\alpha i} \qquad \text{for } j = 1, 2, \ldots, p.}$$

From $\psi_{\alpha i}$ and $u_{\alpha j}$, the following relation is deduced:

$$\boxed{\psi_{\alpha i} = \lambda_\alpha^{-1/2} \sum_{j=1}^{p} f_j^i \phi_{\alpha j} \qquad \text{for } i = 1, 2, \ldots, n.}$$

The two previous relations are known as the transition relations. Thus, any coordinate of the point i of $N(I)$ is a pseudo-center of gravity for the coordinates of the points j of $N(J)$, and is provided with masses that are the components of the profile $f^i_J = \{f^i_1, f^i_2, \ldots, f^i_j, \ldots, f^i_p\}$; it is called a pseudo-center because of the value $\lambda_\alpha^{-1/2}$. The equivalent relation holds for any coordinate of the point j of $N(J)$.

Note 1. *Property*: $\lambda_\alpha \leq 1$ *for any value of* α.

Proof. From the transition relation, the following is deduced:

$$\phi_{\alpha j}\lambda_\alpha^{1/2} = \sum_{i \in I} f^i_j \psi_{\alpha i}.$$

Thus, $\phi_{\alpha j}\lambda_\alpha^{-1/2}$ is the center of gravity of the points i with coordinates $\psi_{\alpha i}$ and masses f^i_j. Thus,

$$\min_{i \in I} \psi_{\alpha i} \leq \phi_{\alpha j}\lambda_\alpha^{1/2} \leq \max_{i \in I} \psi_{\alpha i};$$

hence

$$\max_{j \in J} \phi_{\alpha j}\lambda_\alpha^{1/2} \leq \max_{i \in I} \psi_{\alpha i}.$$

Because of the symmetry between I and J, it follows that

$$\max_{i \in I} \psi_{\alpha i}\lambda_\alpha^{1/2} \leq \max_{j \in J} \phi_{\alpha j}.$$

As $\phi_\alpha \geq 0$, from the two previous relations it follows that

$$\lambda_\alpha \leq 1.$$

Note 2. *Interest in the transition relation.*

Proof. Due to the transition relations, it is easy to deduce from one set (I or J) the relations and the coordinates of the other set (I or J). Thus, the set that has the lowest dimension is diagonalized, and then the coordinates of the other set are computed. From a computational point of view, this is rather important, because one set (often I) has generally a larger dimension than the other set (often J).

4.9. Reconstruction of the Basic Correspondence Data Set

Based on the general analysis (*cf.* Chapter 6), the values of $Y = \{y_{ij} = f_{ij}/f_i f_j;\ i \in I;\ j \in J\}$ can be reconstructed from the following formula:

$$Y = \sum_{\alpha=1}^{p} \lambda_\alpha^{1/2} v_\alpha u'_\alpha;$$

4. Geometric Representation of the Sets I and J

v_α and u'_α are replaced by their values $\{v_{\alpha i}; i \in I\}$ and $\{u'_{\alpha j}; j \in J\}$ to give $\{y_{ij}; i \in I; j \in J\}$:

$$v_{\alpha i} = f_i^{1/2} \lambda_\alpha^{-1/2} \psi_{\alpha i} \qquad \text{for } i = 1, 2, \ldots, n;$$

$$u_{\alpha j} = f_j^{1/2} \lambda_\alpha^{-1/2} \phi_{\alpha j} \qquad \text{for } j = 1, 2, \ldots, p;$$

$$y_{ij} = \sum_{\alpha=1}^{p} \lambda_\alpha^{1/2} (\lambda_\alpha^{-1/2} f_i^{1/2} \lambda_\alpha^{-1/2} f_j^{1/2}) \psi_{\alpha i} \phi_{\alpha j} \qquad \begin{array}{l} \text{for } i = 1, 2, \ldots, n \\ \text{and } j = 1, 2, \ldots, p \end{array}$$

$$= \sum_{\alpha=1}^{p} \lambda_\alpha^{-1/2} \psi_{\alpha i} \phi_{\alpha j} f_i^{1/2} f_j^{1/2} \qquad \begin{array}{l} \text{for } i = 1, 2, \ldots, n \\ \text{and } j = 1, 2, \ldots, p. \end{array}$$

Since

$$y_{ij} = f_{ij} f_i^{-1/2} f_j^{-1/2} \qquad \text{for } i = 1, 2, \ldots, n \text{ and } j = 1, 2, \ldots, p,$$

it follows that

$$(f_{ij}/f_i f_j) = \sum_{\alpha=1}^{p} \lambda_\alpha^{-1/2} \psi_{\alpha i} \phi_{\alpha j} \qquad \text{for } i = 1, 2, \ldots, n \text{ and } j = 1, 2, \ldots, p.$$

Since, for $\alpha = 1$, $\lambda_\alpha = 1$,

$$u_{1j} = \sqrt{f_j}; \qquad \phi_{1j} = 1;$$
$$v_{1i} = \sqrt{f_i}; \qquad \psi_{1i} = 1,$$

we finally have

$$\boxed{f_{ij} = f_i f_j \left(1 + \sum_{\alpha=2}^{p} \lambda_\alpha^{-1/2} \psi_{\alpha i} \phi_{\alpha j}\right) \qquad \text{for } i = 1, 2, \ldots, n \text{ and } j = 1, 2, \ldots, p.}$$

Remarks. 1. If $(f_{ij}/f_i f_j) = 1$ for $i = 1, 2, \ldots, n$ and $j = 1, 2, \ldots, p$, then this represents the independence of I and J ($f_{IJ} = f_I \cdot f_J$).

Therefore, we have

$$f_{ij}/f_i f_j = 1 + \underbrace{\sum_{\alpha=2}^{p} \lambda_\alpha^{-1/2} \psi_{\alpha i} \phi_{\alpha j}}_{\text{represents the deviation from the independence}}.$$

2. If $\lambda_\alpha = 0$ for $\alpha \geq 2$, then $f_{ij} = f_i f_j$ for $i = 1, 2, \ldots, n$ and $j = 1, 2, \ldots, p$. This then implies that the sets I and J are independent.

This reconstruction formula shows how correspondence analysis provides a way of studying both the dependence and the independence of the two sets I and J. From the reconstruction formula, the following exact formula is deduced:

$$\boxed{k_{ij} = (k_i k_j / k)\left(1 + \sum_{\alpha=2}^{p} \lambda_\alpha^{-1/2} \psi_{\alpha i} \phi_{\alpha j}\right) \quad \begin{array}{l} \text{for } i = 1, 2, \ldots, n \\ \text{and } j = 1, 2, \ldots, p. \end{array}}$$

In practice, only s factor coordinates are used (with $s \leq p$). This gives the following approximate formula:

$$\boxed{k_{ij} = (k_i k_j / k)\left(1 + \sum_{\alpha=2}^{s} \lambda_\alpha^{-1/2} \psi_{\alpha i} \phi_{\alpha j}\right) \quad \begin{array}{l} \text{for } i = 1, 2, \ldots, n \\ \text{and } j = 1, 2, \ldots, p. \end{array}}$$

4.10. Reconstruction of the Distributional Distances

From Chapter 6, we have

$$d^2(i, i') = d^2(f_J^i, f_J^{i'}) = \sum_{j=1}^{p} (f_j^i - f_j^{i'})^2 / f_j.$$

Now, f_j^i and $f_j^{i'}$ are replaced by their respective values in the reconstruction formula to give

$$d^2(i, i') = \sum_{j=1}^{p} \frac{\left(f_j\left(1 + \sum_{\alpha=2}^{p} \lambda_\alpha^{-1/2} \psi_{\alpha i} \phi_{\alpha j}\right) - f_j\left(1 + \sum_{\alpha=2}^{p} \lambda_\alpha^{-1/2} \psi_{\alpha i'} \phi_{\alpha j}\right)\right)^2}{f_j}$$

$$= \sum_{\alpha=2}^{p} (\psi_{\alpha i} - \psi_{\alpha i'})^2 \quad \left(\text{because } \sum_{j \in J} f_j \phi_{\alpha j}^2 = \lambda_\alpha\right).$$

The distributional distances are reconstructed from the factor coordinates only:

$$d^2(i, i') = \sum_{\alpha=2}^{p} (\psi_{\alpha i} - \psi_{\alpha i'})^2.$$

Conversely,

$$d^2(j, j') = \sum_{\alpha=2}^{p} (\phi_{\alpha j} - \phi_{\alpha j'})^2.$$

5. Interpretation of the 2-D Correspondence Analysis

5.1. What Does "To Interpret" Mean?

To interpret an analysis means to give a specific meaning to each axis and to the proximities between points when projected onto the factor axes. It is not sufficient to compute only factor coordinates nor to draw graphics. The next sections describe how to provide interpretations of the results from a correspondence analysis program based on the patents registration data set (*cf.* Table 8.7).

5.2. Data Set Analyzed

TABLE 8.7. Data set from patents registration for the telecommunications sector during the years 1980–1986.

I \ J	1986	1985	1984	1983	1982	1981	1980
USA	986	774	711	591	467	404	258
Japan	653	552	361	307	195	128	43
FRG	405	357	347	254	293	319	208
France	189	158	200	171	153	184	147
Great Britain	204	182	158	137	92	86	67
Italy	31	28	28	21	22	29	15
Holland	64	59	61	64	33	30	11
Sweden	25	19	31	25	15	12	13
Switzerland	23	34	19	30	17	30	15

5.3. Eigenvalues and Associated Parameters

Recall that eigenvalues represent the dispersion of each factor axis. Table 8.8 contains the nonequal zero eigenvalues, the proportions of variance of factor axes, and the cumulative proportions of variance of factor spaces.

TABLE 8.8. Eigenvalues data table and associated parameters:
num = serial number of eigenvalues in decreasing order ($\alpha = 2, \ldots, p$);
iter = number of iterations to obtain the factor axes, for each factor axis;
eigenvalue = λ_α for $\alpha = 2, \ldots, p$; for $\alpha = 1$, $\lambda_\alpha = 1$; λ_α are the eigenvalues of the diagonalized matrix;
percent = $\tau_\alpha = \lambda_\alpha/M^2(J)$, with $M^2(J) = M^2(I) = \sum_{\alpha=2}^{p} \lambda_\alpha$;
the τ_α are the percentage of variance of each factor axis;
cumul. = cumulative percentage of variance of factor axes considering successively 1, then 2, ..., then p factor axes;
$M^2(I) = M^2(J)$ = total variance of the cloud of points; here, $M^2(J) = 0.0461241$.

num	iter	eigenvalue	percent	cumul.
2	0	0.04104792	88.998	88.998
3	1	0.00276522	5.995	94.993
4	2	0.00145948	3.164	98.158
5	2	0.00068186	1.478	99.636
6	2	0.00012750	0.276	99.912
7	5	0.00004043	0.088	100.000
		0.04612241		

$\uparrow$
$M^2(I)$

5.4. Factor Coordinates of the Variable-points of N(J)

The coordinates $\{\phi_{\alpha j}; \alpha = 2, \ldots p; j = 1, 2, \ldots, p\}$ are given by the following formula:

$$\phi_{\alpha j} = \sum_{i=1}^{n} f_i^j f_i^{-1/2} v_{\alpha i},$$

where $v_{\alpha i}$ is the ith component of the eigenvector v_α of the matrix $S^* = YY'$. The results are given in Table 8.9.

5.5. Rules for Interpreting Factor Axes by the Variable-points of N(J)

5.5.1. Explicative Points

The *explicative points of an α-axis* are the variable-points j of $N(J)$ whose contributions are higher than those of the other points of $N(J)$. The factor axes α can be interpreted based on the explicative points j of $N(J)$. How are these points chosen? Each of the factor axes depends on the

5. Interpretation of the 2-D Correspondence Analysis

TABLE 8.9. Factor coordinates and associated parameters for J (all the numerical values are multiplied by 1000).

					1st factor $\alpha = 2$			2nd factor $\alpha = 3$			3rd factor $\alpha = 4$		
n^a	j^b	qlt_s^c	weig.^{d}	inr^e	ϕ_1^f	cor_1^g	ctr_1^h	ϕ_2	cor_2	ctr_2	ϕ_3	cor_3	ctr_3
1	1986	1000	223	183	190	958	197	−28	23	69	13	4	26
2	1985	1000	187	135	171	882	133	−35	40	90	−48	71	302
3	1984	1000	166	17	10	20	0	44	428	118	47	479	251
4	1983	999	138	40	26	50	2	107	858	569	−26	53	67
5	1982	999	111	51	−127	779	45	−34	57	48	56	148	239
6	1981	1000	106	240	−315	955	258	−51	26	103	−7	1	5
7	1980	999	68	335	−468	969	365	9	0	2	−47	10	110
				1000			1000			1000			1000

$^a n$ = serial number associated with j
$^b j$ = label associated with j
$^c qlt_s(j) = \sum_{\alpha=1}^{s} cor_\alpha(j)$ = quality of explanation of j in the factor space $(1, s)$ (here, $s = 6$)
$^d f_j$ = weight associated with j
$^e f_j \rho^2(J)/M^2(J)$ = percentage of variance of j related to the total variance $M^2(J)$
$^f \phi_{\alpha j}$ = α-coordinate of the variable j
$^g cor_\alpha(j) = \phi_{\alpha j}^2/\rho^2(j)$ = cosine squared of the angle formed by j and α-axis = relative contribution of α-axis to the eccentricity of j
$^h ctr_\alpha(j) = f_j \phi_{\alpha j}^2/\lambda_\alpha$ = relative contribution of the variable j to the variance of the axis α

points j that contribute the most to the total variance of the axis, according to the following formula:

$$\sum_{j \in J} f_j \phi_{\alpha j}^2 = \lambda_\alpha \quad \text{for } \alpha = 2, \ldots p.$$

Thus, the explicative points of an α-axis are selected when their contributions, denoted by $ctr_\alpha(j)$, are higher than the average of the whole contributions (*cf.* The rules of selection in Section 6). The set of explicative points is then divided into two parts; the left part contains explicative points whose factor coordinates are negative, and the right part contains explicative points whose factor coordinates are positive (*cf.* Table 8.10).

The points selected are those on which interpretation must be based, if we are to express the real meaning of the factor axes.

5.5.2. Explained Points

The *explained points by an α-axis* are those points j of $N(J)$ whose contributions to the eccentricity are higher than those of the other points

TABLE 8.10. Explicative points for j of $N(J)$ (the values given in parentheses are the relative contributions to the variance).

explicative points negative coordinates	explicative points positive coordinates
Axis 1	
1980 (365) 1981 (258)	1986 (197)
Axis 2	
	1983 (569)
Axis 3	
1985 (302)	1984 (251) 1982 (239)

of $N(J)$. They must not be confused with explicative points. A point j can be an explained point (by an α-axis) without being an explicative point (of the same α-axis), because of the value of f_j in the previous formula. The contributions to the eccentricity are similar to a squared coefficient of correlation. As with the explicative points, the explained points can be selected according to the highest values of $\text{cor}_\alpha(j)$ compared to a given threshold (cf. the rules of selection in Section 6). Then, the set of explained points is divided into two parts; the left part contains the explained points whose factor coordinates are negative; the right part contains the explained points whose factor coordinates are positive (cf. Table 8.11).

5.5.3. Basic Interpretation

The first factor axis reflects the importance of three years; the years 1980 and 1981, and in opposition, the year 1986. The second factor axis reflects the importance of the year 1983. The third factor axis reflects the importance of the year 1985 on one hand, and of 1984 and 1985 on the other hand. This shows that some event happened during these years within the telecommunications sector. Graphics associated with the factor coordinates would highlight the facts given by these computation.

5. Interpretation of the 2-D Correspondence Analysis

TABLE 8.11. Explained points for j of $N(J)$ (the values given in parentheses are the relative contributions to the eccentricity).

explained points negative coordinates	explained points positive coordinates
Axis 1	
1980 (969)	1986 (858)
1981 (955)	1985 (882)
Axis 2	
none	1983 (858)
Axis 3	
none	none

5.6. Factor Coordinates of the Individual-points of $N(I)$

The coordinates $\{\psi_{\alpha i}; \alpha = 2, \ldots, p; i = 1, 2, \ldots, n\}$ are given by the following formula:

$$\psi_{\alpha i} = \sum_{j=1}^{p} f_j^i f_j^{-1/2} u_{\alpha j},$$

where $u_{\alpha j}$ is the jth component of the eigenvector u of the matrix $T^* = YY'$. These results are given in Table 8.12.

5.7. Rules for Interpreting Factor Axes by the Individual-points of $N(I)$

5.7.1. Explicative Points

The *explicative points of an α-axis* are the individual-points i of $N(I)$ whose contributons are higher than those of the other points of $N(I)$. The factor axes α can be interpreted based on the explicative points. How are these points chosen? Each factor axis depends on the points i that contribute the most to the total variance of the axis, according to the following formula:

$$\sum_{i \in I} f_i \psi_{\alpha i}^2 = \lambda_\alpha \quad \text{for } \alpha = 2, \ldots, p.$$

Thus, the explicative points of an α-axis are selected when their contributions, denoted by $\text{ctr}_\alpha(i)$, are higher than the average of the whole contributions (*cf.* the rules of selection in Section 6). The set of

194   **2-D Correspondence Analysis**

TABLE 8.12. Factor coordinates and associated parameters (all the numerical values are multiplied by 1000).

n^a	i^b	$qlt_s{}^c$	weig.^{d}	inr^e	1st factor $\alpha=2$			2nd factor $\alpha=3$			3rd factor $\alpha=4$		
					$\psi_1{}^f$	$cor_1{}^g$	$ctr_1{}^h$	ψ_2	cor_2	ctr_2	ψ_3	cor_3	ctr_3
1	U.S.A.	1000	363	18	39	681	14	12	65	19	20	183	103
2	Japan	1000	194	431	318	988	478	−28	8	60	−19	4	54
3	FRG	1000	189	183	−198	891	183	−65	98	299	15	5	30
4	France	1000	104	232	−317	983	256	36	12	48	−8	1	6
5	Great Britain	1000	80	8	24	134	1	40	365	47	−33	259	64
6	Italy	1000	15	19	−212	798	17	−63	71	22	10	2	1
7	Holland	1000	28	31	86	144	5	172	571	298	16	5	5
8	Sweden	999	12	17	−100	153	3	216	702	205	57	49	27
9	Switzerland	1001	15	61	−338	627	43	22	3	3	−258	365	709
				1000			1000			1000			1000

$^a n$ = serial number associated with i
$^b i$ = label associated with i
$^c qlt_s(i) = \sum_{\alpha=1}^{s} cor_\alpha(i)$ = quality of explanation of i in the factor space $(1,s)$ (here $s=6$)
$^d f_i$ = weight associated with i
$^e inr(i) = f_i \rho^2(i)/M^2(I)$ = percentage of variance of i related to the total variance $M^2(I)$.
$^f \psi_{\alpha i}$ = α-coordinate of the individual i
$^g cor_\alpha(i) = \psi_{\alpha i}^2/\rho^2(i)$ = relative contribution of α-axis to the eccentricity of i = cosine squared of the angle formed by i and the α-axis
$^h ctr_\alpha(i) = f_i \psi_{\alpha i}^2/\lambda_\alpha$ = relative contribution of the individual i to the variance of the α-axis

TABLE 8.13. Explicative points for i of $N(I)$ (the values given in parentheses are the relative contributions to the variance).

explicative points negative coordinates	explicative points positive coordinates
Axis 1	
France (256) West Germany (183)	Japan (478)
Axis 2	
West Germany (299)	The Netherlands (298) Sweden (205)
Axis 3	
Switzerland (709)	

5. Interpretation of the 2-D Correspondence Analysis

explicative points is divided into two parts; the left part contains explicative points whose factor coordinates are negative; the right part contains explicative points whose factor coordinates are positive. The results are given in Table 8.13.

The points chosen are those on which interpretation must be based if we are to express the meaning of the factor axes.

5.7.2. Explained Points

The *explained points by an α-axis* are those points i of $N(I)$ whose contributions to the eccentricity are higher than those of other points of $N(I)$. They should not be confused with explicative points. A point i can be an explained point without being an explicative point (*cf.* Section 5.5.2). The contributions to the eccentricity are similar to squared coefficients of correlation. As with the explicative points, the explained points can be selected according to the values of $\text{cor}_\alpha(i)$ higher than a given threshold (*cf.* the rules of selection in Section 6). Then, the set of explained points is divided into two parts; the left part contains the explained points whose factor coordinates are negative; the right part contains the explained points whose factor coordinates are positive (*cf.* Table 8.14).

TABLE 8.14. Explained points for i of $N(I)$ (the values given in parentheses are the contributions to the eccentricity).

explained points negative coordinates	explained points positive coordinates
Axis 1	
France (983) West Germany (891) Italy (798) Switzerland (627)	Japan (988) USA (681)
Axis 2	
	The Netherlands (571) Sweden (702)

5.7.3. Basic Interpretation

The first axis characterizes the high growth in the telecommunication sector of the USA and Japan in oppositon to West Germany and France, whose growth is weak. The second axis characterizes the intensity of the growth of the Netherlands and Sweden during the period 1980–1986. This basic interpretation of $N(I)$ must be related to that of $N(J)$, in order to allow a simultaneous interpretation of the points of $N(I)$ and $N(J)$.

5.8. Joint Interpretation of Factor Axes by the Points of $N(I)$ and $N(J)$

All the previous information is gathered in the Table 8.15 for the explicative points, and in Table 8.16 for the explained points.

5.9. Remark on the Contributions

The relative contributions of each factor axis α are given by a percentage of variance of the points i of $N(I)$ or j of $N(J)$, according the following

TABLE 8.15. Explicative points for $N(I)$ and $N(J)$ (the values given in parentheses are $\text{ctr}_\alpha(i)$ or $\text{ctr}_\alpha(j)$).

Points with negative coordinates	Points with positive coordinates
Axis 1 ($\lambda_1 = 0.041$, $\tau_1 = 89\%$)	
1980 (365) 1981 (258)	1986 (197)
France (256) West Germany (183)	Japan (478)
Axis 2 ($\lambda_2 = 0.0028$; $\tau_2 = 6\%$)	
none	1983 (569)
West Germany (299)	The Netherlands (298) Sweden (205)
Axis 3 ($\lambda_3 = 0.0015$; $\tau_3 = 3.16\%$)	
1985 (302)	1984 (251) 1982 (239)
Switzerland (709)	none

5. Interpretation of the 2-D Correspondence Analysis

TABLE 8.16. Explained points for $N(I)$ and $N(J)$ (the values given in parentheses are $\text{cor}_\alpha(i)$ or $\text{cor}_\alpha(j)$).

points with negative coordinates	points with positive coordinates
Axis 1 ($\lambda_1 = 0.041$; $\tau_1 = 89\%$)	
1980 (969)	1986 (958)
1981 (955)	1985 (882)
France (983)	
West Germany (891)	Japan (988)
Italy (798)	USA (681)
Switzerland (627)	
Axis 2 ($\lambda_2 = 0.0028$; $\tau_2 = 6\%$)	
none	1983 (853)
none	The Netherlands (571)
	Sweden (702)
Axis 3 ($\lambda_3 = 0.0015$; $\tau_3 = 3.16\%$)	
none	none

formula:

$$\text{ctr}_\alpha(i) = \frac{f_i \psi_{\alpha i}^2}{\lambda_\alpha} \quad \text{for } \alpha = 2, \ldots, p \text{ and } i = 1, 2, \ldots, n;$$

$$\text{ctr}_\alpha(j) = \frac{f_j \phi_{\alpha j}^2}{\lambda_\alpha} \quad \text{for } \alpha = 2, \ldots, p \text{ and } j = 1, 2, \ldots, p.$$

However, the values λ_α can have a large variation, making it difficult to compare the values of $\text{ctr}_\alpha(i)$ for different values of α. In this case, it is preferable to compute the absolute contributions as follows:

$$\text{cta}_\alpha(i) = f_i \psi_{\alpha i}^2 \quad \text{for } \alpha = 2, \ldots, p \text{ and } i = 1, 2, \ldots, n;$$
$$\text{cta}_\alpha(j) = f_j \phi_{\alpha j}^2 \quad \text{for } \alpha = 2, \ldots, p \text{ and } j = 1, 2, \ldots, p;$$

from the following formulas:

$$\sum_{i \in I} \sum_{\alpha=2}^{p} f_i \psi_{\alpha i}^2 = M^2(N(I));$$

$$\sum_{j \in J} \sum_{\alpha=2}^{p} f_j \phi_{\alpha j}^2 = M^2(N(J)).$$

TABLE 8.17. Absolute contributions for I.

![Table 8.17: rows indexed by INDIVIDUALS I (1 to n), columns are FACTORS Λ with $\alpha = 2, \ldots, \alpha, \ldots, p$; cell shows $f_i\psi^2_{\alpha i}$; right column I with $\mathrm{In}(i)$; bottom row Λ with $\lambda_2, \ldots, \lambda_\alpha, \ldots, \lambda_p$ and $M^2(N(I))$.]

In order to compare the absolute contributions, the following data tables must be computed (*cf.* Tables 8.17 and 8.18).

By definition, $\sum_{i \in I} f_i \psi^2_{\alpha i} = \lambda_\alpha$ and $\sum_{\alpha=2}^{p} f_i \psi^2_{\alpha i} = \mathrm{In}(i)$ ($\mathrm{In}(i)$ is the variance of the point i).

The same table is done for $N(J)$ as follows:

TABLE 8.18. Absolute contributions for J.

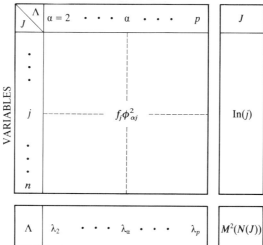

5. Interpretation of the 2-D Correspondence Analysis

By definition, $\sum_{j \in J} f_j \phi_{\alpha j}^2 = \lambda_\alpha$ and $\sum_{\alpha=2}^{p} f_i \phi_{\alpha j}^2 = \text{In}(j)$ ($\text{In}(j)$ is the variance of the point j).

These previous data tables may be standardized as follows:

$$\text{Table 1} = \{(i, \alpha); f_i \psi_{\alpha i}^2 / M^2(N(I)) \text{ for } i \in I \text{ and } \alpha \in \Lambda\},$$

$$\text{Table 2} = \{(j, \alpha); f_j \phi_{\alpha j}^2 / M^2(N(J)) \text{ for } j \in J \text{ and } \alpha \in \Lambda\},$$

where Λ is the space of factors.

5.10. Quality of Explanation

To finish the basic interpretation of correspondence analysis from a numerical point of view, the quality of explanation of factor axes needs to be studied, as well as the quality of explanation of the points i of $N(I)$ and j of $N(J)$ by the factor axes.

5.10.1. Global Quality of Explanation

The global quality of explanation, denoted by qge_s, is given by the sum of the percentages of variance of the factor axes considered as being *interpretable* (i.e., significant): let $s \leq p$ be the number of these factor axes. Then,

$$\text{qge}_s = \sum_{\alpha=2}^{s} \tau_\alpha = \frac{\sum_{\alpha=2}^{s} \lambda_\alpha}{\sum_{\alpha=2}^{p} \lambda_\alpha}.$$

For the example of patents registration the following results hold:

For one factor axis, $\text{qge}_1 = 89\%$;

For two factor axes, $\text{qge}_2 = 96\%$;

For three factor axes, $\text{qge}_3 = 98\%$;

this means that the three factor axes account for 98% of the total variance.

5.10.2. Local Quality of Explanation

For each point i of $N(I)$ and j of $N(J)$, the quality of explanation, denoted by $\text{qlt}_s(i)$ and $\text{qlt}_s(j)$ can be computed in a given factor space with rank s:

$$\text{qlt}_s(i) = \sum_{\alpha=2}^{s} \text{cor}_\alpha(i);$$

$$\text{qlt}_s(j) = \sum_{\alpha=2}^{s} \text{cor}_\alpha(j).$$

TABLE 8.19. Quality of explanation for $N(I)$ and $N(J)$ for the three first factor axes (the values are multiplied by 1000).

I	Quality of explanation qlt	qlt$_1$	qlt$_2$	qlt$_3$	J	Quality of explanation qlt	qlt$_1$	qlt$_2$	qlt$_3$
USA		681	746	929	1986		958	981	985
Japan		988	996	1000	1985		882	922	993
FRG		891	989	994	1984		20	448	927
France		983	995	996	1983		50	908	961
Great Britain		134	499	758	1982		779	836	984
Italy		798	869	871	1981		995	981	982
Holland		144	715	720	1980		969	969	979
Sweden		153	855	904					
Switzerland		627	630	995					

The results for the example of patents registration are given in Table 8.19.

6. Factor Graphics

6.1. Introduction

The purpose of correspondence analysis is to represent graphically the points of $N(I)$ and $N(J)$ in a geometric space whose dimension is as low as possible (i.e., it means having the maximum of variance for the cloud of points). This is done in order to highlight relationships that are impossible to show only in numerical data sets. As representation in a s-D space is not easy for the human mind to grasp, the s-D space is studied by considering as significant or interpretable every projection plane related to the s-D space (this means: planes $(1, 1)$, $(1, 2)$, $(3, 3)$, $(2, 3)$, $(2, 4)$, ... ,$(s - 1, s)$, (s, s)). We now give some examples of graphics based on the example of patents registration.

6.2. Graphics Associated with the Points of $N(J)$

The graphical representations of $N(J)$ according to the factor planes $(1, 2)$ and $(1, 3)$ are given in Figs. 8.4 and 8.5. The factor coordinates are from the set $\{\phi_{\alpha j}; \alpha = 2, 3, 4; j = 1, 2, \ldots, p\}$. Note that the plane $(1, 2)$

6. Factor Graphics

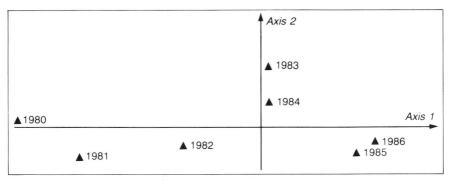

FIGURE 8.4. Representation of the points j of $N(J)$, denoted by ▲, in the factor space $(1, 2)$.

is from $\alpha = 2$ and 3, because the value $\alpha = 1$ is a trivial value in correspondence analysis (i.e., noninterpretable).

6.3. Graphics Associated with the Points of $N(I)$

The graphical representations of $N(I)$ according to the factor planes $(1, 2)$ and $(1, 3)$ are given in Figs. 8.6 and 8.7. The factor coordinates are from the set $\{\psi_{\alpha i}; \alpha = 2, 3, 4; i = 1, 2, \ldots, n\}$. Note that the plane $(1, 2)$ corresponds to the values $\alpha = 2$ and 3, because the value $\alpha = 1$ is a trivial value in correspondence analysis (i.e., noninterpretable).

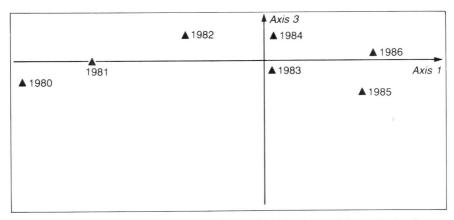

FIGURE 8.5. Representation of the points j of $N(J)$, denoted by ▲, in the factor space $(1, 3)$.

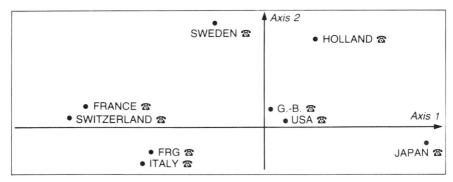

FIGURE 8.6. Representation of the points i of $N(I)$, denoted by ●, in the factor space $(1, 2)$.

6.4. Graphics Associated with the Points of $N(I)$ and $N(J)$ Simultaneously

The graphical representations of $N(I)$ and $N(J)$ on the same graphic allow the user to interpret each set (I or J) by the other set (I or J). The simultaneous representations are given for the factor planes $(1, 2)$ and $(1, 3)$ in Figs. 8.8 and 8.9.

6.5. Graphics Associated with the Explicative Points of $N(I)$ and $N(J)$ of the Factor Axes

Since factor axes are explained by the contribution values of variables, it is useful to transfer that information onto graphics in order to highlight

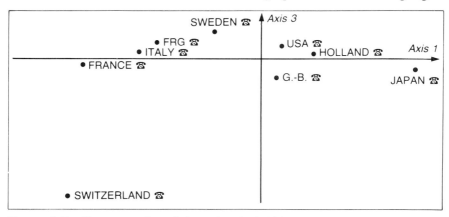

FIGURE 8.7. Representation of the points i of $N(I)$, denoted by ●, in the factor space $(1, 3)$.

7. Classifying Supplementary Points into Graphics

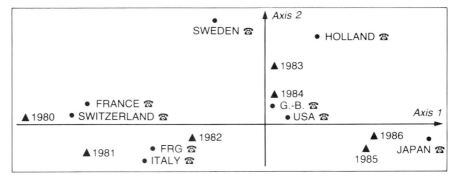

FIGURE 8.8. Representation of the points i of $N(I)$ (●) and j of $N(J)$ (▲) in the factor space $(1, 2)$.

the points that are the most explicative of the factor axes (Figs. 8.10 and 8.11). The contributions associated with each point ($i \in I$ or $j \in J$) are denoted by a rectangle whose sides are proportional to the contribution values.

7. Classifying Supplementary Points into Graphics

7.1. Introduction

The most fruitful part of correspondence analysis techniques is to introduce supplementary elements onto factor graphics. The principle is to add points that are not involved in the basic data table. These points

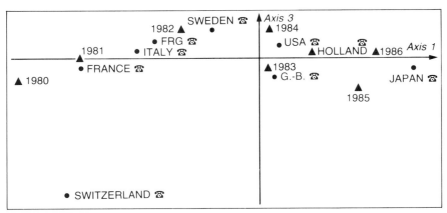

FIGURE 8.9. Representation of the points i of $N(I)$ (●) and j of $N(J)$ (▲) in the factor space $(1, 3)$.

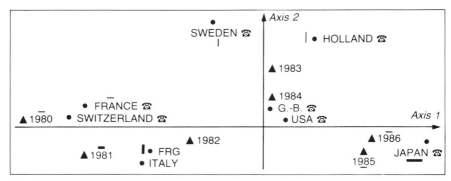

FIGURE 8.10. Representation of the points i of $N(I)$ (●) and j of $N(J)$ (▲) in the factor space $(1, 2)$. The values of the contributions are denoted by rectangles.

are passive (i.e., they have no influence on the computation of factor axes); their factor coordinates can be computed from the transition formula, in order to enrich the interpretation of graphics. The process is done by first performing a standard correspondence analysis, and then computing the coordinates of any supplementary element i_s or j_s, in order to represent them in the correspondence analysis graphics.

7.2. Areas of Application

The current applications of classifying supplementary elements are the following:

- To suppress a particular point in a factor analysis graphic and then to reintroduce it as a supplementary element. This is done when a

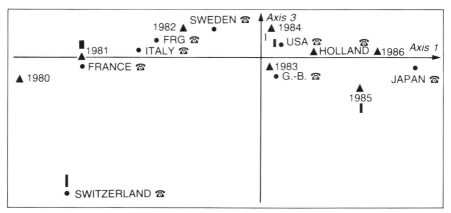

FIGURE 8.11. Representation of the points i of $N(I)$ (●) and j of $N(J)$ (▲) in the factor space $(1, 3)$. The values of contributions are denoted by rectangles.

particular pair (i, j) of $(N(I), N(J))$ has a contribution value close to the total variance while the other values are all smaller, so that only the pair (i, j) explains the factor axis; this is not significant for analysis, and it prevents other relationships being studied on the graphics. A new principal analysis is made therefore with the sets $I - \{i\}$ and $J - \{j\}$ (principal means that the factor axes are computed from $I - \{i\}$ and $J - \{j\}$). Then, the elements i and j considered as dummy elements are computed from the previous factor analysis and projected onto the graphics. These elements cannot be considered as *explicative elements* because they did not contribute to a factor axis.

- To check the stability of factor axes. This can be done by suppressing points of I or J and studying the variation of factor axes.
- To classify elements whose description, in terms of profiles, is missing or incomplete. In this case, the data values are estimated, and then these points are introduced as supplementary elements into graphics.
- To illustrate graphics by passive elements representing, for example, centers of gravity of classes, or elements of the same set but taken at different periods, etc. These illustrations help to give correspondence analysis results a better interpretation. In what follows, we give computations of supplementary elements, illustrated by the example of patents registration (*cf.* Appendix 2, §3).

7.3. *Principal and Supplementary Data Sets*

Consider the statistics of patents registration by industrial countries during the years 1980–1986, for telecommunications on one hand, and for all the economic sectors on the other hand (*cf.* Table 8.20). The situation of the telecommunications sector in each country is compared to the global economic situation. To do this, the procedure is the following:

(1) Correspondence analysis is done on $I \times J$, in order to study the global economic situation.
(2) To the telecommunications sector in each country, a supplementary element is assigned; a profile on the set of years is assigned to each telecommunications sector.
(3) Each previous element is used as a supplementary element in the correspondence analysis of $I \times J$; its coordinates are computed, and then it is represented on the resulting graphics correspondence analysis graphics of $I \times J$.

TABLE 8.20. Patents registration data sets during the period 1980–1986. $I \times J$: all the economic sectors mixed (principal data set); $IS \times J$: telecommunications sector (supplementary data set).

YEARS

	J	1986	1985	1984	1983	1982	1981	1980
	I							
COUNTRIES	USA	11126	9762	8969	7486	6243	5820	3574
	Japan	6259	5384	4192	3728	2980	2026	853
	FRG	8710	7892	6604	6360	6002	5747	4087
	France	3304	2891	2797	2735	2252	2371	1634
	Great Britain	3396	3076	2944	2652	2366	2146	1853
	Italy	1180	969	863	748	616	490	206
	Holland	905	779	732	677	545	542	268
	Sweden	843	737	764	659	551	478	302
	Switzerland	1193	1220	1059	1047	912	947	699

YEARS

	J	1986	1985	1984	1983	1982	1981	1980
	IS							
COUNTRIES	USA	986	774	711	591	467	404	258
	Japan	653	552	361	307	195	128	43
	FRG	405	357	347	251	293	319	208
	France	189	158	200	171	153	184	147
	Great Britain	204	182	158	137	92	86	67
	Italy	31	28	28	21	22	29	15
	Holland	64	59	61	64	33	30	11
	Sweden	25	19	31	25	15	12	13
	Switzerland	23	34	19	30	17	30	15
	CNET	130	107	76	81	72	70	81

7.4. Coordinates of Supplementary Points

7.4.1. Coordinates of Supplementary Elements s of IS

Let us consider the principal data set k_{IJ} and the supplementary data $k_{IS,J}$ as presented in Table 8.21. For each supplementary element s, its profile can be computed as follows:

$$f_j^s = \left\{ \frac{k_{sj}}{k_s} = \frac{f_{sj}}{f_s} ; j \in J \right\}.$$

7. Classifying Supplementary Points into Graphics

TABLE 8.21. Principal and supplementary data sets k_{IJ} and $k_{IS,J}$

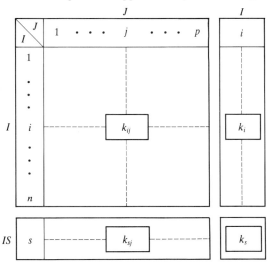

To compute the coordinates $\psi_{\alpha s}$, the transition formula is used by replacing i by s as follows:

$$\psi_{\alpha i} = \lambda_\alpha^{-1/2} \sum_{j=1}^{p} f_j^i \phi_{\alpha j} \quad \text{for } \alpha = 2, \ldots, p \text{ and } i = 1, 2, \ldots, n;$$

thus,

$$\psi_{\alpha s} = \lambda_\alpha^{-1/2} \sum_{j=1}^{p} f_j^s \phi_{\alpha j} \quad \text{for } \alpha = 2, \ldots, p.$$

For the example of patents registration, the following results were produced (*cf.* Tables 8.22–8.25).

7.4.2. Coordinates of Supplementary Elements t of JS

Let us consider the principal data set k_{IJ} and the supplementary data set $k_{I,JS}$ as presented in Table 8.26. Since there is complete symmetry between I and J, the supplementary elements t are computed as i the previous section with the transition formula:

$$\phi_{\alpha t} = \lambda_\alpha^{-1/2} \sum_{i=1}^{n} f_i^t \psi_{\alpha i} \quad \text{for } \alpha = 2, \ldots, p.$$

There is no supplementary data for the example of patents registration but it is easy to imagine. Every year these statistics of patents registration are gathered. Without redoing the complete correspondence analysis of

208 2-D Correspondence Analysis

TABLE 8.22. Eigenvalues data table and associated parameters with the correspondence analysis of $I \times J$:
num = serial number of eigenvalues for $\alpha \geq 2$;
iter = number of iterations giving the result;
eigenvalue = λ_α = variance of the α-axis;
percent = τ_α = percentage of variance of the α-axis;
cumul. = $\sum_{\alpha'=2}^{\alpha} \tau_{\alpha'}$ = cumulative percentage of variance of the α-axis.

num	iter	eigenvalue	percent	cumul.
2	0	0.01203015	91.863	91.863
3	0	0.00052446	4.005	95.868
4	2	0.00032154	2.455	98.323
5	2	0.00013383	1.022	99.345
6	5	0.00005384	0.411	99.757
7	2	0.00003187	0.243	100.000

TABLE 8.23. Factor coordinates of I and associated contributions resulting from correspondence analysis of $I \times J$.

					1st factor			2nd factor			3rd factor		
n	i	qlt_7	weig	inr	ψ_1	cor_1	ctr_1	ψ_2	cor_2	ctr_2	ψ_3	cor_3	ctr_3
1	USA	999	292	44	37	691	33	20	203	224	3	5	9
2	Japan	1000	140	507	216	988	545	−20	9	118	5	1	12
3	FRG	999	251	146	−81	871	139	−29	119	437	−5	4	24
4	France	999	99	86	−100	899	84	19	31	66	−21	42	148
5	Great Britain	1000	102	96	−101	842	88	9	7	16	41	139	542
6	Italy	1000	28	50	149	953	52	10	4	5	−23	25	50
7	Holland	1000	25	8	25	153	1	25	159	30	−51	664	207
8	Sweden	1000	24	7	5	8	0	48	619	104	−3	4	1
9	Switzerland	1000	39	57	−133	940	58	−3	1	1	−6	3	6
				1000			1000			1000			1000

TABLE 8.24. Factor coordinates of J and associated contributions resulting from correspondence analysis of $I \times J$.

					1st factor			2nd factor			3rd factor		
n	j	qlt_7	weig	inr	ϕ_1	cor_1	ctr_1	ϕ_2	cor_2	ctr_2	ϕ_3	cor_3	ctr_3
1	1986	1000	204	169	103	977	180	−10	11	46	2	0	2
2	1985	999	181	93	78	907	92	−18	55	128	7	8	31
3	1984	1000	160	47	40	413	21	46	548	638	10	28	53
4	1983	1000	144	9	3	13	0	11	138	32	−10	138	53
5	1982	1000	124	23	−38	625	16	−24	252	147	−1	2	2
6	1981	1000	114	180	−137	918	180	4	1	3	−38	72	524
7	1980	1000	74	479	−287	982	512	−6	1	6	38	17	336
				1000			1000			1000			1000

7. Classifying Supplementary Points into Graphics

TABLE 8.25. Factor coordinates of IS and associated contributions resulting from correspondence analysis of $I \times J$. The values $\text{ctr}_\alpha(s)$ have no meaning.

					1st factor			2nd factor			3rd factor		
n	s	qlt_7	weig	inr	ψ_1	cor_1	ctr_1	ψ_2	cor_2	ctr_2	ψ_3	cor_3	ctr_3
1	USA	1000	23	21	96	782	*	15	19	*	23	46	*
2	Japan	1000	12	134	359	905	*	−49	18	*	49	17	*
3	FRG	1000	12	23	−129	670	*	−3	1	*	−17	14	*
4	France	1000	7	35	−252	927	*	65	61	*	−2	0	*
5	Great Britain	999	5	5	71	410	*	26	57	*	55	245	*
6	Italy	1000	1	3	−136	522	*	23	15	*	−84	201	*
7	Holland	1000	2	8	148	396	*	117	247	*	−54	53	*
8	Sweden	999	1	4	−39	26	*	198	651	*	95	149	*
9	Switzerland	1000	1	11	−312	694	*	−46	16	*	−22	4	*

the whole set of years, the coordinates of the supplementary years can be computed and classified as supplementary elements in correspondence analysis graphics.

TABLE 8.26. Principal and supplementary data sets k_{IJ} and $k_{I,JS}$.

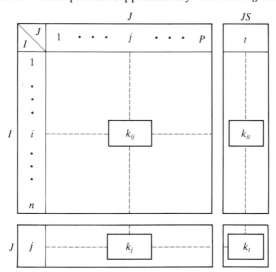

7.5. Graphics with Supplementary Points

Recall that the aim of correspondence analysis is to represent *graphically* the relationships between all the elements of two sets *put into correspondence*, to highlight and suggest relationships that cannot be seen by studying numerical data tables. With the procedure of classifying supplementary elements, two more sets are involved: *IS*, the set of supplementary points of *I*, and *JS*, the set of supplementary points of *J*.

Therefore, correspondence analysis graphics involve combining the following sets:

- the principal sets *I* and *J*;
- the explicative subsets from *I* and *J* formed by the explicative points of *I* and *J*;
- the explained subsets from *I* and *J* formed by the explained points of *I* and *J*;
- the supplementary sets *IS* and *JS*;
- the explained subsets from *IS* and *JS* formed by the explained points of *IS* and *JS* (remember that there is no explicative points for *IS* and *JS* because ctr_α has no meaning for any supplementary points);
- the set of interpretable factor axes $\alpha = 2, \ldots, s$.

Thus, any correspondence analysis software must be able to manage all of these previous sets; otherwise, it cannot be considered an efficient one. An example of classifying supplementary elements is given based on the patents registration data sets (*cf.* Figs. 8.12 and 8.13).

8. Rules for Selecting Significant Axes and Elements

8.1. Introduction

After twenty years of experience in applying correspondence analysis and other factor analysis models, it appears clear that rules must be specified to help in selecting significant axes and elements of *I* or *J* because of the size of the sets involved. Twenty years ago, a data set whose size was (50, 50), was considered large, so it was easy to select "by hand" each explained point and each explicative point. The data sets involved now are larger; a standard questionnaire can have 2000 to 3000 individuals and can involve 100 variables that can give 300 to 400 forms (or modalities). Because of computer power, these sets can be studied directly without any aggregation. However, selection of the points, the axes, or the graphics must be made *automatically*. This explains the purpose of this section, which is devoted to helping the user apply correspondence analysis.

8. Rules for Selecting Significant Axes and Elements

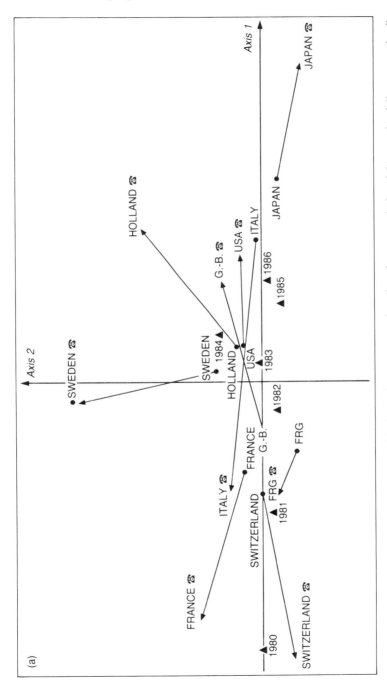

FIGURE 8.12. Correspondence analysis graphic from the patents registration data set: ● industrial countries (all sectors mixed) considered as principal points; industrial countries (telecommunications sectors) considered as supplementary points; ▲ years considered as principal points. Representation according to the factor axes 1 and 2.

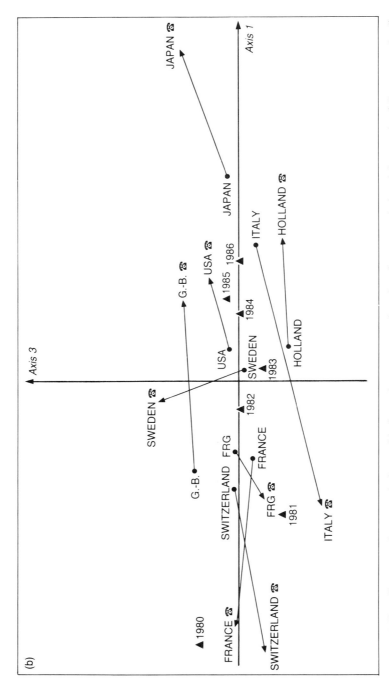

FIGURE 8.13. Correspondence analysis graphic from the patents registration data set: ● industrial countries (all sectors mixed) considered as principal points; industrial countries (telecommunications sectors) considered as supplementary points; ▲ years considered as principal points. Representation according to the factor axes 1 and 3.

8.2. Rules for Selecting Significant Axes

Here, significant means "necessary to study in detail"; the significance is not based on statistical tests, which are difficult to apply. Two types of factor axes are considered, known as first order factor axes and second order factor axes; the first order factor axes are considered as the principal ones, based on the contributions to the total variance; the second order factor axes are based on the contributions to the eccentricity.

8.2.1. Rules for Selecting First Order Factor Axes

Let N be the number of axes to be studied, determined by one of the following rules:

Rule 1. N is chosen such that $\sum \{\tau_\alpha ; \alpha \leq N\} \geq p$, where p is a given percentage of variance; in general, $p = 80\%$. This percentage can be modified according to the type of data (contingency data sets, measurements, marks, logical data sets)

Rule 2. N is chosen such that $\tau_N \geq 100/(p-1)$, where p is the minimum of the sizes of I and J.

Rule 3. Consider the following values: $\{f_i \psi_{\alpha i}^2 : \alpha = 2, \ldots, p; \; i = 1, 2, \ldots, n\}$ and $\{f_j \phi_{\alpha j}^2 ; \alpha = 2, \ldots, p; j = 1, \ldots, p\}$

Suppose these values are ordered by decreasing order, and denoted by $O(f_i \psi_{\alpha i}^2)$ and $O(f_j \phi_{\alpha j}^2)$.

We then seek solutions to the following two inequalities:

$$\sum_{i \in I} \sum_{\alpha \in \Lambda} \left\{ O\left(\frac{f_i \psi_{\alpha i}^2}{M^2(I)}\right) \right\} \geq p,$$

$$\sum_{j \in J} \sum_{\alpha \in \Lambda} \left\{ O\left(\frac{f_j \phi_{\alpha j}^2}{M^2(J)}\right) \right\} \geq p,$$

where p is a given percentage of variance (in general, $p = 80\%$) and Λ is the factor space.

Denoted by α and α' the minimum values of $\alpha \in \Lambda$ such that the inequalities are satisfied when the inner sums are truncated at α and α', respectively. Then $N = \sup(\alpha, \alpha')$.

These rules do not necessarily give the same results, but they indicate an order of magnitude of the number of axes to be studied.

8.2.2. Rules for Selecting Second Order Factor Axes

Let N' be the number of factors axes to be studied. These axes are those for which the percentage of variance τ_α is small, but which have some characteristic points with high contributions to the eccentricity. This produces the following rule:

Rule 4. Let n be the rank of the factor axis for which a point i of $N(I)$ or j of $N(J)$ exists such that:

$$\left.\begin{array}{l} \psi_{ni}^2/\rho^2(i) \geq k, \\ \text{or} \\ \phi_{nj}^2/\rho^2(j) \geq k, \end{array}\right\} \quad \text{where } k \text{ is a given value, similar to a squared coefficient of correlation (in general, } k = 0.5\text{).}$$

The value of k can be modified according to the type of data. The number of second order facror axes N' is $n - N$. Now, $N + N'$ factors can be considered for further studies.

8.3. Rules for Selecting Explicative Points of Factor Axes

8.3.1. Selection of Explicative Points of First Order Factor Axes

Recall that an explicative point is a point whose contributons $\text{ctr}_\alpha(i)$ (for $i \in I$) or $\text{ctr}_\alpha(j)$ (for $j \in J$) are distinctly higher than the contributions of other points; to understand the meaning of "distinctly higher," the following rules are proposed:

Rule 5. The points $i \in I$ whose contributons are higher than the average are considered as "explicative points." The following set of explicative points of I can thus be selected:

$$\left\{ i \in I; \left(\frac{f_i \psi_{\alpha i}^2}{\lambda_\alpha} = \text{ctr}_\alpha(i) \geq \frac{1}{n_I} \right) \right\},$$

where n_I is the number of elements of I.

The same is done for J. The following set of explicative points of J can thus be selected:

$$\left\{ j \in J; \left(\frac{f_j \phi_{\alpha j}^2}{\lambda_\alpha} = \text{ctr}_\alpha(j) > \frac{1}{n_J} \right) \right\},$$

where n_J is the number of elements of J.

8. Rules for Selecting Significant Axes and Elements

Rule 6. The points $i \in I$ are ordered by their contribution to $\text{ctr}_\alpha(i)$, in decreasing order. Then, the sum

$$\left\{ \sum_{i \in I} \frac{f_i \psi_{\alpha i}^2}{\lambda_\alpha} \geq p \right\}$$

is truncated at the lowest value $i_0 \in I$ such that the truncated sum is $\geq p$. The set $\{i \in I \mid i \leq i_0\}$ is the set of explicative points. The same is done for J.

$$\left\{ j \in J; \sum_{j \in J} \frac{f_j \phi_{\alpha j}^2}{\lambda_\alpha} \geq p \right\}.$$

8.3.2. Selection of Explicative Points of Second Order Factor Axes

Recall that the second order factor axes are determined according to the contribution to eccentricity, based on the following rule:

Rule 7. The sets of explicative points of second order factor axes are selected as follows:

$$\left\{ i \in I; \frac{\psi_{\alpha i}^2}{\rho^2(i)} \geq k \right\},$$

$$\left\{ j \in J; \frac{\phi_{\alpha j}^2}{\rho^2(j)} \leq k \right\},$$

where k is a given coefficient, similar to a squared coefficient of correlation (in general, $k = 0.5$). This coefficient can be modified according to the type of data.

8.4. Selection of Explained Points by Factor Axes

The explained points $i \in I$ or $j \in J$ are those which have a high value of $\text{cor}_\alpha(i)$ or $\text{cor}_\alpha(j)$, respectively, resulting in the following rule;

Rule 8. The sets of explained points by factor axes (first order factor axis or second order factor axis) are selected as follows:

$$\left\{ i \in I; \frac{\psi_{\alpha i}^2}{\rho^2(i)} \geq k \right\},$$

$$\left\{ j \in J; \frac{\phi_{\alpha j}^2}{\rho^2(j)} \geq k \right\},$$

where k is a given coefficient, similar to a squared coefficient of correlation (in general, $k = 0.5$). This coefficient can be modified according to the type of data.

Notice that rule 8 is similar to Rule 7. They are based on the same principle, but they do not play the same role in interpreting correspondence analysis. The first rule is given to interpret the axes, the second rule is given to explain the position of the points according to axes previously interpreted.

8.5. Conclusions

With these rules, it is easy to study large data sets, because only qualified points (explained or explicative) and significant axes are retained. As with many rules, a few exceptions arise:

- When λ_2 is significantly higher than the other eigenvalues, it is necessary to study more factor axes than indicated by the rules.
- When the shape of the projected cloud of points is a continuum, the contributions may not be significant. The best policy then is to interpret the results in terms of ordering rather than in terms of opposition between variables.

Notice that some rules are dependent on two given values: p and k. This means that the way of interpreting data by correspondence analysis must be *adjusted* progressively through experience and learning. Correspondence analysis is a learning machine for exploring and understanding data.

9. 2-D Correspondence Analysis Formulas

9.1. Basic Data Sets and Associated Clouds

$$k_{IJ} = \{k_{ij}; i \in I; j \in J\}, \quad k_{ij} \geq 0, \quad \forall i, j \in I \times J;$$

$$k_i = \sum_{j \in J} k_{ij}; k_j = \sum_{i \in I} k_i; \quad k = \sum_{i,j \in I \times J} k_{ij} = \sum_{j \in J} k_j = \sum_{i \in I} k_i;$$

$$f_{IJ} = \{f_{ij} = k_{ij}/k; i \in I; j \in J\};$$

9. 2-D Correspondence Analysis Formulas

$$f_I = \{f_i; i \in I\};$$
$$f_J = \{f_j; j \in J\};$$
$$f^i_J = \{f^i_j; f_i \neq 0; j \in J\};$$
$$f^j_I = \{f^j_i; f_j \neq 0; i \in I\};$$
$$f^I_J = \{f^i_J; i \in I\};$$
$$f^J_I = \{f^j_I; j \in J\};$$
$$N(I) = \{(f^i_J, f_i); i \in I\} \subset \mathcal{P}(J) \subset R_J;$$
$$N(J) = \{(f^j_I, f_i); j \in J\} \subset \mathcal{P}(I) \subset R_I.$$

9.2. Distributional Distances and Variance

$$d^2(i, i') = d^2(f^i_J, f^{i'}_J) = \sum_{j \in J} (f^i_j - f^{i'}_j)^2 / f_j;$$

$$d^2(j, j') = d^2(f^j_I, f^{j'}_I) = \sum_{i \in I} (f^j_i - f^{j'}_i)^2 / f_i;$$

$$\rho^2(i) = d^2(f^i_J - f_J) = \sum_{j \in J} (f^i_j - f_j)^2 / f_j;$$

$$\rho^2(j) = d^2(f^j_I - f_I) = \sum_{i \in I} (f^j_i - f^{j'}_i)^2 / f_i;$$

$$M^2(N(I)) = M^2(N(J)) = \sum_{i \in I} f_i \rho^2(i) = \sum_{j \in J} f_j \rho^2(j)$$
$$= \sum_{i,j \in I \times J} (f_{ij} - f_i f_j)^2 / f_i f_j.$$

9.3. The Matrix to Be Diagonalized

$$Y = y_{ij} = \frac{f_{ij}}{\sqrt{f_i f_j}}.$$

9.4. Factor Axes and Factors

The factors are ordered functions on I or J, denoted by ψ for I and ϕ for J, with which is associated a real number λ_α, between 0 and 1. Thus, we have:

$$\phi_1 = \{\phi_{1j}; j \in J\}, \qquad \psi_1 = \{\psi_{1i}; i \in I\},$$
$$\phi_2 = \{\phi_{2j}; j \in J\}, \qquad \psi_2 = \{\psi_{2i}; i \in I\},$$
$$\vdots \qquad\qquad\qquad\qquad \vdots$$
$$\phi_\alpha = \{\phi_{\alpha j}; j \in J\}, \qquad \psi_\alpha = \{\psi_{\alpha i}; i \in I\},$$
$$\vdots \qquad\qquad\qquad\qquad \vdots$$
$$\phi_N = \{\phi_{Nj}; j \in J\}, \qquad \psi_N = \{\psi_{Ni}; i \in I\};$$

$N = \text{minimum}(n, p); \lambda_1 \geq \lambda_2 \ldots \geq \lambda_N;$
$\phi_{\alpha j} = \alpha$-coordinate of the point j of $N(J)$;
$\psi_{\alpha i} = \alpha$-coordinate of the point i of $N(I)$.

Properties of α-coordinates:

$$\sum_{i \in I} f_i \psi_{\alpha i} = 0; \qquad \sum_{j \in J} f_j \phi_{\alpha j} = 0;$$
$$\sum_{i \in I} f_i \psi_{\alpha i}^2 = \lambda_\alpha; \qquad \sum_{j \in J} f_j \phi_{\alpha j}^2 = \lambda_\alpha;$$
$$\sum_{\substack{i \in I \\ \alpha \neq \beta}} f_i \psi_{\alpha i} \psi_{\beta i} = 0; \qquad \sum_{\substack{j \in J \\ \alpha \neq \beta}} f_j \phi_{\alpha j} \phi_{\beta j} = 0.$$

9.5. Contribution Formulas

9.5.1. Contributions to $N(J)$

$M^2(N(J))$: total variance of the cloud $N(J) = \sum_{\alpha=2}^{N} \lambda_\alpha$;

λ_α: absolute contribution of the α-axis to the total variance of the cloud, $M^2(N(J)); 0 \leq \lambda_\alpha \leq 1$;

$\tau_\alpha = \dfrac{\lambda_\alpha}{M^2(N(J))}$: relative contribution of the α-axis to the total variance of the cloud, $M^2(N(J))$;

$\phi_{\alpha j}$: α-coordinate of the point j of J;

$f_j \phi_{\alpha j}^2$: absolute contribution of the point j of J to the variance of the α-axis, λ_α;

: absolute contribution of the pair (α, j) to the total variance of the cloud, $M^2(N(J))$;

$\text{ctr}_\alpha(j) = \dfrac{f_j \phi_{\alpha j}^2}{\lambda_\alpha}$: relative contribution of the point j of J to the variance of the α-axis, λ_α;

9. 2-D Correspondence Analysis Formulas

$\dfrac{f_j \phi_{\alpha j}^2}{M^2(N(J))}$: relative contribution of the pair (α, j) to the total variance, $M^2(N(J))$;

$\rho^2(j) = d^2(j, G)$: squared distance from $j \in J$ to G, center of gravity of the cloud $N(J)$ = eccentricity of $j \in J = \sum_{\alpha=2}^{N} \phi_{\alpha j}^2$;

$\phi_{\alpha j}^2$: absolute contribution of the point j of J to the squared distance $d^2(j, G)$ = absolute contribution to the eccentricity $\rho^2(j)$;

$\dfrac{\phi_{\alpha j}^2}{\rho^2(j)}$: relative contribution of the point j of J to the squared distance $d^2(j, G)$ = relative contribution to the eccentricity $\rho^2(j)$;

: squared cosine of the angle formed by the radius vector of $j \in J$ and the α-axis, similar to a squared coefficient of correlation = $\text{cor}_\alpha(j)$ (cf. Fig. 8.14);

$\text{qlt}_s(j) = \sum_{\alpha=1}^{s} \dfrac{\phi_{\alpha j}^2}{\rho^2(j)}$: quality of explanation of the point j of J in the factor space formed by s factor axes;

$\text{inr}(j) = f_j \rho^2(j)$: absolute contribution of the point j of J to the total variance of the cloud, $M^2(N(J))$;

$\dfrac{\text{inr}(j)}{M^2(N(J))}$: relative contribution of the point j of J to the total variance of the cloud, $M^2(N(J))$.

9.5.2. Contributions to $N(I)$

$M^2(N(I))$: total variance of the cloud $N(I) = \sum_{\alpha=2}^{N} \lambda_\alpha$

λ_α: absolute contribution of the α-axis to the total variance of the cloud, $M^2(N(I)); 0 \leq \lambda_\alpha \leq 1$;

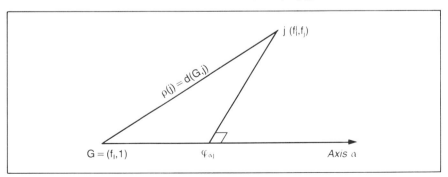

FIGURE 8.14. Geometric representation of the point j of J relative to the α-axis.

$\tau_\alpha = \dfrac{\lambda_\alpha}{M^2(N(I))}$: relative contribution of the α-axis to the total variance of the cloud, $M^2(N(I))$;

$\psi_{\alpha i}$: α-coordinate of the point i of I;

$f_i \psi_{\alpha i}^2$: absolute contribution of the point i of I to the variance of the α-axis;

: absolute contribution of the pair (α, i) to the total variance of the cloud, $M^2(N(I))$;

$\dfrac{f_i \psi_{\alpha i}^2}{\lambda_\alpha}$: relative contribution of the point i of I to the variance of the α-axis;

$\dfrac{f_i \psi_{\alpha i}^2}{M^2(N(I))}$: relative contribution of the pair (α, i) to the total variance, $M^2(N(I))$;

$\rho^2(i) = d^2(i, G)$: squared distance from $i \in I$ to G, center of gravity of the cloud $N(I)$ = eccentricity of the point $i \in I = \sum_{\alpha=2}^{N} \psi_{\alpha i}^2$;

$\psi_{\alpha i}^2$: absolute contribution of the point i of I to the squared distance $d^2(i, G)$ = absolute contribution to the eccentricity $\rho^2(i)$;

$\dfrac{\psi_{\alpha i}^2}{\rho^2(i)}$: relative contribution of the point i of I to the squared distance $d^2(i, G)$ = relative contribution to the eccentricity $\rho^2(i)$;

: squared cosine of the angle formed by the radius vector of $i \in I$ and the α-axis, similar to a squared coefficient of correlation = $\operatorname{cor}_\alpha(j)$ (*cf*. Fig. 8.15).

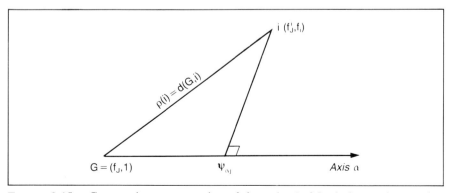

FIGURE 8.15. Geometric representation of the point i of I relative to the α-axis.

$\mathrm{qlt}_s(i) = \sum_{\alpha=2}^{s} \dfrac{\psi_{\alpha i}^2}{\rho^2(i)}$: quality of explanation of the point i of I in the factor space formed by s factor axes;

$\mathrm{inr}(i) = f_i \rho^2(i)$: absolute contribution of the point i of I to the total variance of the cloud, $M^2(N(I))$;

$\dfrac{\mathrm{inr}(i)}{M^2(N(I))}$: relative contribution of the point i of I to the total variance of the cloud $M^2(N(I))$.

9.6. Transition Formulas

$$\psi_{\alpha i} = \lambda_\alpha^{-1/2} \sum_{j \in J} f_j^i \phi_{\alpha j}, \qquad \alpha = 1, 2, \ldots, N, i \in I;$$

$$\phi_{\alpha j} = \lambda_\alpha^{-1/2} \sum_{i \in I} f_i^j \psi_{\alpha i}, \qquad \alpha = 1, 2, \ldots, N, j \in J.$$

9.7. Reconstruction Formulas

$$d^2(i, i') = \sum_{\alpha=1}^{N} (\psi_{\alpha i} - \psi_{\alpha i'})^2;$$

$$d^2(j, j') = \sum_{\alpha=1}^{N} (\phi_{\alpha j} - \phi_{\alpha j'})^2;$$

$$\rho^2(i) = \sum_{\alpha=1}^{N} \psi_{\alpha i}^2;$$

$$\rho^2(j) = \sum_{\alpha=1}^{N} \phi_{\alpha j}^2;$$

$$f_{ij} = f_i f_j \left(1 + \sum_{\alpha \geq 2}^{N} \lambda_\alpha^{-1/2} \psi_{\alpha i} \phi_{\alpha j}\right) \qquad \text{for } i = 1, 2, \ldots, n; j = 1, 2, \ldots, p.$$

10. Patterns of Clouds of Points

Computation of factors and the examination of the associated contributions are not sufficient enough for understanding data. Graphics are also essential in interpreting data. From time to time, typical shapes of clouds can arise in factor graphics and, sometimes, these patterns are related to standard types of data sets. Some typical patterns are given in what follows.

10.1. Elliptic Pattern

The elliptic pattern is related to Laplace–Gauss variables. Consider two normal (quantitative) variables, which are recoded into two dummy qualitative variables. The contingency data set associated with these two qualitative variables (the dispersion data table) is then built. The resulting graphic associated with such a data set shows an elliptic cloud (*cf.* Fig. 8.16).

10.2. Partition Pattern

After factor analysis, a typical data set containing the two sets $N(I)$ and $N(J)$ can be represented graphically as two subclouds (*cf.* Fig. 8.17). From I, there are two distinct subsets I_1 and I_2; from J, there are two distinct subsets J_1 and J_2. The elements of I are each assigned to one of the sets I_1 or I_2, and the same is done for J; thus, k_{IJ} is transformed into $k_{I_1 I_2 J_1 J_2}$. In this situation, the shape of k_{IJ} is as follows: There are two diagonal hachured subsets containing the highest values of the original data set, and the two off-diagonal subsets contain values close to zero.

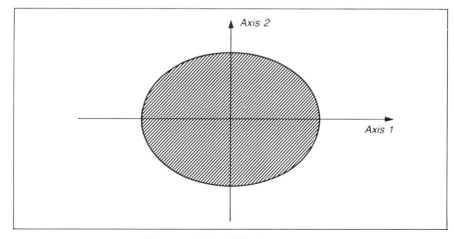

FIGURE 8.16. Elliptic pattern.

10. Patterns of Clouds of Points   223

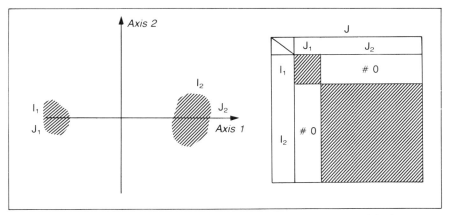

FIGURE 8.17. Data set and resulting factor graphic associated with the representation of a partition.

There are two different cases for the analyst:

1. If the number of elements in one subset is small, then the analysis must be redone with the other data set and then the elements of the first subset are classified into the second analysis as supplementary elements.

2. If the number of elements is approximately the same for the two subsets of I and J, then the analysis can be redone with $I_1 \times J_1$ and $I_2 \times J_2$ separately and each previous subset is classified as a supplementary element of the other one. When the two subsets are mathematically disjoint, corresponding to an exact partition, the first nontrivial eigenvalue is equal to 1; when there are n disjoint subsets from the original set $I \times J$ (instead of two as in the previous discussion), there are $(n-1)$ eigenvalues equal to 1. Conversely, if there are $(n-1)$ eigenvalues equal to 1, then there are n disjoint subsets of $I \times J$, and therefore there are two exact partitions of I and J into n classes.

10.3. Parabolic Pattern

Suppose that a graphical representation has a parabolic shaped cloud of points (cf. Fig. 8.18). If the points i of I and j of J are reordered according to their projections onto the first factor axis, and put into the data table k_{IJ} according to this order, then, this table has a certain form, as seen in Fig. 8.18.

On the diagonal, and on each side of it, the highest values occur, with the lowest ones being off the diagonal. The first factor axis is interpreted as an ordering of the sets I and J, and the other factor axes are

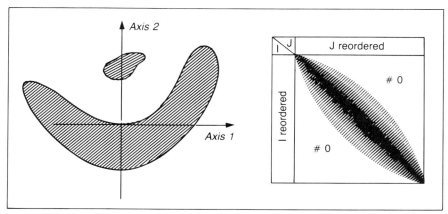

FIGURE 8.18. Data set and resulting factor graphic associated with the horseshoe effect.

polynomial functions of the first one. This effect is known as the Guttmann effect or horseshoe effect. In this case, a point i (or j) located at the inner boundary of the parabola corresponds to a row (or a column) whose profile f_j^i (or f_i^j) is "flat" in the data set k_{IJ}. Conversely, a point i (or j) located at the outer boundary of the parabola corresponds to a row (or a column) whose profile f_j^i (or f_i^j) is acute. The points located inside and at the center of the parabola have no specific meaning in the ordering.

10.4. Triangular Pattern

A cloud of points frequently has a triangular shape in correspondence analysis. This clearly means that, "for a constant value of the first factor," there is a variation on the other factor axis (cf. Fig. 8.19).

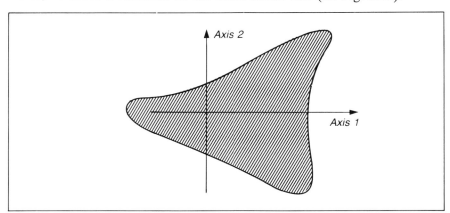

FIGURE 8.19. Triangular pattern.

10.5. General Case

When there is no specific structure, all the points must be studied using their contributions, the possibilities of classifying supplementary elements, and the graphical presentation of data.

11. Patterns of Acceptable Data Sets

11.1. Introduction

First applied to contingency data sets, correspondence analysis has been applied since to many types of data sets whose row-profiles and column-profiles were defined, but not necessarily in terms of conditional probabilities. During the 1965–1970 period, different types of data were studied: marks data sets, measurement data sets, logical data sets, questionnaires, and ranks data sets. In the following, different situations are studied, and the procedures that are needed to perform data analysis using correspondence analysis are examined. Some data sets are analyzed as they are; other are first recoded using dummy data sets.

11.2. Data Sets Analyzed as Such

11.2.1. Contingency Data Sets

A frequency data set, denoted by k_{IJ}, is a two-way data table containing the associations between the elements of I and J as follows: k_{ij}, an element of k_{IJ}, is the number of times that i is associated with j. This occurs when two qualitative variables derived from a questionnaire or a campaign of measurements are crossed (*cf.* Appendix 2, §5). Another example is the patents registration data set; the set I is the set of developed countries; the set J is the set of years; k_{ij} is the number of patents related to the year j for the country i (*cf.* Appendix 2, §3). Another example is the statistics of transmission errors. Two qualitative variables were selected: the type of switching units (I) and the category of errors (J) (number of errors grouped into categories); k_{ij} is the number of times where the category of errors j was associated with the type of switching unit i. A frequency data set can also be given by qualitative variables with multiple forms. Consider the crossing of one qualitative variable (J) having multiple forms with itself. The data set k_{JJ} is formed, where the element $k_{jj'}$ is the number of times that the form j is associated with the form j'. In addition, $k_{jj'} = k_{j'j}$, with $j \neq j'$. For the diagonal of the symmetric data set, there are two possibilities: either $k_{jj} = 0$ (the form j cannot be associated with itself) or $k_{jj} = n_j =$ number of occurrences of j (cf. Appendix 2, §5). Using qualitative variables with multiple forms gives other frequency data sets: a data set crossing one qualitative variable and another one having multiple forms; the data table crossing two qualitative variables having multiple forms (cf. Appendix 2, §5).

11.2.2. Extended Contingency Data Sets

Extended contingency data sets are the data sets k_{IJ} where k_{ij} is not associated with counting, but results from an association between i and j given in the same unit system (e.g., currency, volume, measure, duration, weight) (cf. Appendix 2, §10).

11.2.3. Measurements Data Sets

Measurements data sets can be analyzed as such if the variables involved (quantitative variables) represent a set of homogeneous measurements (such as sometimes occurs in geology, biology, or quantitative taxonomy). In general, however, it is preferable to analyze such data sets after creating a recoded dummy data set (cf. Section 11.3).

11.2.4. Marks and Ratios Data Sets

A number of data sets contain the following type of information: Consider a set of students I and a set of subjects J (cf. Appendix 2, §2), where k_{ij} is the mark obtained by the student i of I for the subject j of J. This type of data set exists in many types of studies, such as quality control based on indices, production indices, economic indices, tests of equipment, psychometric tests, etc. It also occurs in data sets produced by economists to study the investment risks as presented in Appendix 2, §9. Such data sets can be analyzed as such by principal components analysis, or by correspondence analysis, but it is generally preferable to study them using a recoded dummy data set (cf. Section 11.3). The same holds for ratio data sets.

11.2.5. Logical Data Sets

Logical data sets are the data sets k_{IJ} where the element $k_{ij} = 0$ or 1. In general, 1 corresponds to an association between i and j; 0 corresponds to no association. Such data sets occur in quantitative taxonomy (k_{ij} is the presence of an attribute i in a species j), in ecology (k_{ij} is the presence of a plant i in an area j), in archeology (k_{ij} is the presence of decoration j on a vase i), etc. They can be analyzed by correspondence analysis even if

11. Patterns of Acceptable Data Sets

they are not contingency data sets. However, a probabilistic interpretation of the relations cannot be given; only a geometric interpretation can be given. Sometimes, it is more relevant to study reverse dummy data sets, in which the value 1 is replaced by the value 0, and conversely, because the absence of association can be more significant than its presence. Sometimes, absence and presence have the same importance, so that it is necessary to analyze dummy recoded data sets as described in Section 11.3.2.

11.2.6. Logical Data Sets from Multiple Forms Variables

Consider a qualitative variable with multiple forms observed on a population I. Let J be the set of the possible forms; $k_{ij} \in k_{IJ}$ has value 1 for the elements j of J having a positive answer, otherwise 0 (*cf.* Table 8.27).

This data set can be analyzed as it is, using correspondence analysis, or it can be analyzed in a recoded form as indicated in Section 11.3.

11.3. Data Sets Recoded and Analyzed

11.3.1. Recoding a Data Set

Data sets must represent faithfully the relationships between elements of I and J. The framework of observation is usually precise, so that the data set is well determined, but this is not always so. In the latter situation, another dummy data sets must be created by the analyst, so that more

TABLE 8.27. Representation of a data set from a variable with multiple forms (k positive responses among p possible ones).

I \ J	1	...	j	...	p	k_I
1						
.						
.						
.						
i	0	0	1	0	1 1	$k_i = k$
.						
.						
.						
n						

(INDIVIDUALS on vertical axis, VARIABLES on horizontal axis)

than one data set may be involved. The meaning of recoding a data set is to find a data set that represents mathematically the relationships between elements. This operation is described by Benzecri (1982) as follows:

> "The essence of recoding data is to translate precisely the relationships observed by mathematical relations, in such a way that the mathematical structure chosen to represent the reality and synthesized by computer gives a simplified image open to intuition and interpretation with the guarantee of mathematical verification."

This shows that numerous data sets must be built by the analyst. In what follows, we present several model cases of recoded data sets.

11.3.2. Dummy Doubled Data Sets

11.3.2.1. Marks Data Sets. Consider the following example, entitled "economic data set concerning investments abroad" (*cf.* Appendix 2, §9); 43 countries and 15 confidence criteria were selected. The marks given to each criteria varied between 0 and 4 (0 for the worst and 4 for the best), and k_{ij} is the mark given by the expert to the confidence variable j for the country i. To give a marginal mark that is equal for each country, a dummy data set denoted by k_{IJ^-} is built as follows:

$$k_{ij^-} = (\text{maximum value of } j) - k_{ij} \quad \text{for } i \in I \text{ and } j \in J.$$

A representation of such a data set is given in Table 8.28. Studying the

TABLE 8.28. Doubled data set. k_{IJ^-} is the dummy doubled data set.

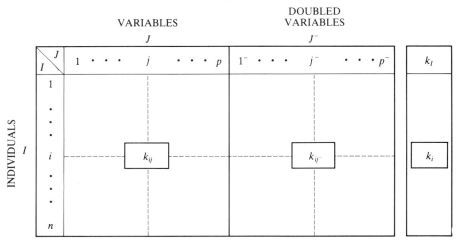

11. Patterns of Acceptable Data Sets

data by correspondence analysis can take place as follows:

- analyze k_{IJ} as such;
- analyze $k_{IJ \cup J^-}$ (the data set J^- is associated with the data set J);
- analyze k_{IJ^-} (if the mistrust is analyzed instead of the confidence).

11.3.2.2. Questionnaire Data Sets. In the case of a questionnaire with two possible replies for each question (1 for yes and 0 for no), the doubling operation can be done as follows: a dummy question, denoted by q^-, is associated with each question such that

$$k_{iq^-} = 1 - k_{iq} \quad \text{for } i \in I \text{ and } q \in J.$$

A representation of such a data set is given in Table 8.29.

The basic data set and the dummy one, denoted by $k_{IJ \cup J^-}$, can be analyzed together by correspondence analysis. The interest of such a data set is the following: On graphics, the points q and q^- will correspond to the two possible replies (yes and no).

11.3.2.3. Preference Data Sets. Consider the data set formed by students (set I) and subjects (set J); k_{ij} is the rank of preference given by the student i to the subject j (1 is the best rank; k_j is the worst rank). A dummy data set denoted by k_{IJ^-} is formed as follows:

$$k_{ij^-} = 1 + \max(k_{ij}) - k_{ij} + \quad \text{for } i \in I \text{ and } j \in J.$$

A representation of such a data set is given in Table 8.30.

TABLE 8.29. Representation of a data set involving questions with two possible replies. p is the number of questions.

		VARIABLES J			DOUBLED VARIABLES J^-			k_I
		1 ...	j	... p	1^- ...	j^-	... p^-	
INDIVIDUALS	1	1 1 1 0 1			0 0 0 1 0			p p . . .
I	i	1			0			p
	. . n	0 1 1 0 1			1 0 0 1 0			. . p

TABLE 8.30. Representation of a data set involving ranks or preferences.

	RANK VARIABLES J					DOUBLED RANK VARIABLES J^-					
I $\diagdown$ J	1	$\cdots$	j	$\cdots$	p	1^-	$\cdots$	j^-	$\cdots$	p^-	k_I
1	1	5	3	4	2	5	1	3	2	4	
$\vdots$											
i			r_{ij}					r_{ij^-}			k_i
$\vdots$											
n											

(INDIVIDUALS I)

Processing data analysis takes place as follows:

- analyze k_{IJ} as such;
- analyze $k_{IJ \cup J^-}$ (the data set J^- is associated with J);
- analyze k_{IJ^-} (if the non-preference is preferred to preference).

11.3.2.4. Questionnaire with Multiple Forms Data Sets. Consider a variable with multiple forms observed on a set I. There are k forms chosen among p possible ones (with $k \leq p$); k_{ij} is equal to 1 if i has given a positive reply to j and 0 if i has given a negative reply to j; this gives a logical data set. A dummy data set denoted by k_{IJ^-} is formed as follows: k_{ij^-} is equal to $1 - k_{ij}$. A representation of these two previous data sets is given in Table 8.31.

Processing this by correspondence analysis takes place as follows:

- analyze k_{IJ} as such;
- analyze k_{IJ^-};
- analyze $k_{IJ \cup J^-}$;
- build and analyze the crossed data set from k_{IJ}, denoted by k_{JJ^-};
- build and analyze the crossed data set from k_{JJ^-}, denoted by $k_{J^-J^-}$;
- build and analyze the crossed data set from $k_{IJ \cup J^-}$, denoted by $k_{J \cup J^- J \cup J^-}$.

11. Patterns of Acceptable Data Sets

TABLE 8.31. Representation of a data set from a qualitative variable with multiple forms. $k_i = p$ = number of possible forms.

	VARIABLES J					DOUBLED VARIABLES J^-						
$I \backslash J$	1 ...	j	... p			1^- ...	j^-	... p^-				k_I
INDIVIDUALS I												
i	0 1	0 1	0 1	0		1 0	1 0	1 0	1			k_i
n												

The final choice by the analyst depends on what needs to be highlighted in the data.

11.3.3. Dummy Disjunctive Form Data Sets

Dummy disjunctive form data sets are dummy data sets associated with qualitative variables. Since they are connected with multiple correspondence analysis, their mathematical description will be given in the next chapter. Note that disjunctive from data sets can be applied in various important fields, such as for questionnaires, panels, measurement campaigns, data bases, and information systems (*cf.* Chapter 9).

11.3.4. Chronological Data Sets

All the previous data sets can be considered as representing situations on a *given date*. The same data sets can be studied using a new qualitative variable, whose forms are time intervals. Formally, the chronological data sets can represented as follows:

$$k_{IJT} = \{k_{ijt}, \text{ where } i \in I; j \in J; t \in T\}.$$

Correspondence analysis can be applied satisfactorily to binary data sets, denoted k_{IJ}, but not to the sets k_{IJT}. How can correspondence analysis be applied here? The procedure for classifying supplementary elements can be used as follows: let k_{IJ} be such that:

$$k_{IJ} = \sum_{t_0}^{t_n} k_{IJt};$$

k_{IJ} is the cumulative data set associated with the set of time intervals. Now, K_{IJ} is considered as the reference data set on which correspondence analysis is done. The other data sets $k_{IJt_0}, \ldots, k_{IJt_n}$ are classified as supplementary elements. This classifying is made "in rows" and "in columns," and the point i is represented as many times as there are time intervals $(i(t_0), i(t_1), \ldots, i(t_n), i(k_{IJ}))$, and the point j is represented as many times as there are time intervals $(j(t_0), j(t_1), \ldots, j(k_{IJ}))$. This allows the analyst to follow the trajectory of each point i of I or j of J. The data sets are arranged as presented in Table 8.32.

12. Case Studies

12.1. Contingency Data Sets

The type of contingency data set described in Section 11.2 that is come across most frequently is generally based on the crossing of two categorical (or qualitative) variables, or on the crossing of two quantita-

TABLE 8.32. Chronological data sets ordered for application of correspondence analysis.

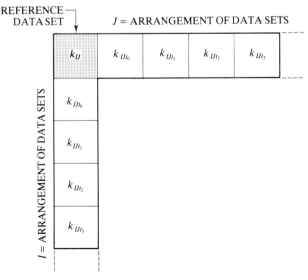

12. Case Studies

tive variables via the intermediary of a correlation data table. An example of correspondence analysis of a contingency data set is given below based on the crossing of two qualitative variables concerning the survey on Minitel services and the electronic directory (cf. Appendix 2, §5).

Question 1. How much time do you spend each week using the Minitel services?

    Code 1 = <5 min;
    Code 2 = from 5 to 10 min;
    Code 3 = from 11 to 15 min;
    Code 4 = from 16 to 30 min;
    Code 5 = from 31 to 45 min;
    Code 6 = from 46 to 60 min;
    Code 7 = from 14 to 24 min;
    Code 8 = from 2 to 34 min;
    Code 9 = ≥34 min.

Question 2. Are you satisfied by the Minitel services?

    Code 1 = yes, very satisfied;
    Code 2 = yes, rather satisfied;
    Code 3 = no, rather unsatisfied;
    Code 4 = no, very unsatisfied.

Correspondence analysis results are given in Fig. 8.20.

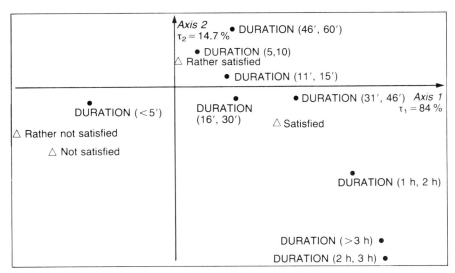

FIGURE 8.20. Analysis of the contingency data set from the questionnaire on the Minitel services. Representation in the factor space (1, 2).

12.2. Contingency Data Sets from Multiple Forms Variables

Contingency data sets from multiple forms variables can be formed from questionnaires. The crossing of two variables with multiple forms can give a contingency data set as follows: consider two questions from the survey on the Minitel services (*cf.* Appendix 2 §5).

Question 1. What are your main wishes relating to the Minitel services?
        Code 1 = faster display;
        Code 2 = faster input;
        Code 3 = color display;
        Code 4 = less bulky;
        Code 5 = to use a microcomputer;
        Code 6 = evolution to a microcomputer;
        Code 7 = no wishes;
        Code 8 = other responses;
        Code 9 = more aesthetic;
        Code 10 = ergonomical input;
        Code 11 = telephone set included.

Question 2. What are the reasons for using the Minitel services?
        Code 1 = for information;
        Code 2 = for curiosity;
        Code 3 = for fun;
        Code 4 = to learn the use of it;
        Code 5 = to show to people;
        Code 6 = it is fast;
        Code 7 = it is useful;
        Code 8 = no response.

Correspondence analysis results are given in Fig. 8.21.

From a questionnaire involving variables with multiple forms, other contingency data sets can be built as follows: each variable is crossed with itself, to give a contingency data set that can be analyzed by correspondence analysis. For example, one of the previous questions related to the reasons for using the Minitel services is taken (*cf.* Question 2). The correspondence analysis results are given in Fig. 8.22.

12.3. Confusion Data Sets

Confusion data sets occur in learning processes in pattern recognition. On one hand, there is a set of stimuli, denoted by I, and on the other

12. Case Studies

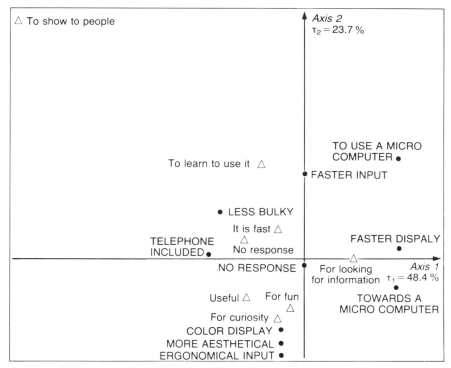

FIGURE 8.21. Analysis of the contingency data set from two qualitative variables with multiple forms from the survey on the Minitel services. Representation in the factor space (1, 2).

hand, there is a set of responses given to the stimuli, denoted by J. I and J have the same size and the same elements; the element n_{ij} of the data set n_{IJ} is the number of times that the stimulus i is associated with the response j. Consider Shepard's experiment: The stimuli are eight circles whose radii are marked as given in Fig. 8.23. The subjects undergoing the experiment learn to associate a letter (D, H, K, M, O, R, S, W) with each stimulus; n_{ij} is the number of times that the stimuli i is associated with responses corresponding to the stimulus j. The contingency data set n_{IJ} is called a confusion data set. If $n_{ij} = 0$ (with $i \neq j$), then the learning process is almost perfect. The errors in the learning process are above and below the diagonal. Correspondence analysis can be done to study the structure of the errors in the learning process.

12.4. Flow Data Sets

Flow data sets are also contingency data sets. On one hand, there is a set of individuals, denoted by I, and on the other hand, there is the same set

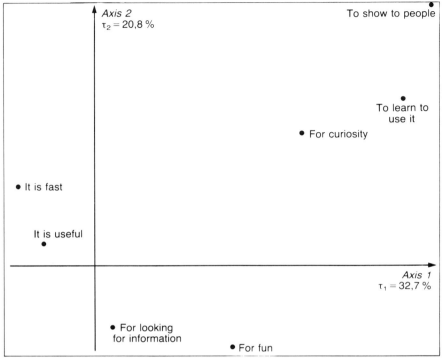

FIGURE 8.22. Analysis of the contingency data set given by the crossing of a qualitative variable with itself. Representation in the factor space (1, 2).

of individuals, also denoted by I. I is crossed with itself. The element n_{ij} of n_{II} is the number of times (or amount of currency, number of durations,) that i is associated with j. Such data sets occur in economics (exchange matrix), in traffic, and in demography, and can be analyzed by correspondence analysis as such. The value of n_{ii} represents the inner exchanges.

12.5. Marks Data Sets

Consider the following data set formed by the trade experts to advice their patrons on investment. Forty-three countries were judged according to 15 criteria of confidence (politic, economic, financial) (*cf.* Appendix 2, §9). The marks given by the experts were between 0 and 4. To each criterion denoted by j^+, is assigned a dummy criterion, denoted by j^-,

12. Case Studies

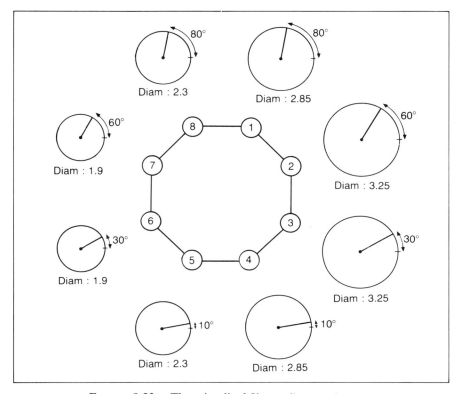

FIGURE 8.23. The stimuli of Shepard's experiment.

determined as follows:

$$k_{ij^-} = 4 - k_{ij^+} \quad \text{for } i \in I \text{ and } j \in J.$$

Correspondence analysis was applied on the doubled data set $k_{IJ \cup J^-}$, giving the results in Fig. 8.24.

Marks data sets are involved in any study concerning quality, choice, confidence, or indicators. The reader is invited to study the graphics in order to learn how to use that information efficiently.

12.6. Preferences Data Sets

Such data sets are similar to the previous ones, though instead of giving marks, the individuals interviewed must indicate an order for some given

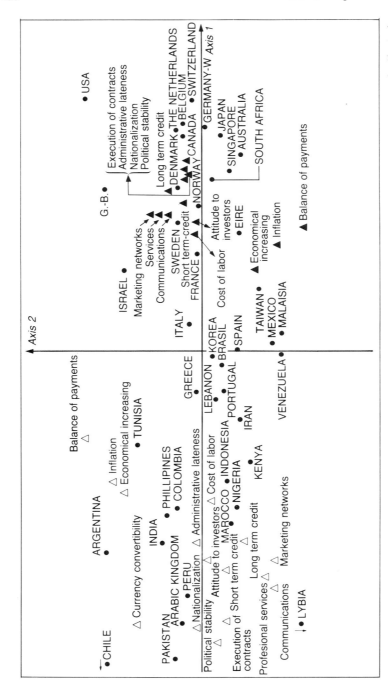

FIGURE 8.24. Analysis of a marks data set concerning the risk of investment. Representation in the factor space $(1, 2) \cdot j^-$ is denoted by $\triangle$; j^+ is denoted by ▲.

12. Case Studies

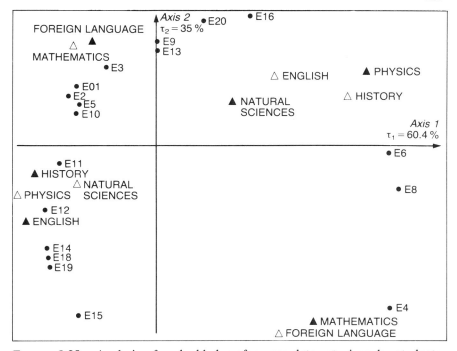

FIGURE 8.25. Analysis of a doubled preferences data set given by students. Representation in the factor space (1, 2). j^+ are denoted by △; j^- are denoted by ▲.

criteria. The given example is based on the preferences of students for certain subjects.

The complete data set is given in Appendix 2, §4. This data set can be analyzed as such, but it is preferable to use the doubled procedure as such described in Section 11.3.2. Correspondence analysis results of the doubled data set are given in Fig. 8.25.

Chapter 9

N-D Correspondence Analysis

1. Introduction

The method of N-D correspondence analysis is not actually new, but is derived from 2-D correspondence analysis. It is widely applied to data in many fields, such as economics, marketing, business sciences, psychometrics, biometrics, geology, social sciences, etc., where the data can originate from questionnaires, statistical files, data bases, panels, opinion samples, or measurements campaigns. It is the essential tool for any specialist in data analysis. Sometimes, N-D correspondence analysis is known as "multiple correspondence analysis" because correspondence analysis originally involved the crossing of two qualitative variables, giving 2-D correspondence analysis, and then was extended to a disjunctive form data set with N qualitative variables, giving N-D correspondence analysis. Since the mathematical descriptions of 2-D and N-D correspondence analysis are very similar, only the additional mathematical formulas useful for N-D correspondence analysis will be given here. Now, we describe the contents of this chapter. First, the basic data sets that can be analyzed by N-D correspondence analysis are studied, followed by the mathematical model, the rules for interpretation illustrated by an actual example using graphical outputs and, finally, some real case studies in order to demonstrate the advantages of the method. The references of this chapter are from Benzecri (1973, 1980, 1984), Cazes (1980, 1982), Greenacre (1984), Jambu (1978, 1983), Lebart (1979, 1984), and from numerous concrete applications published in the journal *"Les cahiers de l'analyse des données."*

2. Basic Data Sets

2.1. Data Sets with Qualitative and Quantitative Variables

Consider a questionnaire involving two sets, denoted by I (the individuals) and by J (the variables).

• The set I of individuals can be: plants (natural sciences, taxonomy), animals (zoology), minerals (geology), cells (biology), pieces or objects (archeology), exchanges (finances), regions (geography), countries (politics, geography), stars (astronomy), components (industry), points of observations (meteorology), points of networks (telecommunications), enterprises (economics), patients (medicine), people interviewed (marketing), etc. This nonexhaustive list shows that the term "individuals" can cover numerous types of sets. This term is used to name any "statistical unit." It also shows that individuals can be different from human populations, which was not the case initially since N-D correspondence analysis was applied only to survey data.

• The set J of variables can be: responses to questions (categorical variables), attributes (archeology), characteristics (taxonomy), measurements (geology, biology), descriptors (medicine), indices (quality of service, economics, finances), ratios (geology, economics), etc. Whatever name is given, the set J is a set of qualitative and/or quantitative variables.

• The element of V_{IJ} denoted by v_{IJ} is the information given by the variable j of J on the unit i on I, as presented in Table 9.1. This table shows the quasi-universal model of a data set generally used in data analysis. From it, various subsets can be formed such as contingency data sets (two-way table formed by the crossing of two qualitative variables), logical data sets formed by 1/0 data sets, measurements data sets formed by quantitative variables, etc.

2.2. Data Sets Associated with the Basic Data Sets

Correspondence analysis cannot be applied to the previous basic data sets as such. To use the mathematical model correctly, dummy data sets need to be formed and then N-D correspondence analysis is applied to these. First, we study qualitative data sets, followed by quantitative data sets, and finally data sets mixing qualitative and quantitative data sets.

2. Basic Data Sets

TABLE 9.1. Pattern of basic data set involving qualitative and quantitative variables.

	QUALITATIVE VARIABLES			QUANTITATIVE VARIABLES			
I \ J	V_1 V_2	...		... V_j		... V_p	
1							
2							
⋮							
i	v_{i1} v_{i2}	-- -- --	-- --	v_{ij}	-- -- -- -- -- --	v_{ip}	values of the variables $V_1, V_2, ..., V_p$ for the individual i
⋮							
n							

(INDIVIDUALS)

2.2.1. Disjunctive Form Data Sets Associated with Qualitative Variables

Consider the basic data set, denoted by k_{IQ}, where I is the set of individuals and Q is the set of qualitative variables. With k_{IQ} is associated a dummy data set, denoted by k_{IJ}, where I is the set of individuals and J is the set of the whole forms (or modalities), denoted by J_q, associated with the qualitative variables q of Q. Therefore,

$$J = \bigcup \{ \{J_q\} ; q \in Q \}.$$

For any individual i of I and q of Q, it is clear that the individual has given one and only one response to the qualitative variable q of Q. Therefore, any two forms j and j' from a qualitative variable q are mutually exclusive. Therefore,

$$\forall i \in I, \forall j \in J_q; \quad k_{ij} = 1, \quad \text{if } i \text{ has been given the form } j \text{ of } J_q,$$

$$k_{ij} = 0 \quad \text{if not;}$$

then

$$k_{ij} = 1 \text{ or } 0 \quad \forall i, j \in I \times J.$$

k_{IJ} is called the disjunctive form data set associated with k_{IQ}.

The following relations hold:

$$\sum_{j \in J_q} k_{ij} = 1;$$

$$k_i = \sum_{j \in J} k_{ij} = \sum_{q \in Q} \left(\sum_{j \in Q} k_{ij} \right) = n_Q,$$

where n_Q is the number of qualitative variables;

$$k_j = \sum_{i \in I} k_{ij} = \text{number of individuals having been given the form } j.$$

Therefore,

$$\sum_{j \in J_q} k_j = n_I,$$

where n_I is the number of individuals of I;

$$k = \sum_{i \in I} \left(\sum_{j \in J} k_{ij} \right) = n_I \cdot n_Q$$

Finally, $k_{IJ} = k_I(J_1 J_2 \cdots J_q)$. The basic data set and its dummy data set put into disjunctive form are presented in Table 9.2.

The previous dummy data set can be formed from data sets associated with questionnaires.

TABLE 9.2. Model of qualitative variables data set and its associated disjunctive form data set.

	QUALITATIVE VARIABLES					QUALITATIVE VARIABLES AND ASSOCIATED FORMS														
	J_1	J_2	J_3	J_4	J_5		J_1			J_2				J_3				J_4		J_5
I	1	2	3	4	5	I	1	2	3	1	2	3	4	1	2	3	4	1	2	3	1	2	3

	QUALITATIVE VARIABLES					QUALITATIVE VARIABLES AND ASSOCIATED FORMS																	
	J_1	J_2	J_3	J_4	J_5		J_1			J_2				J_3				J_4			J_5		
I	1	2	3	4	5	I	1	2	3	1	2	3	4	1	2	3	4	1	2	3	1	2	3
1	1	1	4	2	1	1	1			1								1			1		
2	2	1	4	2	2	2		1		1								1				1	
3	2	1	3	2	3	3		1		1						1		1					1
4	2	2	3	1	3	4		1			1					1		1					1
5	1	2	2	3	3	5	1				1				1					1			1
6	3	2	1	1	3	6			1		1			1				1					1
7	3	3	2	2	3	7			1			1			1				1				1
8	3	3	3	2	2	8			1			1				1			1			1	
9	1	4	3	3	1	9	1						1			1				1	1		
10	1	3	3	1	1	10	1					1				1		1			1		
11	2	4	2	3	1	11		1					1		1					1	1		
12	2	3	2	3	2	12		1				1			1					1		1	

2. Basic Data Sets

2.2.2. Disjunctive Form Data Sets Associated with Quantitative Variables

In a set of quantitative variables, each quantitative variable is associated with a dummy qualitative variable as follows: The range of a quantitative variable is divided into a given number of intervals. These intervals determine classes of values, and a form is associated with each of the intervals. Thus, a dummy qualitative variable is formed for each interval. For any i and any j, the value v_{ij} will fall within a certain interval; a form is then associated with the interval as for any qualitative variable. Then these dummy qualitative variables form a disjunctive form data set. The process of obtaining a disjunctive form data set from quantitative variables is summarized in Table 9.3.

In conclusion, any quantitative data set can be transformed in a qualitative data set using the previous procedure. How should the intervals for the classes be chosen? This question has no theoretical response, but the user will eventually gain experience to help him in his choice. Nevertheless, some guidelines can be obtained through the use of univariate analysis, which gives information on the distribution of quantitative variables. According to the distribution shape, either the values can be divided into equally ranged classes, the values can be divided into classes with equal frequency, or the class intervals can be based on specific values. The choice of one way or the other has relatively little influence on the results given by N-D correspondence analysis. It is of course possible to verify this, choosing several different

TABLE 9.3. Model of quantitative variables data set and its associated disjunctive form data set.

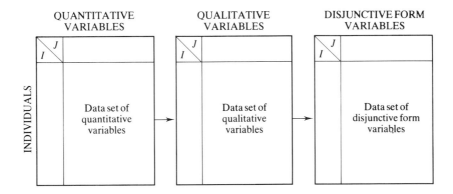

intervals, applying correspondence analysis to them, and then comparing graphically the results.

2.2.3. Disjunctive Form Data Sets Associated with Qualitative and Quantitative Variables

From the two previous sections, it is clear that disjunctive form data sets can be associated with qualitative and quantitative variables in a process described in Table 9.4.

2.2.4. N-D Contingency Data Sets from Qualitative and Quantitative Variables

Consider any disjunctive form data set as described in the previous section, denoted by k_{IJ} (I = set of individuals; J = set of forms), derived from qualitative and quantitative variables data sets. An N-D contingency data set, denoted by b_{JJ}, can be associated with k_{IJ} as follows:

$$\forall j, j' \in J, \ b_{jj'} = \sum_{i \in I} k_{ij} k_{ij'} \qquad (b_{jj'} \text{ is an element of } b_{JJ}).$$

Therefore,

$$\forall j, j' \in J_q, \quad b_{jj'} = 0 \quad \text{if } j \neq j',$$
$$= k_j \quad \text{if } j = j',$$

$$b_j = \sum_{j \in J} b_{jj'} = k_j n_Q,$$

$$b = \sum_{j,j \in J} b_{jj'} = n_I n_Q^2.$$

The N-D contingency data set is an arrangement of the 2-D contingency data sets formed by all pairs of qualitative variables (or dummy qualitative variables associated with quantitative variables). The diagonal 2-D contingency data sets are filled with zeros except on their diagonals, which contain the marginal frequencies of the forms ($b_{jj} = k_j$). The N-D contingency data set is also known as the multiple contingency data set (cf. Table 9.5).

2.2.5. Example of Data Transformations

The previous transformations are illustrated by the example of the family timetables data set (*cf*. Table 9.6). Each quantitative variable, associated with a daily activity, is divided into three classes of activity of equal frequency. Then, the disjunctive form data set is formed, followed by the N-D contingency data set (*cf*. Tables 9.7 and 9.8).

2. Basic Data Sets

TABLE 9.4. Models of qualitative and quantitative variables data sets and their associated disjunctive form data sets, which are then gathered in a unique disjunctive form data set (3).

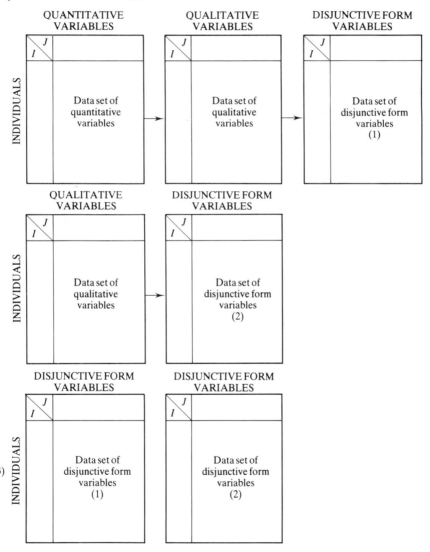

TABLE 9.5. Model of an N-D contingency data set.

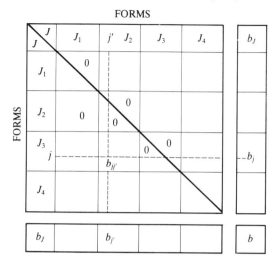

2.2.6. Conclusion

The principle of N-D correspondence analysis is based on the application of 2-D correspondence analysis to disjunctive form data sets obtained from qualitative or quantitative variables. The results obtained may be interpreted mathematically since there is a mathematical equivalence between the 2-D correspondence analysis of a disjunctive form data set and the 2-D correspondence analysis of its associated N-D contingency data set; this is proved in the following section.

3. Equivalence between the Analyses of b_{JJ} and k_{IJ}

3.1. General Case

Consider the correspondence data set $k_{IJ} = \{k_{ij}; i \in I; j \in J\}$, from which a new data set, denoted by g_{JJ}, is formed as follows:

$$g_{jj'} = \sum_{i \in I} \frac{k_{ij} k_{ij'}}{k_i} \quad \text{for } j, j' \in J.$$

3. **Equivalence between the Analyses of b_{JJ} and k_{IJ}**

TABLE 9.6. Family timetables data set (cf. Appendix 2 §10).

I \ J	WORK	TRANSPORT	HOUSEHOLD	CHILDREN	SHOPPING	PERSONAL CARE	MEALS	SLEEP	TELEVISION	LEISURE
EM. USA	610	140	60	10	120	95	115	760	175	315
EW. USA	475	90	250	30	140	120	100	775	115	305
UW. USA	10		495	110	170	110	130	875	160	430
MM. USA	615	140	65	10	115	90	115	765	180	305
MW. USA	179	29	421	87	161	112	119	776	143	373
SM. USA	585	115	50		150	105	100	760	150	385
SW. USA	482	94	196	18	141	130	96	775	132	336
EM. WEST	653	100	95	7	57	85	150	808	115	330
EW. WEST	511	70	307	30	80	95	142	816	87	262
UW. WEST	20	7	568	87	112	90	180	843	125	368
MM. WEST	655	97	97	10	52	85	152	807	122	320
MW. WEST	168	22	528	69	102	83	174	824	119	311
SM. WEST	643	105	72		62	77	140	813	100	388
SW. WEST	429	34	262	14	92	97	147	849	84	392
EM. EAST	650	142	122	22	76	94	100	764	96	334
EW. EAST	578	106	338	42	106	94	92	752	64	228
UW. EAST	24	8	594	72	158	92	128	840	86	398
MM. EAST	652	133	134	22	68	94	102	763	122	310
MW. EAST	436	79	433	60	119	90	107	772	73	231
SM. EAST	627	148	68		88	92	86	770	58	463
SW. EAST	434	86	297	21	129	102	94	799	58	380
EM. YUG	630	140	120	15	85	90	105	760	70	365
EW. YUG	560	105	375	45	90	90	95	745	60	235
UW. YUG	10	10	710	55	145	85	130	815	60	380
MM. YUG	650	145	112	15	85	90	105	760	80	358
MW. YUG	260	52	576	59	116	85	117	775	65	295
SM. YUG	615	125	95		115	90	85	760	40	475
SW. YUG	433	89	318	23	112	96	102	774	45	408

INDIVIDUALS

TABLE 9.7. Disjunctive form data set associated with the family timetables data set (each original variable is divided into three dummy logical variables).

FORMS / INDIVIDUALS

I \ J	WORK$_1$	WORK$_2$	WORK$_3$	TRANSPORT$_1$	TRANSPORT$_2$	TRANSPORT$_3$	HOUSEHOLD$_1$	HOUSEHOLD$_2$	HOUSEHOLD$_3$	CHILDREN$_1$	CHILDREN$_2$	CHILDREN$_3$	SHOPPING$_1$	SHOPPING$_2$	SHOPPING$_3$	PERSONAL CARE$_1$	PERSONAL CARE$_2$	PERSONAL CARE$_3$	MEALS$_1$	MEALS$_2$	MEALS$_3$	SLEEP$_1$	SLEEP$_2$	SLEEP$_3$	TV$_1$	TV$_2$	TV$_3$	LEISURE$_1$	LEISURE$_2$	LEISURE$_3$
EM. USA		1				1	1			1					1		1		1			1					1	1		
EW. USA		1				1	1			1					1	1			1			1					1	1		
UW. USA	1				1		1							1		1				1		1				1				1
MM. USA	1				1			1		1				1		1				1			1			1			1	
MW. USA		1				1			1		1			1		1				1			1		1				1	
SM. USA	1			1				1					1			1			1				1		1			1		
SW. USA	1			1				1			1		1			1			1				1		1			1		
EM. WEST			1			1			1			1			1			1			1			1			1			1
EW. WEST			1			1			1			1			1			1			1			1			1			1
UW. WEST		1			1			1						1			1			1				1		1			1	
MM. WEST		1			1			1			1			1			1			1				1		1			1	
MW. WEST		1			1			1			1			1			1			1				1		1			1	
SM. WEST	1			1			1						1			1			1					1	1			1		
SW. WEST	1			1			1				1		1			1			1					1	1			1		
EM. EAST			1			1			1						1			1			1			1			1			1
EW. EAST			1			1			1						1			1			1			1			1			1

3. Equivalence between the Analyses of b_{IJ} and k_{IJ}

TABLE 9.8. *N-D* contingency data set from the family timetables data set given in Table 9.6.

FORMS

J \ J	WORK₁	WORK₂	WORK₃	TRANSPORT₁	TRANSPORT₂	TRANSPORT₃	HOUSEHOLD₁	HOUSEHOLD₂	HOUSEHOLD₃	CHILDREN₁	CHILDREN₂	CHILDREN₃	SHOPPING₁	SHOPPING₂	SHOPPING₃	PERSONAL CARE₁	PERSONAL CARE₂	PERSONAL CARE₃	MEALS₁	MEALS₂	MEALS₃	SLEEP₁	SLEEP₂	SLEEP₃	TV₁	TV₂	TV₃	LEISURE₁	LEISURE₂	LEISURE₃
WORK₁	10	0	0	8	2	2	0	3	7	1	2	7	0	5	5	4	1	5	2	3	5	0	5	5	4	3	3	2	4	4
WORK₂	0	10	0	2	4	4	4	4	2	4	4	2	2	4	4	4	3	3	6	3	1	6	3	1	4	2	4	7	1	2
WORK₃	0	0	8	0	3	5	6	2	0	6	2	0	8	0	0	5	3	0	3	2	3	4	1	3	2	4	2	1	5	2
TRAN₁	8	8	0	10	0	0	0	2	8	1	1	8	1	5	4	5	2	3	0	4	6	0	4	6	3	4	3	4	3	3
TRAN₂	2	4	3	0	9	0	3	3	1	3	5	1	4	2	3	4	1	4	6	0	3	2	4	3	4	3	2	3	4	2
TRAN₃	0	4	5	0	0	9	7	2	0	7	2	0	5	2	2	4	4	1	5	4	0	8	1	0	3	2	4	3	3	3
HOUS₁	0	4	6	0	3	7	10	0	0	10	0	0	6	5	2	7	2	1	3	4	3	6	1	3	3	3	4	2	4	4
HOUS₂	3	4	2	2	5	2	0	9	0	1	8	0	3	3	3	0	4	5	7	0	2	3	4	2	3	4	2	4	4	2
HOUS₃	7	2	0	8	1	0	0	0	9	0	0	9	1	4	4	6	1	2	1	4	4	1	4	4	4	2	3	4	3	2
CHIL₁	1	4	6	1	3	7	10	1	0	11	0	0	4	5	2	7	2	2	3	4	4	6	1	4	3	4	4	2	4	5
CHIL₂	2	4	2	1	5	2	0	8	0	0	8	0	3	2	3	0	4	4	7	0	1	3	4	1	3	3	2	4	3	1
CHIL₃	7	2	9	8	1	0	0	0	9	0	0	9	1	4	4	6	1	2	1	4	4	1	4	4	4	2	3	4	3	2
SHOP₁	0	2	8	1	4	5	6	5	1	6	3	1	10	0	6	6	4	0	4	2	4	5	1	4	3	5	2	3	5	2
SHOP₂	5	4	0	5	2	2	2	3	4	3	2	4	9	9	0	6	1	3	3	3	3	3	3	3	5	2	2	5	1	3
SHOP₃	5	4	0	4	3	2	3	3	4	2	3	4	0	0	9	1	2	6	4	3	2	2	5	2	2	2	5	2	4	3
CARE₁	4	4	5	5	4	4	7	0	6	7	0	6	6	6	1	13	0	0	3	5	6	5	2	6	6	4	3	3	6	2

FORMS

3. Equivalence between the Analyses of b_{JJ} and k_{IJ}

The following relations exist:

$$g_j = \sum_{j' \in J} g_{jj'} = \sum_{j' \in J} \sum_{i \in I} \frac{k_{ij} k_{ij'}}{k_i}$$

$$= \sum_{i \in I} \frac{k_{ij}}{k} \sum_{j' \in J} k_{ij'}$$

$$= \sum_{i \in I} k_{ij}$$

$$= k_j \, ;$$

$$g = \sum_{j,j' \in J} g_{jj'} = \sum_{j \in J} g_j = \sum_{j \in J} k_j = k.$$

Correspondence analysis of g_{JJ} amounts to diagonalizing the matrices (UU') and $(U'U)$, U being the matrix whose term $U_{jj'}$ is determined as follows:

$$U_{jj'} = \frac{\left(g_{jj'} - \dfrac{g_j g_{j'}}{g}\right)}{g_j g_{j'}} \quad \text{for } j, j' \in J.$$

As $g_j = k_j$, $g_{j'} = k_{j'}$, and $g = k$,

$$U_{jj'} = \sum_{i \in I} \frac{\dfrac{k_{ij} k_{ij'}}{k_i} - \dfrac{k_j k_{j'}}{k}}{k_j k_{j'}},$$

$U_{jj'} = t_{jj'}$,   term of the matrix T diagonalized by 2-D correspondence analysis.

Then,

$$U = U' = T.$$

Thus, correspondence analysis of g_{JJ} amounts to diagonalizing the matrix T^2, so we have the following results:

- the factors of J with unit variance from analysis of g_{JJ} are identical to those with unit variance from analysis of k_{IJ};
- the eigenvalues from analysis of g_{JJ} are the squares of the eigenvalues from analysis of k_{IJ}.

3. Equivalence between the Analyses of b_{JJ} and k_{IJ}

3.2. Case of Disjunctive Form Data Sets

Consider the data set $k_{IJ} = \{k_{ij}; k_{ij} = 0 \text{ or } 1; i \in I; j \in J\}$, the set put into disjunctive form and its associated N-D contingency data set $b_{JJ} = \{b_{jj'}; j, j' \in J\}$. Then,

$$b_{jj'} = n_Q \sum_{i \in I} \frac{k_{ij}k_{ij'}}{k_i} = n_Q g_{jj'} \quad \text{for } j, j' \in J.$$

Since b_{JJ} is proportional to g_{JJ}, the correspondence analyses of b_{JJ} and g_{JJ} are equivalent (same factor units; eigenvalues of b_{JJ} are equal to the squares of eigenvalues from analysis of k_{IJ}).

3.3. Properties of Factors from the Analyses of k_{IJ} and b_{JJ}

Consider the pair of factors, denoted by $(\phi_{\alpha'}, \phi_\alpha)$, with unit variance associated with the correspondence analysis of k_{IJ}. It has been shown that $(\phi_{\alpha'}, \phi_\alpha)$ is also a pair of factors with unit variance associated with the correspondence analysis of b_{JJ}.

Consider now the couple of factors with variance λ_α, denoted by (F_α, G_α), associated with the correspondence analysis of k_{IJ}, and the pair of factors with variance $\lambda_\alpha^2 = \mu_\alpha$, denoted by (F'_α, G'_α), associated with the correspondence analysis of b_{JJ}. The following relations hold:

$$F'_\alpha(j) = G'_\alpha(j) = \sqrt{\mu_\alpha}\,\phi_\alpha(j) = \lambda_\alpha \phi_\alpha(j) = \sqrt{\lambda_\alpha}\,G_\alpha(j);$$

$$F_\alpha(j) = \sum_{j \in J} \frac{f_{ij}}{f_j} \phi_\alpha(j) \quad \text{for } i \in I$$

$$= \sum_{j \in J} \frac{k_{ij}}{k_j} \frac{G_\alpha(j)}{\sqrt{\lambda_\alpha}}$$

$$= \frac{\sum_{i \in I} k_{ij}\phi_\alpha(j)}{n_Q}$$

$$= \sum_{q \in Q} \frac{\sum_{j \in J_q} k_{ij}\phi_\alpha(j)}{n_Q}.$$

As $k_{ij} = 0$ or 1,

$$F_\alpha(i) = \frac{\sum_{q \in Q} \phi_\alpha(j)}{n_Q}.$$

Since $\phi_\alpha(j)$ is the α-coordinate of the point j of J where a positive response exists for i of I, it is deduced that $F_\alpha(i)$ is the average of the α-coordinates corresponding to the forms taken by the individuals i of I. On the other hand,

$$\sum_{j \in j_q} f_j \phi_\alpha(j) = 0,$$

$$\sum_{j \in j_q} f_j G_\alpha(j) = 0.$$

It is deduced that the center of gravity of the points j of J_q is equal to the center of gravity of the whole set J.

3.4. Contribution Computations in N-D Correspondence Analysis Based on k_{IJ}

3.4.1. Introduction

Computations of contributions are based on the same principles in both 2-D and N-D correspondence analysis. In the latter, new features exist because of the particular structure of the disjunctive form data set k_{IJ}. Presented in the following are two ways for computing contributions; the first is based on the forms j of J and the individuals i of I, which are interpreted in the same way as in 2-D correspondence analysis. The second is based on questions q of Q that help to improve the interpretation of the factorial results.

3.4.2. Preliminary Computations of the Total Variance Associated with k_{IJ}

The following relations exist for the data set k_{IJ}:

$\forall i \in I$:

$k_i = n_Q =$ number of questions;

$f_i = \dfrac{1}{n}$ with $n =$ number of individuals;

$k_j =$ number of individuals giving

the form j for the question q;

$p_j = \dfrac{k_j}{n} =$ proportion of individuals i giving

the form j for the question q.

3. Equivalence between the Analyses of b_{JJ} and k_{IJ}

Thus,

$$\sum_{j \in J_q} p_j = 1 \quad \text{and} \quad f_j = \frac{p_j}{n_Q},$$

$$\rho^2(j) = d^2(j, G) = \sum_{i \in I} \frac{\left(\frac{f_{ij}}{f_j} - f_i\right)^2}{f_i}$$

$$= n \left(\frac{n_Q}{p_j} f_{ij} - \frac{1}{n}\right)^2.$$

This expression is decomposed according to the values taken by f_{ij}:

$$f_{ij} = 1/nn_Q \quad (np_j \text{ times}) \quad \text{and} \quad 0 \quad (n(1 - p_j) \text{ times});$$

$$\rho^2(j) = p_j \left(\frac{1 - p_j}{p_j}\right)^2 + 1(1 - p_j)$$

$$= (1 - p_j) p_j.$$

Thus,

$$f_j \rho^2(j) = (1 - p_j)/n_Q.$$

Consider

$$M^2(N(J_q)) = \text{total variance of } N(J_q)$$

$$= \sum_{j \in J_q} f_j \rho^2(j)$$

$$= \frac{n_q - 1}{n_Q}$$

and

$$M^2(N(J)) = \sum_{q \in Q} M^2(N(J_q))$$

$$= \sum_{q \in Q} \frac{(n_Q - 1)}{n_Q}$$

$$= \frac{n_J - n_Q}{n_Q},$$

where n_J = total number of forms of J, and n_Q = total number of questions (or qualitative variables).

From the previous computations, the following relations are deduced:

- The contribution of a form j to the total variance of $N(J)$ is higher when the frequency of j is smaller. This implies that the infrequent forms play too important a role in the analysis. Thus, an analysis is more stable when the distribution of $j \in J$ is more regular.
- The contribution of a question q to the total variance of $N(J)$ is a function of its associated forms. This implies that each q must have approximately the same number of forms.

3.4.3. Contributions of Questions and Forms According to the Factors

The total variance of $N(J)$, denoted by $M^2(N(J))$, is decomposed according to the factors as follows:

$$M^2(N(J)) = \sum_{j \in J} f_j \rho^2(j) \quad \text{(definition)}$$

$$= \sum_{q \in Q} \left(\sum_{j \in J_q} f_j \rho^2(j) \right)$$

$$= \sum_{\alpha \in \wedge} \left(\sum_{j \in J_q} f_j G_\alpha^2(j) \right) \quad (\wedge \text{ is the factor space associated with } k_{IJ}).$$

Consider the following relations:

$$\text{cta}_\alpha(j) = f_j G_\alpha^2(j) = \lambda_\alpha \, \text{ctr}_\alpha(j).$$

In the usual notation of correspondence analysis, $\text{ctr}_\alpha(j)$ is the relative contribution of j to the variance of the α-axis; $\text{cta}_\alpha(j)$ is the absolute contribution of j to the variance of the α-axis.

$$\text{cta}_\alpha(J_q) = \sum_{j \in J_q} f_j G_\alpha^2(j)$$

= absolute contribution of J_q to the variance of the α-axis

= mutual contribution of the couple (α, q) to $M^2(N(J))$;

$$\text{ctr}_\alpha(q) = \frac{\text{cta}_\alpha(J_q)}{\lambda_\alpha}$$

$$= \sum_{j \in J_q} \text{ctr}_\alpha(j),$$

where $\text{ctr}_\alpha(j)$ is the (usual) relative contribution of j to the variance of

3. Equivalence between the Analyses of b_{IJ} and k_{IJ}

the α-axis;

$\text{ctr}_\alpha(q)$ = relative contribution of J_q to the variance of the α-axis;

$$\sum_{\alpha \in \Lambda} \text{cta}_\alpha(J_q) = \text{cta}(J_q) = \frac{n_J - n_Q}{n_Q}.$$

Thus,

$$\sum_{q \in Q} \text{cta}_\alpha(J_q) = \lambda_\alpha.$$

Due to the previous definitions and relations, it turns out the following relations similar to those given for 2-D correspondence analysis:

$$\text{cor}_\alpha(q) = \frac{\text{cta}_\alpha(J_q)}{\text{cta}(J_q)};$$

$$\text{inr}(q) = \frac{\text{cta}(J_q)}{M^2(N(J))};$$

$$\text{qlt}_s(q) = \sum_{\alpha \leq s} \frac{\text{cta}_\alpha(J_q)}{\text{cta}(J_q)}$$

$$= \sum_{\alpha \leq s} \text{cor}_\alpha(J_q), \quad \text{where } s \text{ is the number of retained factors.}$$

For each question (or qualitative variable) involved in N-D correspondence analysis, the above computations and indicators are used for interpreting the data. However, $\text{cor}_\alpha(q)$ and $\text{qlt}_s(q)$ are not similar to square cosines (or a sum of square cosines), and therefore they are not geometrically interpretable as in 2-D correspondence analysis.

3.4.4. Factor Coordinates of Questions

Consider the following relations:

$$f_q = \sum_{j \in J_q} f_j = \frac{1}{n_Q};$$

$$G_\alpha^2(q) = \frac{\text{cta}_\alpha(q)}{f_q};$$

$$M^2(N(J)) = \sum_{q \in Q} \left(\sum_{\alpha \in \Lambda} f_q G_\alpha^2(q) \right).$$

By analogy with 2-D correspondence analysis, this suggests that $G_\alpha(q)$ should be used as an indicator representing the question q in the factor space. By extension, $G_\alpha(q)$ is known as the α-coordinate of the question q.

$$G_\alpha(q) = \sqrt{\frac{\mathrm{cta}_\alpha(q)}{f_q}}$$

$$= \sqrt{n_Q \, \mathrm{cta}_\alpha(q)}.$$

3.5. Contributions Computations in N-D Correspondence Analysis Based on b_{JJ}

3.5.1. Introduction

Contributions computations are the same for the analysis of b_{JJ} as for that of k_{JJ} because of their equivalence. However, in practice, N-D correspondence analysis of k_{JJ} is preferred, so that the contributions from the analysis of b_{JJ} are little used. Nevertheless, the details of contributions computations from b_{JJ} are given in the next section.

3.5.2. Preliminary Computations of the Total Variance Associated with b_{JJ}

Consider the variance associated with the crossing of two questions (or qualitative variables) q and q', denoted by $\chi^2_{J_q J_{q'}}$. The total variance associated with b_{JJ}, denoted by $\chi^2_{b_{JJ}}$, can be written as follows:

$$\chi^2_{b_{JJ}} = \frac{\sum_{q \in Q, q' \in Q} \chi^2_{J_q J_{q'}}}{n_Q^2},$$

where Q is the set of questions and J_q is the set associated with the question q. Since

$$\chi^2_{J_q J_q} = (n_q - 1),$$

we have

$$\chi^2_{b_{JJ}} = \frac{(n_J - n_Q) + \sum_{q,q' \in Q; q \neq q'} \chi^2_{J_q J_{q'}}}{n_Q^2}.$$

The absolute contribution of a question q to the total variance $\chi^2_{b_{JJ}}$,

3. Equivalence between the Analyses of b_{JJ} and k_{IJ}

denoted by $\text{ctab}(q)$, can be written as follows:

$$\text{ctab}(q) = \frac{\sum_{q' \in Q} \chi^2_{J_q J_{q'}}}{n_Q^2}$$

$$= \frac{(n_q - 1) + \sum_{q' \in Q; q' \neq q} \chi^2_{J_q J_{q'}}}{n_Q^2}.$$

3.5.3. Contributions of the Questions According to the Factors

The previous quantity $\text{ctab}(q)$ can be decomposed according to the factors, so that the contributions of questions can be obtained as a function of the variance of the axes α. By definition,

$$\text{ctab}_\alpha(q) = \sum_{j \in J_q} \frac{b_j}{b} Gb_\alpha^2(j),$$

where $Gb_\alpha(j)$ is the α-coordinate of the point j from the analysis of b_{JJ}. Since

$$\frac{b_j}{b} = \frac{k_j}{k} = \frac{k_j}{n \cdot n_Q},$$

we have

$$\text{ctab}_\alpha(q) = \frac{1}{n_Q} \left\{ \sum_{j \in J_q} \frac{k_j}{n} Gb_\alpha^2(j) \right\},$$

$$\text{ctab}_\alpha(q) = \frac{1}{n_Q} \left(\sum_{j \in J_q} p_j Gb_\alpha^2(j) \right),$$

$$\text{ctab}_\alpha(q) = \lambda_\alpha \, \text{cta}_\alpha(q).$$

3.6. Conclusions on the N-D Correspondence Analysis of k_{IJ} and b_{JJ}

• The factor axes are the same for the N-D correspondence analysis of k_{IJ} and b_{JJ} (only the variances are different). This shows that analysis of a logical data set k_{IJ} is equivalent to that of a N-D contingency data set b_{JJ}.

This is one of the given reasons why 2-D correspondence analysis is applied to a logical data set k_{IJ}.

- As a corollary, in practice, it is not useful to compute b_{JJ} to obtain the results of the analysis of b_{JJ} and k_{IJ} (because of the equivalence). Only the standard basic computer program for 2-D correspondence analysis is therefore applied to the disjunctive form data set k_{IJ} to obtain an N-D correspondence analysis. When applying correspondence analysis to b_{JJ}, the individuals i of I do not appear; when applying correspondence analysis to k_{IJ}, the individuals i of I appear on graphics. The only usefulness of b_{JJ} is in the data set, which gives all the crossings of n_Q questions or qualitative variables.

4. Interpretation of N-D Correspondence Analysis

4.1. Introduction

Once again "to interpret" an analysis means not only to compute factors and to plot graphics, but to give a meaning to the factors using graphics. This is done using contributions and graphics. The only problem is to extract the relevant information. In the following, we give a step-by-step account of the interpretation process based on the output from the standard 2-D correspondence analysis computer program applied to the disjunctive form data set recoded from the family timetables data set (*cf.* Appendix 2, §10).

4.2. The Basic Data Set Analyzed

The data set described in Appendix 2, §10 is from a survey on the activities timetables of families for population groups based on the variables sex, employed/unemployed, country, and marital status. Ten main activities were finally retained. As was shown in Section 2.2.5, three dummy qualitative variables (logical values) have been associated with each activity. These dummy variables have been chosen such that their marginal distributions are approximately equal. They have the following labels:

WORK	gives:	$WORK_1$, $WORK_2$, $WORK_3$;
TRANSPORT	gives:	$TRANSPORT_1$, $TRANSPORT_2$, $TRANSPORT_3$;
HOUSEHOLD	gives:	$HOUSEHOLD_1$, $HOUSEHOLD_2$, $HOUSEHOLD_3$;

4. Interpretation of N-D Correspondence Analysis

CHILDREN	gives:	CHILDREN$_1$, CHILDREN$_2$, CHILDREN$_3$;
SHOPPING	gives:	SHOPPING$_1$, SHOPPING$_2$, SHOPPING$_3$;
PERSONAL CARE	gives:	PERSONAL CARE$_1$, PERSONAL CARE$_2$, PERSONAL CARE$_3$;
MEALS	gives:	MEALS$_1$, MEALS$_2$, MEALS$_3$;
SLEEP	gives:	SLEEP$_1$, SLEEP$_2$, SLEEP$_3$;
TELEVISION	gives:	TELEVISION$_1$, TELEVISION$_2$, TELEVISION$_3$;
LEISURE	gives:	LEISURE$_1$, LEISURE$_2$, LEISURE$_3$.

The labels for the individuals are the same as in the basic data set (*cf.* Appendix 2, §10).

4.3. Eigenvalues, Variances, Rates of Variance

The results given by the computer program are given in Table 9.9. According to the specific structure of k_{IJ} (the disjunctive form data set), $(n_J - n_Q)$ nontrivial factor axes are obtained. There are $(n_Q - 1)$ factors related to the zero eigenvalue, and one factor related to the unit eigenvalue. As $M^2(N(J))$ is equal to $(n_J - n_Q)/n_Q$, the average of an eigenvalue is $1/n_Q$. This value is taken as an *a priori* threshold for the number of factors to be studied. Notice that the rates of variance of the axes α are underestimated by the N-D correspondence analysis of k_{IJ}; it is preferable to use the rates of variance given by the analysis of b_{JJ}. A summary of all these results is given in Table 9.9.

4.4. Factor Coordinates of the Points of N(J), N(I), and N(Q).

The factor coordinates of the points j of $N(J)$, are computed using the same formulas as those used in 2-D correspondence analysis. Similarly, the relative contributions to the variance of the α-axes and to the eccentricity are computed. These results can be interpreted in the same manner as for 2-D correspondence analysis. The factor coordinates of j and the contributions from the family timetables data set are given in Table 9.10.

The factor coordinates and the associated contributions of i of $N(I)$ are computed using the same formulas as those used in 2-D correspondence analysis. These results can be interpreted in the same manner as for 2-D correspondence analysis. The factor coordinates of i and the contributions from the family timetables data sets are given in Table 9.11.

TABLE 9.9. Eigenvalues and the associated rates of variance: num = serial number of α ($\alpha = 2, \ldots, 21$); iter = number of iterations; eigenvalues (k_{IJ}) = eigenvalues from the analysis of k_{IJ}; percentage (k_{IJ}) = rates of variance of the α-axis from the analysis of k_{IJ}; cumul (k_{IJ}) = cumulative rates of variance from the analysis of k_{IJ}; eigenvalues (b_{JJ}) = eigenvalues from the analysis of b_{JJ}; percentage (b_{JJ}) = rates of variance of the α-axis from the analysis of b_{JJ}; cumul (b_{JJ}) = cumulative rates of variance from the analysis of b_{JJ}.

Num	iter	eigenvalues (k_{IJ})	percentage (k_{IJ})	cumul (k_{IJ})	eigenvalues (b_{JJ})	percentage (b_{JJ})	cumul (b_{JJ})
2	0	0.429	21.47	21.474	0.184	38.09	38.09
3	1	0.348	17.43	38.908	0.121	25.05	63.14
4	1	0.251	12.574	51.482	0.063	13.04	76.18
5	1	0.195	9.754	61.237	0.038	7.86	84.04
6	2	0.152	7.613	68.850	0.023	4.76	88.80
7	1	0.142	7.137	75.987	0.020	4.14	92.94
8	2	0.119	5.987	81.974	0.014	2.89	95.83
9	1	0.079	3.957	85.931	0.006	1.24	97.07
10	2	0.077	3.899	89.830	0.006	1.24	98.31
11	1	0.057	2.895	92.725	0.003	0.06	98.37
12	1	0.044	2.244	94.970	0.002	0.05	98.42
13	1	0.032	1.601	96.570	0.001	0.02	98.44
14	2	0.022	1.131	97.702	0.001	0.02	98.46
15	2	0.019	0.969	98.670	0.001	0.02	100.00
16	1	0.009	0.492	99.162	0.000	0.00	100.00
17	2	0.007	0.396	99.558	0.000	0.00	100.00
18	3	0.002	0.285	99.849	0.000	0.00	100.00
19	2	0.002	0.107	99.950	0.000	0.00	100.00
20	2	0.001	0.050	100.00	0.000	0.00	100.00
21	0	0.0003	0.000	100.00	0.000	0.00	100.00

$N(Q)$ is the cloud of questions (or qualitative variables). The factor coordinates and the associated contributions of q of $N(Q)$ are computed using the specific formulas given in Section 3.4. The factor coordinates of q and the contributions from the family timetables data set are given in Table 9.12.

4.5. Rules for Interpreting Factor Axes by the Points j of $N(J)$, i of $N(I)$, and q of $N(Q)$

4.5.1. Explicative Points

As in 2-D correspondence analysis, the explicative points of the α-axis are those points j of $N(J)$ (respectively, i of $N(I)$ and q of $N(Q)$) whose

4. Interpretation of N-D Correspondence Analysis

contributions to the variance of the α-axis, denoted by $\text{ctr}_\alpha(j)$ (respectively, $\text{ctr}_\alpha(i)$ and $\text{ctr}_\alpha(q)$), are much higher than the contributions of the other elements. A general rule for selecting these explicative points is given in Section 7, and the associated results are summarized in Table 9.13.

4.5.2. Explained Points

As in 2-D correspondence analysis, the explained points by the α-axis are those points j of $N(J)$ (respectively, i of $N(I)$ and q of $N(Q)$) whose contributions to the eccentricity, denoted by $\text{cor}_\alpha(j)$ (respectively, $\text{cor}_\alpha(i)$ and $\text{cor}_\alpha(q)$), and much higher than a given threshold that is similar to a squared coefficient of correlation. A general rule for selecting this threshold is given in Section 7. The explained points are interpreted in the same way here as in 2-D correspondence analysis.

4.6. Quality of Explanation

To achieve the interpretation of results, it is necessary to study the quality of explanation of factor axes, and the quality of explanation of the points i of $N(I)$, j of $N(J)$, and q of $N(Q)$ related to those factor axes retained as interpretable.

4.6.1. Global Quality of Explanation

The global quality of explanation, denoted by qge_s, is the sum of the rates of variance of the axes considered as interpretable:

$$\text{qge}_s = \sum_{\alpha=2}^{s} \tau_\alpha.$$

In the particular case of N-D correspondence analysis, there are n_Q trivial factor axes, so only $(n_J - n_Q)$ factor axes are studied. As $M^2(N(J))$ is equal to $(n_J/n_Q - 1)$, $M^2(N(J))$ increases with the number of forms. This implies that the rates of variance of the axes decrease (since $\lambda_\alpha \leq 1$). Thus,

$$\tau_\alpha \leq \frac{1}{\sum_{\alpha \in \Lambda} \lambda_\alpha}, \quad \tau_\alpha \leq \frac{n_Q}{n_J - n_Q}.$$

TABLE 9.10. Factor coordinates and contributions associated with the forms j of J (all the numerical values are multiplied by 1000).

p^a	j^b	qlt_5^c	weig^d	inr^e	1st factor			2nd factor			3rd factor			4th factor			5th factor		
					ϕ_1^f	cor_1^g	ctr_1^h	ϕ_2	cor_2	ctr_2	ϕ_3	cor_3	ctr_3	ϕ_4	cor_4	ctr_4	ϕ_5	cor_5	ctr_5
1	WORK$_1$	825	36	32	1123	701	105	−31	1	0	158	14	4	−220	27	9	1	0	0
2	WORK$_2$	813	36	32	−315	56	8	−506	143	26	−616	211	54	261	38	12	52	2	1
3	WORK$_3$	902	29	36	−1008	407	68	673	181	37	574	132	37	−49	1	0	−66	2	1
4	TRANSPORT$_1$	923	36	32	1165	754	113	306	52	10	−32	1	0	208	24	8	361	72	31
5	TRANSPORT$_2$	783	32	34	228	25	4	−541	139	27	648	199	54	24	0	0	−752	269	120
6	TRANSPORT$_3$	886	32	34	−1065	538	85	203	19	4	−611	178	48	−254	31	11	352	59	26
7	HOUSEHOLD$_1$	934	36	32	−889	440	66	703	274	51	−177	18	4	−504	142	47	−169	16	7
8	HOUSEHOLD$_2$	873	32	34	−219	23	4	−1105	580	113	612	177	48	415	81	28	145	10	4
9	HOUSEHOLD$_3$	891	32	34	1209	693	109	326	50	10	−413	81	22	147	10	4	44	1	0
10	CHILDREN$_1$	927	39	30	−741	356	50	657	279	49	−40	1	0	−471	144	45	−132	12	5
11	CHILDREN$_2$	927	29	36	−340	46	8	−1268	644	132	323	110	31	484	94	34	134	7	3
12	CHILDREN$_3$	691	32	34	1209	693	109	326	50	10	−413	81	22	147	10	4	44	1	0
13	SHOPPING$_1$	821	36	32	−815	370	55	500	139	26	480	132	34	294	48	16	−18	0	0
14	SHOPPING$_2$	678	32	34	434	98	13	10	0	0	−484	111	30	510	123	43	−532	134	60
15	SHOPPING$_3$	762	32	34	452	97	15	−656	152	30	−36	2	0	−835	331	115	554	145	65
16	PERSONAL CARE$_1$	938	46	27	53	2	0	780	527	81	−201	35	8	190	31	9	−549	262	92
17	PERSONAL CARE$_2$	809	25	38	−580	113	20	−313	33	7	1	0	0	713	170	65	1041	361	178

4. Interpretation of N-D Correspondence Analysis

p	j	qlt																	
18	PERSONAL CARE₃	885	29	36	423	71	12	−989	392	80	327	43	12	−933	349	128	−17	0	0
19	MEALS₁	797	39	30	−510	169	24	−925	555	97	2	0	0	15	0	0	−272	48	19
20	MEALS₂	610	29	36	113	3	1	469	88	18	−993	397	113	−142	8	3	325	42	20
21	MEALS₃	856	32	34	524	130	21	715	242	47	883	369	100	109	6	2	44	1	0
22	SLEEP₁	742	36	32	−907	458	69	07	0	0	−673	254	65	138	11	3	180	18	8
23	SLEEP₂	679	32	34	324	130	21	−798	303	59	−128	8	2	−448	96	33	−366	64	28
24	SLEEP₃	900	34	34	486	112	18	792	297	38	880	367	99	296	41	14	167	13	6
25	TELEVISION₁	824	36	32	52	2	0	−251	35	7	−479	128	33	245	33	11	−918	469	198
26	TELEVISION₂	595	32	34	−34	1	0	339	54	11	831	327	88	434	89	31	413	81	36
27	TELEVISION₃	672	32	34	−2	0	0	−58	2	0	−298	42	11	−706	237	82	608	175	78
28	LEISURE₁	823	36	32	17	0	0	−248	34	6	−590	194	50	917	467	154	193	21	9
29	LEISURE₂	761	36	32	−47	1	0	253	36	7	436	105	27	−425	101	33	−120	8	3
30	LEISURE₃	894	29	36	39	1	0	−4	0	0	194	15	4	−613	151	55	−89	3	2
		1000					1000				1000				1000				1000

[a] p = serial number associated with the forms

[b] j = labels of forms

[c] $\text{qlt}_s(j) = \sum_{\alpha=2}^{s} \text{cor}_\alpha(j)$ = quality of explanation of the form j in the factor space formed by s factors (here, $s=7$)

[d] f_j = weight associated with the form j (marginal frequency of j)

[e] $f_j \rho^2(j)/M^2(J)$ = percentage of variance of the form j related to the total variance of $N(J)$

[f] $\phi_{\alpha j}$ = α-coordinate of the form j

[g] $\text{cor}_\alpha(j)$ = cosine squared of the angle formed by j and the α-axis = relative contribution of the α-axis to the eccentricity of the form j

[h] $\text{ctr}_\alpha(j) = f_j \phi_{\alpha j}^2 / \lambda_\alpha$ = relative contribution of the form j to the variance of the α-axis

TABLE 9.11. Factor coordinates and contributions associated with the individuals i of I (all the numerical values are multiplied by 1000).

n^a	i^b	qlt_5^c	$weig^d$	inr^e	1st factor			2nd factor			3rd factor			4th factor			5th factor		
					ψ_1^f	cor_1^g	ctr_1^h	ψ_2	cor_2	ctr_2	ψ_3	cor_3	ctr_3	ψ_4	cor_4	ctr_4	ψ_5	cor_5	ctr_5
1	EM. USA	865	36	37	−598	174	30	58	2	0	−810	319	93	−200	20	7	769	288	139
2	EW. USA	740	36	37	−38	1	0	1119	606	129	309	46	14	75	3	1	20	0	0
3	UW. USA	796	36	38	1017	486	86	−131	8	2	43	1	0	−712	239	93	303	43	22
4	MM. USA	876	36	33	−501	134	21	341	62	12	−936	468	125	−14	0	0	83	4	2
5	MW. USA	947	36	37	941	423	74	−129	8	2	−283	38	11	−727	253	97	367	64	32
6	SM. USA	880	36	35	−537	146	24	−250	32	6	−389	77	22	−884	395	143	272	37	17
7	SW. USA	941	36	37	−40	1	0	1102	587	125	288	40	12	−486	115	43	−9	0	0
8	EM. WEST	950	36	34	−415	91	14	825	357	70	861	389	105	−23	0	0	−304	49	22
9	EW. WEST	824	36	38	−20	0	0	−134	9	2	594	167	50	935	414	160	649	199	99
10	UW. WEST	688	36	34	942	462	74	579	174	34	102	5	1	57	2	1	17	0	0
11	MM. WEST	861	36	34	407	87	14	757	301	59	636	212	57	−282	42	15	−254	34	15
12	MW. WEST	878	36	34	944	463	74	561	164	32	122	8	2	620	200	70	48	1	1
13	SM. WEST	848	36	35	−402	82	13	781	309	62	813	335	94	−66	2	1	−296	45	21
14	SW. WEST	783	36	37	488	115	20	116	7	1	663	231	62	−60	2	2	92	4	2
15	EM. EAST	883	36	34	802	342	54	676	243	47	−336	60	16	−222	26	9	−286	44	19
16	EW. EAST	505	36	32	−36	1	0	−90	5	1	−449	112	29	538	161	53	−511	146	62

4. Interpretation of N-D Correspondence Analysis

17	UW. EAST	666	36	34	950	477	75	448	106	21	151	12	3	-31	1	0	-95	5	2
18	MM. EAST	776	36	34	-818	351	56	776	315	62	-74	3	1	-179	17	6	54	2	1
19	MW. EAST	897	36	34	903	423	68	150	12	2	-714	265	73	351	64	23	-357	67	30
20	SM. EAST	849	36	32	-584	188	28	113	7	1	-616	209	34	-109	7	2	-533	157	67
21	SW. EAST	897	36	38	201	19	3	-1000	475	103	273	35	11	-118	7	3	-684	222	110
22	EM. YUG	855	36	38	-847	338	60	-277	36	8	435	89	27	399	75	29	458	98	49
23	EW. YUG	849	36	37	-393	76	13	-871	370	78	-211	22	6	843	345	130	-187	17	8
24	UW. YUG	716	36	40	787	280	52	278	35	8	30	0	0	30	0	0	733	243	126
25	MM. YUG	730	36	38	-829	324	57	-430	87	19	5	0	0	445	93	36	588	162	81
26	MW. YUG	815	36	34	684	242	39	69	2	0	-869	392	107	460	110	39	-344	62	28
27	SM. YUG	728	36	37	-762	281	48	74	3	1	-35	1	0	-243	29	11	-164	13	6
28	SW. YUG	892	36	36	187	17	3	-1054	546	114	407	81	24	-380	71	27	-414	84	40
			1000	1000			1000			1000			1000			1000			1000

[a] n = serial number associated with individuals
[b] i = labels of individuals i
[c] $\text{qlt}_s(i) = \sum_{\alpha=2}^{s} \text{cor}_\alpha(i)$ = quality of explanation of i in the factor space formed by s factor axes (here, $s = 7$)
[d] f_i = weight associated with i (marginal frequency of i)
[e] $f_i \rho^2(i)/M^2(I)$ = percentage of variance of i related to the total variance of $N(I)$
[f] $\psi_{\alpha i}$ = α-coordinate of i
[g] $\text{cor}_\alpha(i) = \psi_{\alpha i}^2/\rho^2(i)$ = cosine squared of the angle formed by i and the α-axis = relative contribution of the α-axis to the eccentricity of i
[h] $\text{ctr}_\alpha(i) = f_i \psi_{\alpha i}^2/\lambda_\alpha$ = relative contribution of i to the variance of the α-axis

TABLE 9.12. Factor coordinates and contributions associated with the questions q of Q (all the numerical values are multiplied by 1000).

						1st factor			2nd factor			3rd factor			4th factor			5th factor		
q^a	q^b	qlt_s^c	$weig^d$	inr^e	ϕ_1^f	cor_1^g	ctr_1^h	ϕ_2	cor_2	ctr_2	ϕ_3	cor_3	ctr_3	ϕ_4	cor_4	ctr_4	ϕ_5	cor_5	ctr_5
1	WORK	247	100	100	882	389	181	469	110	63	489	119	95	202	20	21	55	2	2
2	TRANSPORT	865	100	100	931	434	202	378	71	41	506	128	102	193	19	19	519	135	177
3	HOUSEHOLD	900	100	100	877	384	179	779	303	174	431	93	74	393	77	79	129	8	11
4	CHILDREN	914	100	100	847	359	167	816	333	191	865	67	53	402	81	83	110	6	8
5	SHOPPING	752	100	100	604	183	85	442	98	56	401	80	64	583	170	174	436	95	125
6	PERSONAL CARE	870	100	100	371	69	32	765	293	168	224	25	20	628	197	202	641	206	270
7	MEALS	754	100	100	445	99	46	752	282	162	732	268	213	99	5	0	244	30	39
8	SLEEP	776	100	100	681	232	108	639	204	117	646	209	166	312	49	50	253	32	42
9	TELEVISION	694	100	100	0	0	0	251	31	18	576	166	132	492	121	124	659	238	312
10	LEISURE	827	100	100	0	0	0	213	23	13	451	102	81	687	236	242	146	11	14
			1000	1000			1000			1000			1000			1000			1000

[a] q = serial number of the question q
[b] q = labels of the question q
[c] $qlt_s(q) = \sum_{\alpha=2}^{s} cor_\alpha(q)$ = quality of explanation of the question q in the factor space formed by s factor axes (here, $s = 5$)
[d] f_q = marginal frequency of the question q
[e] $f_q \rho^2(q)/M^2(Q)$ = percentage of variance of the question q related to the total variance of $N(Q)$
[f] $\phi_{\alpha q}$ = α-coordinate of the question q
[g] $cor_\alpha(q)$ = relative contribution of the α-axis to the eccentricity of q
[h] $ctr_\alpha(q)$ = relative contribution of the question q to the variance of the α-axis

TABLE 9.13. Explicative points i of $N(I)$, j of $N(J)$, and q of $N(Q)$ from the analysis of the family timetables data set for the first two axes.

Points with negative coordinates	Points with positive coordinates
Axis 1	
WORK$_3$ (68) TRANSPORT$_3$ (85) HOUSEHOLD$_1$ (66) CHILDREN$_1$ (50) SHOPPING$_1$ (55) SLEEP$_1$ (69)	WORK$_1$ (105) TRANSPORT$_1$ (113) HOUSEHOLD$_3$ (109) CHILDREN$_3$ (109)
Employed men EAST (50) Married men EAST (55) Employed men YUG (60) Married men YUG (57) Single men YUG (48)	Unemployed women USA (86) Married women USA (74) Unemployed women WEST (74) Married women WEST (74) Unemployed women EAST (75) Married women EAST (68) Unemployed women YUG (52) Married women YUG (59)
	WORK (181) TRANSPORT (202) HOUSEHOLD (179) CHILDREN (167) SLEEP (108)
Axis 2	
HOUSEHOLD$_2$ (113) CHILDREN$_2$ (132) PERSONAL CARE$_3$ (80) MEALS$_1$ (97) SLEEP$_2$ (59)	WORK$_3$ (37) HOUSEHOLD$_1$ (51) CHILDREN$_1$ (49) PERSONAL CARE$_1$ (780) MEALS$_3$ (47) SLEEP$_3$ (58)
Employed women USA (129) Single women USA (125) Single women EAST (103) Single women YUG (114)	Employed men WEST (70) Married men WEST (59) Single men WEST (62) Employed men EAST (47) Married men EAST (62)
	HOUSEHOLD (174) CHILDREN (191) PERSONAL CARE (168) MEALS (162) SLEEP (117)

In N-D correspondence analysis, therefore, the rate of variance is underestimated. To avoid missing some significant factor axes, it is preferable to determine the number of interpretable factor axes by considering the rates of variance of axes from the analysis of b_{JJ}.

The results associated with the family timetables data set are given in Table 9.14.

4.6.2. Local Quality of Explanation

For each point i of $N(I)$, j of $N(J)$, and q of $N(Q)$, the associated quality of explanation can be computed for a given factor space (determined by a given number of axes). Let s be the given number of axes; the following relations exist:

$$\text{qlt}_s(i) = \sum_{\alpha=2}^{s} \text{cor}_\alpha(i);$$

$$\text{qlt}_s(j) = \sum_{\alpha=2}^{s} \text{cor}_\alpha(j);$$

$$\text{qlt}_s(q) = \sum_{\alpha=2}^{s} \text{cor}_\alpha(q).$$

These computations can be interpreted as in 2-D correspondence analysis. The results obtained from the analysis of the family timetables data set for five factors axes are given in Tables 9.15, 9.16, and 9.17.

TABLE 9.14. Global quality of explanation qge_s from the analyses of k_{IJ} and b_{JJ}.

QUALITY OF EXPLANATION

AXES α	qge	from k_{IJ}	from b_{JJ}
	2	21.47	38.09
	3	38.90	63.14
	4	51.42	76.18
	5	61.23	84.04
	6	68.85	88.08

5. Factor Graphics

TABLE 9.15. Quality of explanation of the form-points j of J according to five factor axes.

QUALITY OF EXPLANATION

		qlt	1st factor	2nd factor	3rd factor	4th factor	5th factor
	I						
FORMS OF QUANTITATIVE VARIABLES	WORK$_1$		701	702	716	743	743
	WORK$_2$		56	199	410	448	450
	WORK$_3$		407	588	720	721	723
	TRANSPORT$_1$		754	806	807	731	903
	TRANSPORT$_2$		25	164	363	363	632
	TRANSPORT$_3$		538	557	735	766	825
	HOUSEHOLD$_1$		440	714	732	874	890
	HOUSEHOLD$_2$		23	603	780	861	871
	HOUSEHOLD$_3$		693	743	824	834	835
	CHILDREN$_1$		356	635	636	780	792
	CHILDREN$_2$		44	690	800	894	901
	CHILDREN$_3$		693	743	824	834	835
	SHOPPING$_1$		370	509	641	689	689
	SHOPPING$_2$		98	98	209	332	466
	SHOPPING$_3$		97	249	251	582	727
	PERSONAL CARE$_1$		2	529	564	595	827
	PERSONAL CARE$_2$		113	146	146	316	677
	PERSONAL CARE$_3$		71	463	506	855	855
	MEALS$_1$		169	724	724	724	772
	MEALS$_2$		5	93	490	498	540
	MEALS$_3$		130	372	741	747	748
	SLEEP$_1$		458	452	712	723	741
	SLEEP$_2$		130	433	441	537	601
	SLEEP$_3$		112	409	776	817	830
	TELEVISION$_1$		2	37	165	198	667
	TELEVISION$_2$		1	55	382	471	552
	TELEVISION$_3$		0	2	44	281	456
	LEISURE$_1$		0	34	228	695	716
	LEISURE$_2$		1	37	142	243	251
	LEISURE$_3$		1	0	15	166	169

5. Factor Graphics

5.1. Introduction

Recall that the main purpose of any factor analysis is to present relations between elements $(I, J,$ and $Q)$ in a Euclidean space (called a factor space), whose "dimension is as low as possible and whose information

TABLE 9.16. Quality of explanation of the individual-points i of I according to five factor axes.

QUALITY OF EXPLANATION

	I	qlt	1st factor	2nd factor	3rd factor	4th factor	5th factor
INDIVIDUALS	EM. USA		174	176	495	515	803
	EW. USA		1	607	653	656	656
	UW. USA		486	494	495	734	777
	MM. USA		134	196	664	664	668
	MW. USA		423	431	469	722	786
	SM. USA		146	178	255	650	687
	SW. USA		1	588	628	743	743
	EM. WEST		91	448	837	837	886
	EW. WEST		0	9	176	590	789
	UW. WEST		462	636	641	643	643
	MM. WEST		87	388	600	642	676
	MW. WEST		463	627	635	835	836
	SM. WEST		82	391	726	728	773
	SW. WEST		115	122	335	337	341
	EM. EAST		342	585	645	671	715
	EW. EAST		1	6	118	279	425
	UW. EAST		477	583	595	596	598
	MM. EAST		351	666	669	686	688
	MW. EAST		423	435	700	764	831
	SM. EAST		188	195	404	411	468
	SW. EAST		19	494	529	536	758
	EM. YUG		338	374	463	538	636
	EW. YUG		76	446	468	813	830
	UW. YUG		280	315	315	315	558
	MM. YUG		324	411	411	504	666
	MW. YUG		242	244	636	746	808
	SM. YUG		281	284	285	314	327
	SW. YUG		17	563	644	715	799

kept is as large as possible". To understand and to highlight relationships, graphical representations need to be given factor plane by factor plane, because higher dimensional representations are not easy to use. The problem for the analyst is to select 2-D graphics. Several ways for representing points and presenting graphics are available, according to the following choices:

1. the factor axes;
2. the subsets of qualified points (explained points, explicative points);

5. Factor Graphics

TABLE 9.17. Quality of explanation of the question-points q of Q according to five factor axes.

QUALITY OF EXPLANATION

Q \ qlt	1st factor	2nd factor	3rd factor	4th factor	5th factor
WORK	389	499	618	638	640
TRANSPORT	434	505	633	659	787
HOUSEHOLD	384	687	780	873	881
CHILDREN	359	692	759	840	846
SHOPPING	183	281	361	531	626
PERSONAL CARE	69	362	387	584	790
MEALS	99	381	649	655	685
SLEEP	232	436	645	694	726
TELEVISION	0	31	197	318	556
LEISURE	0	23	125	361	372

(QUANTITATIVE VARIABLES)

3. the set of points I, J, and Q;
4. the set of supplementary points IS, JS (*cf.* Section 6).

In what follows, we give several examples based on the analysis of the family timetables data set.

5.2. Graphical Representations. Examples

To illustrate the previous paragraph, the following representations are given:

$J + Q$ according to the factor space $(1, 2)$ (*cf.* Fig. 9.1);
$J + Q$ according to the factor space $(1, 3)$ (*cf.* Fig. 9.2);
I according to the factor space $(1, 2)$ (*cf.* Fig. 9.3);
I according to the factor space $(1, 3)$ (*cf.* Fig. 9.4);
$I + J + Q$ according to the factor space $(1, 2)$ (*cf.* Fig. 9.5);
$I + J + Q$ according to the factor space $(1, 3)$ (*cf.* Fig. 9.6).

Based on these graphics and the previous contribution results, the following interpretation is proposed:

- Factor space $(1, 2)$.

Joining the forms associated with the same question by a straight line makes the results easier to read. It is seen that the forms associated with

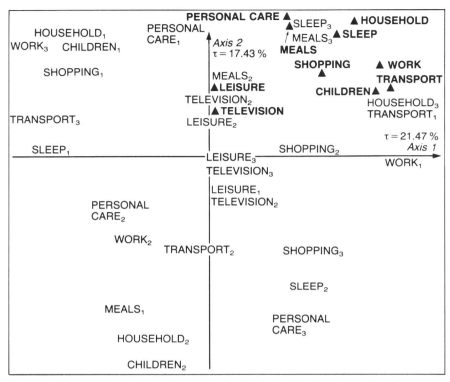

FIGURE 9.1. *N*-D correspondence analysis of the family timetables data set. Representation of the set of questions (denoted by ▲) and the set of forms associated with questions (denoted by indexed labels) in the factor space (1, 2).

the questions HOUSEHOLD and CHILDREN are ordered along the first axis. On the other hand, the forms associated with TRANSPORT and WORK are ordered along the first axis, but in the reverse direction. This highlights the opposition between these two groups of variables. As TRANSPORT and WORK times increase, HOUSEHOLD and CHILDREN times decrease, and conversely. Notice that these four variables contribute the most to the first factor (by forms and by questions). This interpretation is related to the position of the individuals. On the left, the groups with men and employed women are found, whereas on the right, the groups with unemployed and married men are seen. Notice the intermediate position of employed and single women who do both professional work and household work. The second axis characterizes the opposition between the extreme positions of the four previous variables (TRANSPORT, CHILDREN, WORK,

5. Factor Graphics

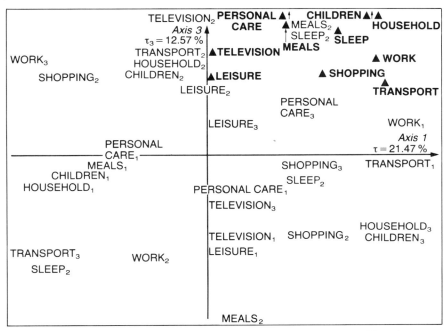

FIGURE 9.2. N-D correspondence analysis of the family timetables data set. Representation of the set of questions (denoted by ▲) and the set of forms associated with questions (denoted by indexed labels) in the factor space (1, 3).

HOUSEHOLD). On one hand, the forms denoted by the indices 1 and 3 are in opposition with the forms denoted by the index 2. On the other hand, the forms associated with MEALS are ordered along the direction given by the second axis, while the forms associated with PERSONAL CARE are ordered along the reverse direction (given by the second axis). These two variables explain the meaning of the second axis. Out along the positive part of the second axis, the individuals denoted by EM. WEST, SM. WEST, and MM. WEST are associated with the forms MEALS$_3$ and SLEEP$_3$. This implies that these three groups spend more time in the MEALS and SLEEP activities than all the other groups. Along the negative part of the second axis, the individuals denoted by EW. USA, EW. WEST, SW. EAST, SW. USA, and SW. YUG are associated with the forms PERSONAL CARE$_3$, SLEEP$_2$, MEALS$_1$, HOUSEHOLD$_2$, and CHILDREN$_2$. This type of interpretation helps to classify the way of life of the groups studied in the survey (in 1965).

- Factor space (1, 3).

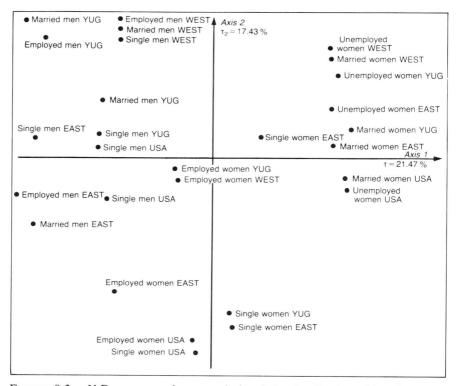

FIGURE 9.3. *N*-D correspondence analysis of the family timetables data set. Representation of the set of individuals (denoted by ●) in the factor space (1, 2).

The reader is invited to interpret for himself the graphics and contributions, and then compare the two analyses done on the same data, i.e., the principle components analysis (*cf.* Chapter 7, Section 4) and the *N*-D correspondence analysis given in this chapter.

6. Classifying Supplementary Points into Graphics

6.1. Introduction

As in 2-D correspondence analysis, it is possible to classify points onto graphics. This is especially useful when large surveys involving numerous individuals and variables are processed. Consider a socio-economic survey where the set I is a set of individuals interviewed and the set J is divided into two subsets: J_1 is a set of forms derived from questions, to be

6. Classifying Supplementary Points into Graphics

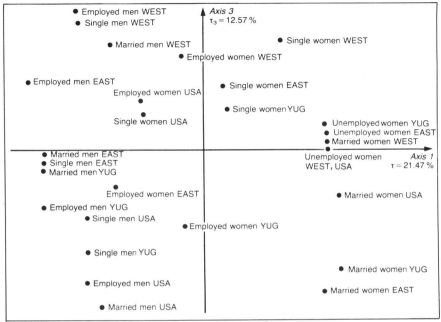

FIGURE 9.4. *N*-D correspondence analysis of the family timetables data set. Representation of the set of individuals (denoted by ●) in the factor space (1, 3).

explained; J_2 is a set of explicative forms; N-D correspondence analysis is performed on J_1, and the set J_2 can be classified as a supplementary set on the graphics associated with the previous analysis.

Another procedure can be proposed to classify points. The center of gravity of any explicative form can be computed and classified as a supplementary element of I on the graphics of the previous analysis. This shows how to use supplementary data sets in N-D correspondence analysis. Different types of situations are presented next.

6.2. Basic and Dummy Supplementary Data Sets

Consider the data sets involved in N-D correspondence analysis from a basic data set involving qualitative variables (to be explained) and qualitative variables (explicative variables), and a set of supplementary individuals, corrresponding to individuals to be classified (*cf.* Table 9.18).

Along with the basic data set, three supplementary data sets can be formed: the dummy supplementary data set involving explicative qualita-

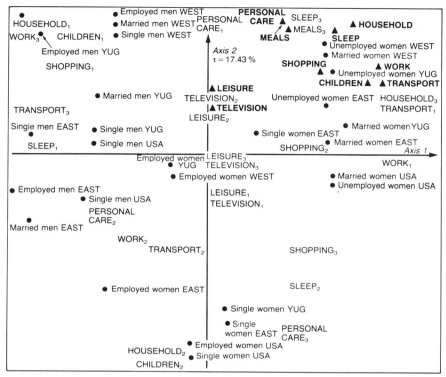

FIGURE 9.5. *N*-D correspondence analysis of the family timetables data set. Representation of the set of individuals (denoted by •), the set of questions (denoted by ▲), and the set of forms associated with questions (denoted by indexed labels), in the factor space (1, 2).

tive variables, and two dummy supplementary data sets of individuals related to explicative variables and variables to be explained. To perform *N*-D correspondence analysis, dummy disjunctive form data sets need to be formed from the previous basic data sets. This gives the arrangement of data sets presented in Table 9.19.

6.3. Coordinates of Supplementary Points

6.3.1. Coordinates of Supplementary Points *t* of *JS*

Consider the disjunctive form data sets k_{IJ} and k_{IJS}, where *JS* is the supplementary set of *J*. *J* is the set of forms associated with the set of basic questions (or qualitative variables to be explained) and *JS* is the set of forms associated with the set of supplementary questions (or qualitative variables to be explained) (*cf.* Table 9.20).

6. Classifying Supplementary Points into Graphics

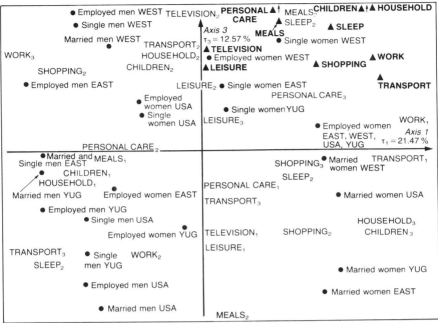

FIGURE 9.6. *N*-D correspondence analysis of the family timetables data set. Representation of the set of individuals (denoted by •), the set of questions (denoted by ▲), and the set of forms associated with questions (denoted by indexed labels), in the factor space (1, 3).

The computation of supplementary elements is based on the transition formula as in 2-D correspondence analysis:

$$\phi_{\alpha t} = \lambda_\alpha^{-1/2} \sum_{i=1}^{n} f_i^t \psi_{\alpha i} \quad \text{for } \alpha = 2, \ldots, p \text{ and any } t \in JS.$$

6.3.2. Coordinates of Supplementary Points *i* of *IS*

Consider the disjunctive form data sets k_{IJ} and k_{ISJ}, where *IS* is the supplementary set of *I*. *J* is the set of forms associated with the set of basic questions (or qualitative variables to be explained). (*cf.* Table 9.21).

The computation of supplementary elements is based on the transition formula as in 2-D correspondence analysis:

$$\psi_{\alpha s} = \lambda_\alpha^{-1/2} \sum_{j=1}^{p} f_j^s \phi_{\alpha j}$$

for $\alpha = 2, \ldots, p$ and any supplementary element *s* of *IS*.

TABLE 9.18. Basic data set and its dummy supplementary data sets.

TABLE 9.19. Dummy disjunctive form data sets from qualitative variable data sets (basic and supplementary).

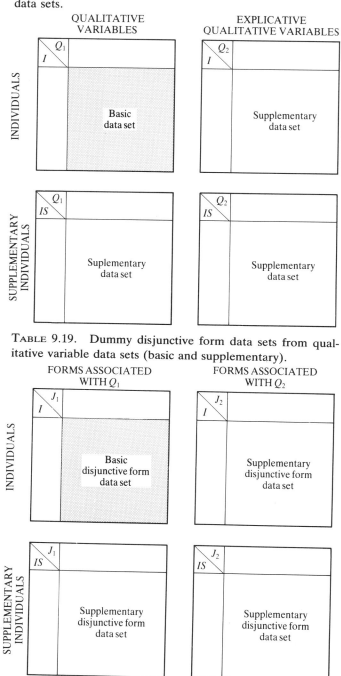

6. Classifying Supplementary Points into Graphics

TABLE 9.20. Model of basic and supplementary data sets associated with principal and supplementary questions.

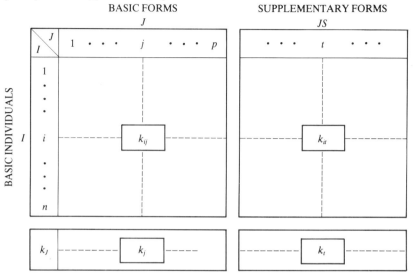

TABLE 9.21. Model of basic and supplementary data sets associated with questions (or qualitative variables).

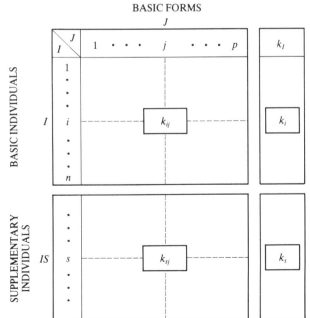

6.3.3. Contributions of Supplementary Elements

It is clear that only the contributions to the eccentricity may be computed and interpreted as a squared coefficient of correlation, as in 2-D correspondence analysis. Contributions of the variance to the α-axis cannot be computed, however, since supplementary elements have no relationship with the variance of the α-axis.

6.4. Graphics with Supplementary Points from the Analysis of k_{IJ}

The graphics involving supplementary elements are the same as those given for 2-D correspondence analysis (*cf.* Section 8). Some examples from a survey are given in Section 10.

6.5. Classifying Supplementary Points from the Analysis of b_{JJ}

After analysis of b_{JJ}, the coordinates and the graphics are computed for only the set J, since the N-D contingency data set does not contain the set I. How, therefore, should the coordinates of I be computed and represented?

Consider the data set b_{JJ}, and the set k_{IJ} considered as a supplementary data set of b_{JJ}, as shown in Table 9.22.

Consider $G'_\alpha(j)$, the α-coordinate of the form $j \in J$ from analysis of b_{JJ}; $\phi_\alpha(j)$, the α-coordinate of the form $j \in J$ from analysis of k_{IJ}; and $\psi_\alpha(i)$, the α-coordinate of $i \in I$ from analysis of k_{IJ}. The following relation holds:

$$G'_\alpha(j) = \sqrt{\lambda_\alpha}\,\phi_\alpha(j)$$

$$= \frac{\sum\limits_{i \in I} k_{ij}\psi_{\alpha i}}{k_j}$$

$$= \frac{\sum\limits_{i \in I, q(i)=j} \psi_{\alpha i}}{k_j}$$

= average of $\psi_{\alpha i}$ associated with the individuals i

that have been given the form j.

Thus, any point $j \in J$ is located at the center of gravity of the points i related to the form j. Nevertheless, from a practical point of view, it is

7. Rules for Selecting Significant Axes and Points

TABLE 9.22. b_{JJ}, and k_{IJ} considered as a supplementary data set of b_{JJ}.

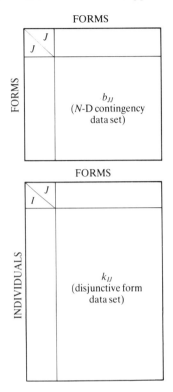

preferable to use the results from k_{IJ} (fewer computations, less data management, and only one computer program with some additional subroutines).

7. Rules for Selecting Significant Axes and Points of $N(I)$, $N(J)$, and $N(Q)$

7.1. Introduction

Due to numerous questions and the many forms associated with them, and the large number of individuals in any survey, it is understandable that data processing and interpretation of surveys are tedious for any analyst. To improve the interpretation process, there are rules for selecting axes and points so that a subset of significant axes and points are studied instead of examining each one of them. These rules are related to those in 2-D correspondence analysis, since N-D correspondence analysis is an extension of 2-D correspondence analysis.

7.2. Rules for Selecting Significant Factor Axes

The rules proposed are the same as those in 2-D correspondence analysis. Because of the structure of k_{IJ}, $M^2(N(J)) = (n_J - n_Q)/n_J$, the average of the eigenvalues is equal to $1/n_Q$, value used as a threshold for determining the number of axes to study. The eigenvalues from the analysis of b_{JJ} can also be used as in 2-D correspondence analysis (*cf.* Chapter 8, Section 8.2).

7.3. Rules for Selecting Explicative Points of Factor Axes

The rules given in Chapter 8, Section 8.3 can be applied as they are, though more rules are added. These concern the selection of the questions q of Q, and are the following:

Rule 9. The points q of Q whose contributions are higher than the average of the contributions are considered as explicative points. The following set of points of Q for each factor axis α is thus selected:

$$\{q \in Q; \text{ctr}_\alpha(q) \geq \text{average of the contributions on the } \alpha\text{-axis}\}.$$

Rule 10. The points q of Q whose contributions are higher than a given percentage of variance, denoted by p, are considered as explicative points. The following set of points of Q for each factor axis is thus selected:

$$\left\{ q \in Q; \sum_q \text{ctr}_\alpha(q) \geq p \right\},$$

where p is a given percentage of variance, and the values $\text{ctr}_\alpha(q)$ are set in decreasing order.

7.4. Rules for Selecting Explained Points by Factor Axes

The rules of selection given in Chapter 8, Section 8.4 can be applied as they are, though another rule needs to be added concerning the questions q of Q, as follows:

Rule 11. The points q of Q whose contributions are higher than a given value k, which is similar to a squared coefficient of correlation, are considered as explained points by the factor axes. The following subset of Q for each factor axis is thus selected:

$$\{q \in Q; \mathrm{cor}_\alpha(q) \geq k\}, \qquad \text{where } k \text{ is a given value.}$$

7.5. Conclusions

In order to perform N-D correspondence analysis for any data survey, we need only a standard 2-D correspondence analysis computer program, to which some subroutines are added: computation of squared eigenvalues, α-coordinates, and contributions of q of Q; graphics with q of Q; selection significant q of Q (*cf.* Chapter 12).

8. *N*-D Correspondence Analysis Formulas

8.1. Reminder

The reader is referred to the formulas of 2-D correspondence analysis given in Chapter 8, since most of them are the same. Only some additional formulas are given here, which are needed because of the particular structure of the disjunctive form data set k_{IJ} and the questions q of Q.

8.2. Basic Data Sets

$$k_{IQ} = \{k_{iq}; i \in I; q \in Q\},$$

where k_{iq} is the value of the response to the question q by the individual i and k_{IQ} is a question data set or a qualitative variable data set;

$$k_{IJ} = \{k_{ij}; i \in I; j \in J\},$$

where $k_{ij} = 0$ or 1; k_{IJ} is a dummy disjunctive form data set associated with k_{IQ};

$$k_i = \sum_{j \in J} k_{ij} = n_Q = \text{number of questions or qualitative variables};$$

$$k = n_I \cdot n_Q;$$

$$M^2(N(J)) = \frac{n_J - n_Q}{n_Q};$$

$$M^2(N(J_q)) = \frac{n_q - 1}{n_Q} = \text{variance associated with the question } q.$$

8.3. Factor Coordinates and Contributions Associated with Questions

$$G_\alpha(q) = \sqrt{n_Q \cdot \lambda_\alpha \cdot \text{ctr}_\alpha(q)}, \qquad \text{with } \text{ctr}_\alpha(q) = \sum_{j \in J_q} \text{ctr}_\alpha(j),$$

$$\text{cta}_\alpha(q) = \sum_{j \in J_q} f_j G_\alpha^2(j),$$

$$\text{ctr}_\alpha(q) = \frac{\text{cta}_\alpha(q)}{\lambda_\alpha},$$

where

$G_\alpha(j) = \alpha$-coordinate of j from the analysis of k_{IJ},

$G_\alpha(q) = \alpha$-coordinate of q.

9. Patterns of Acceptable Data Sets

9.1 Introduction

The 2-D correspondence analysis has been applied to contingency data sets since 1962–1963, and was extended to logical data sets in 1967–1968, and then to different types of data sets not related to contingency data sets. This was the beginning of N-D correspondence analysis, whose mathematical description was given by Benzecri. In what follows, we present different acceptable data sets analyzed by N-D correspondence analysis.

9. Patterns of Acceptable Data Sets

9.2. Data Sets from Qualitative Variables or Questionnaires

These are the most common types of data sets, and many other types of data sets are related to them. The problem is to determine the dummy disjunctive form data set associated with qualitative variable or questionnaire data sets. Consider a data set k_{IQ}, where I is a set of individuals and Q is a set of questions (or qualitative variables); k_{IJ} is a dummy disjunctive form data set associated with k_{IQ} as follows: A logical variable j is associated with each form of q. If the individual has chosen the form j of q, $k_{ij} = 1$; if not, $k_{ij} = 0$. Then,

$$k_{IJ} = \{k_{ij}\,; j = 1 \text{ or } 0; i \in I\}.$$

k_{IJ} is the dummy disjunctive form data set associated with k_{IQ}. k_{IJ} and k_{IQ} are represented in Table 9.23. To compute k_{IJ} from k_{IQ} one needs a standard subroutine (*cf.* Chapter 12).

9.3. Data Sets from Multiple Forms Questionnaires

This situation arises when each individual can give more than one response to each question. Consider the set J of possible responses (or forms) and the set I of individuals; $k_{ij} = 1$ for any form j given by the individual i; $k_{ij} = 0$ if not given. This is represented in Table 9.24.

TABLE 9.23. Model of basic qualitative variables data set and its associated dummy disjunctive form data set.

INDIVIDUALS \ J	J_1	J_2	J_3	J_4	J_5
1	1	1	4	2	1
2	2	1	4	2	2
3	2	1	4	2	3
4	2	2	3	1	3
5	1	2	2	3	3
6	3	2	1	1	3
7	3	3	2	2	3
8	3	3	3	2	2
9	1	4	3	3	1

	J_1			J_2				J_3				J_4			J_5		
INDIVIDUALS \ J	1	2	3	1	2	3	4	1	2	3	4	1	2	3	1	2	3
1	1			1							1		1		1		
2		1		1							1		1			1	
3		1		1							1		1				1
4		1			1					1		1					1
5	1				1				1					1			1
6			1		1			1				1					1
7			1			1			1				1				1
8			1			1				1			1			1	
9	1						1			1				1	1		

TABLE 9.24. Model of data set with two multiple forms questions. The i-margins k_i are not necessarily equal.

There are different strategies for analyzing such data sets:

- correspondence analysis of k_{IJ} as such, indicating that the response 1 is favored by the analyst;
- correspondence analysis of k_{IJ^-}. The data set k_{IJ^-} is the dummy data set of k_{IJ} where $k_{ij} = 0$ for any form j given by the individual i; if not given, $k_{ij} = 1$;
- correspondence analysis of $k_{IJ \cup J^-}$;
- correspondence analysis of b_{JJ} or $b_{J \cup J^- J \cup J^-}$.

9.4. Presence–Absence Data Sets

In numerical taxonomy and in certain domains such as archeology, biology, and psychology, some data sets are formed as follows: consider a set J of attributes (or characteristics) and a set I of individuals. $k_{IJ} = 1$ if an individual i possesses an attribute j; if not, $k_{ij} = 0$.

There are different strategies for analyzing such data sets:

- correspondence analysis of k_{IJ} as such;
- correspondence analysis of k_{IJ^-} (if the absence is considered more important than the presence);
- correspondence analysis of $k_{IJ \cup J^-}$;
- correspondence analysis of b_{JJ} or $b_{J \cup J^- J \cup J^-}$.

9. Patterns of Acceptable Data Sets

9.5. Data Sets from Quantitative Variables

Any set of quantitative variables can be analyzed by N-D correspondence analysis after a transformation into dummy qualitative variables data sets. This is the most fruitful case of an application of correspondence analysis. Its principle is the following:

- each quantitative variable is divided into classes (or categories); to each class, a code or a form is assigned. Thus, the value of the quantitative variable is replaced by the value of its associated form, giving a dummy qualitative variable;
- from each dummy qualitative variable, we form a logical qualitative variable for each form. Therefore, a dummy disjunctive form data set is obtained;
- now, the analysis is the same as that of any qualitative variable data set.

The only problem to solve is to determine the number of classes and how these classes must be defined. There is no theoretical solution because, in practice, this situation is similar to the one in 1-D statistical analysis when a histogram is determined. The recommendations are therefore the same as those given for forming an histogram. Three main procedures are proposed:

1. Division of the quantitative variable into classes with equal range based on the histogram, so that the division approaches the statistical distribution. This is the usual method.
2. Division of the quantitative variable into classes with equal frequencies. This avoids the influence of the marginal weight of the forms on the analysis. There is a disadvantage, however, in that the limit elements can fall into a "wrong" interval.
3. Division of the quantitative variable into classes whose limits are determined by given values (for example, thresholds corresponding to specific values, as is the case in medicine).

The analyst can make as many decisions as wished for his analysis. The evolution and interest of the various divisions can be seen on graphics.

9.6. Data Sets from Quantitative and Qualitative Variables

Generally, any information system contains both quantitative and qualitative variables as described in the previous paragraphs. To analyze these variables, a dummy qualitative variables data set is formed from the quantitative variables data set, and then all the qualitative variables (the basic and dummy ones) are put into disjunctive form. Correspondence analysis is then performed on this new data set.

9.7. Fuzzy Data Sets

When a quantitative variable is divided into intervals, some information is lost since the actual value is replaced by its category. In particular, an individual located at the boundary of two classes can be attached to either of the classes, depending on the sampling. To avoid this, the following rule can be applied: An individual "located at the junction of two classes," will have as values 0.5 for the first class and 0.5 for the second class. More generally, an individual can be assigned any numerical value between 0 and 1, provided that the values for all the forms have a unit sum.

9.8. Data Sets Recoded According to Contributions

The number of forms is not necessarily the same for all the qualitative variables involved in the analysis. This results in the following effect: The contributions to the variance of each variable are not the same, since that variance depends (mathematically) on the number of forms. To have an equal contribution by each qualitative variable (or question) to the total variance, the following transformation of the qualitative variable data set is needed: Mathematically, a value, denoted by $r_q(j)$, is determined, in order to replace the value k_{ij} that is equal to 1 or 0. So:

$$k_{ij} = r_q(j) \qquad \text{for } j \in q \text{ and } 0 \text{ if not.}$$

to compute $r_q(j)$ requires the following:
The contribution of the question q to the total variance is equal to

$$\text{cta}_\alpha(J_q) = \frac{(n_q - 1)r_q}{\sum_{q' \in Q} r(q')}.$$

Then,

$$r_q = \frac{1}{n_q - 1}.$$

9. Patterns of Acceptable Data Sets

Therefore, k_{IJ} has the following values:

$$\forall q \in Q; k_{ij} = \begin{cases} \dfrac{1}{n_q - 1} & \text{for } j \in q, \text{ for } i \in I \text{ that have taken the form } j, \\ 0 & \text{for } j \in q, \text{ for } i \in I \text{ that have not taken the form } j. \end{cases}$$

9.9. Data Sets Derived from N-D Contingency Data Sets

Sometimes, one of the variables involved in qualitative variables data sets can play a particular role compared to the others. This occurs when one variable (qualitative or quantitative) must be *explained* by the other

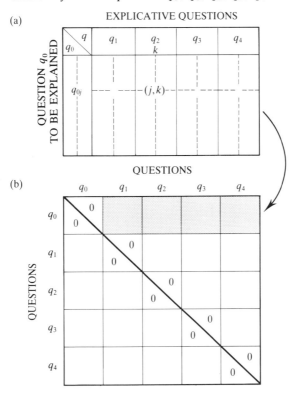

TABLE 9.25. Example of data set crossing one qualitative variable (or question) q_0 and a set of questions $\{q_1, q_2, q_3, q_4\}$. The table shows the position of the frequency data sets ($q_0 \times q_1$, $q_0 \times q_2$, $q_0 \times q_2$, $q_0 \times q_4$) in the N-D contingency data set formed by the five questions q_0, q_1, q_2, q_3, q_4.

variables considered as explicative ones. In this case, an N-D contingency data set is formed by crossing the variable to be explained and the explicative variables (cf. Table 9.25). This data set is a subset of the N-D contingency data set formed from the whole set of variables. This procedure can be used for any quantitative variable that can be transformed into a qualitative variable by a process described in the previous sections. It can be extended to any group of variables to be explained.

9.10. Conclusions

N-D correspondence analysis is actually the most fruitful method in exploratory and multivariate data analysis. It avoids studying the whole set of frequency data sets formed by the crossing of the whole set of variables, two-by-two. This method is also practical since it represents graphically the relations among the whole set of forms. Two examples, from a questionnaire on Minitel telephone services and on telephone quality of service, are presented in the next sections.

10. Case Studies

10.1. Marketing Data Set: Data Processing of a Survey Concerning the Users of Minitel Services

10.1.1. What Are the Minitel Services?

The Minitel is a home terminal connectable to the usual telephone network, which can be used as the usual telephone set, and whose services are the electronic directory, the banking services, the newspapers, the travel agency services, the entertainment services, and many other services. Every quarter, France Telecom undertakes a survey to study the behavior and satisfaction of the users. The questionnaire involves 70 questions (considered as qualitative variables) and 1800 individuals selected according to sociological and geographical criteria.

10. Case Studies

10.1.2. The Variables

Data are divided into two categories; the variables to be explained and the explicative variables:

A. The variables to be explained:

1. Frequency and duration
Q4: How many times do you use the Minitel?
Q5: Which days (during the weekend or the business week)?
Q7: What is the average duration of one call?
Q8: What is the weekly average use?
2. Reasons for using the services
Q11: Reasons for not using services other than the electronic directory service.
Q12: Reasons for using the Minitel.
Q15: Reasons for not using the electronic directory service.
3. Use of the services offered by the Minitel
Q13: A qualitative variable is formed for each service:
Code 1 = does not know the service; Code 2 = knows the service but does not use it; Code 3 = knows the service and uses it.
The services concerned are: electronic directory; bank services; transport services; mail catalogues; lodging; finances; administrative requirements; education; tourism and travel agencies; press.
4. User satisfaction
Q17: Satisfaction in owning the Minitel.
Q24: Satisfaction in using the Minitel.
Q35: Opinion on the utility of the Minitel.
Q36: Opinion on the cost of the Minitel.

TABLE 9.26. Results of numerical computations: eigenvalues and percentages of variance from the analyses of k_{IJ} and b_{JJ}.

α	e.v. (k_{IJ})	perc. (k_{IJ})	cum. (k_{IJ})	e.v. (b_{JJ})	perc. (b_{JJ})	cum. (b_{JJ})
2	0.18	4.08	4.08	0.0324	53.4	53.4
3	0.12	2.70	6.78	0.0144	23.7	77.1
4	0.08	1.84	8.62	0.0064	10.5	87.6
5	0.07	1.60	10.22	0.0049	8.1	95.7
6	0.05	1.24	11.46	0.0025	4.1	99.8

5. User information
Q19: To know how to use the Minitel.
Q20: To be informed of the services.
Q21: Advertising of the services.
Q25: Information on the tariffs.
6. User requests
Q24a: Requests related to the Minitel as equipment.
Q34b: Requests related to the tariff.
Q34c: Requests related to information on the Minitel.
Q34d: Requests related to services given by the Minitel.

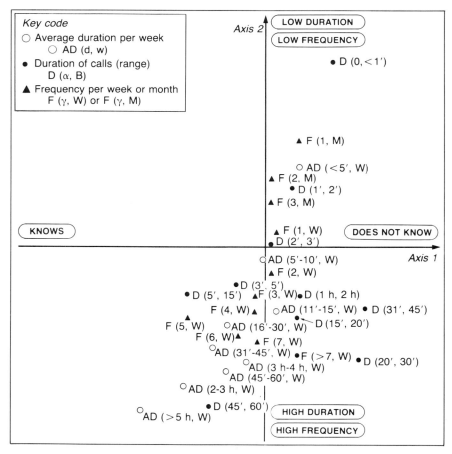

FIGURE 9.7. Graphic 1. Study of the frequency and duration of the Minitel usage. Representation of three variables in the factor space (1, 2).

10. Case Studies

B. The explicative variables:
sex; age; socio-economic category; income class; region.

10.1.3. The Data Set Analyzed

The results of the questionnaire form a qualitative variable data set, which is transformed into a dummy disjunctive form data set. There are 302 logical forms, among which there are 221 forms corresponding to those to be explained and 81 explicative forms. The data set has therefore 302 columns, and 1800 rows corresponding to the individuals. N-D correspondence analysis was performed on the above data set.

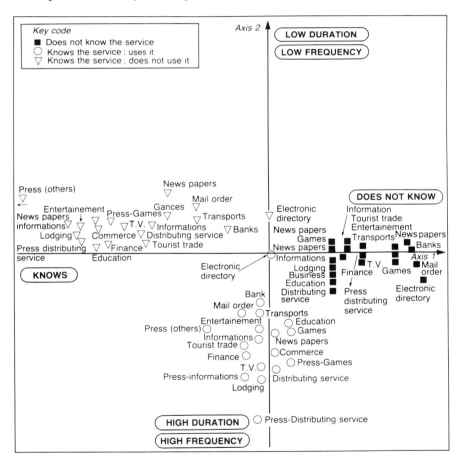

FIGURE 9.8. Graphic 2. Study of the use of the main services offered by Minitel. Representation of three variables in the factor space $(1, 2)$.

10.1.4. Numerical Results

The eigenvalues and the percentage of variance associated with the analysis are given in Table 9.26.

10.1.5. Graphics

To avoid having 302 form-points representing J on the same graphic, different graphics are proposed. Each of them is devoted to a specific set of variables, but may be superimposable. They are easier to understand. The graphics are the following:

- Graphic 1 (*cf.* Fig. 9.7). Study of the frequency and duration of Minitel usage.

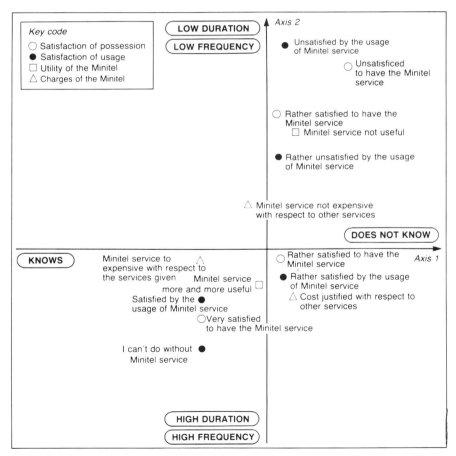

FIGURE 9.9. Graphic 3. Study of user satisfaction. Representation of four variables in the factor space $(1, 2)$.

10. Case Studies

- Graphic 2 (*cf.* Fig. 9.8). Study of the use of the main services offered by Minitel.
- Graphic 3 (*cf.* Fig. 9.9). Study of user satisfaction.
- Graphic 4 (*cf.* Fig. 9.10). Study of user requests.
- Graphic 5 (*cf.* Fig. 9.11). Study of the explicative variables: sex, age, socio-economic categories, income class.
- Graphic 6 (*cf.* Fig. 9.12). Study of the position of the individuals associated with the variable age.

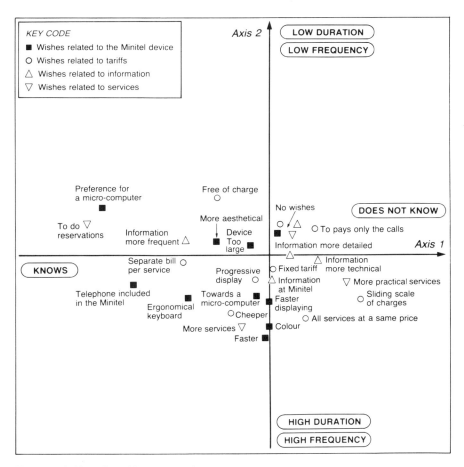

FIGURE 9.10. Graphic 4. Study of user requests. Representation of four variables in the factor space (1, 2).

10.2. Management Data Set: Data Processing of Indicators Related to the Quality of Telephone Service

10.2.1. What Is the Quality of Service?

The management of the France Telecom network is based on technical indicators observed every month and published every year in the annual statistical report of France Telecom. The data, from 1984, concern six main variables (the indicators are quantitative variables) and one composite indicator that is a linear combination of the six previous variables. The data are described in Appendix 2, §14.

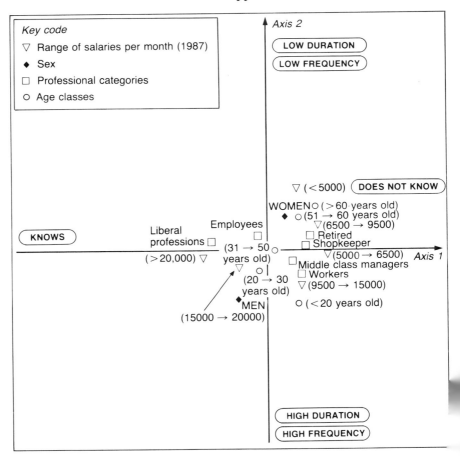

FIGURE 9.11. Graphic 5. Study of the explicative variables. Representation of four variables in the factor space (1, 2).

10. Case Studies

10.2.2. The Data Set Analyzed

To apply N-D correspondence analysis to an indicators data set, a dummy qualitative data set has to be formed. The process chosen involves forming three logical variables with equal marginal frequencies from each of the indicators as follows:

IGQS ⟶ IGS_1, IGS_2, IGS_3;
IZAA ⟶ IZA_1, IZA_2, IZA_3;
EZAA ⟶ EZA_1, EZA_2, EZA_3;
VR2 ⟶ $VR2_1$, $VR2_2$, $VR2_3$;
TCR ⟶ TCR_1, TCR_2, TCR_3;

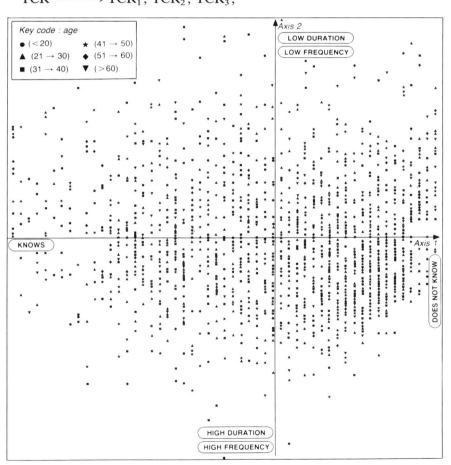

FIGURE 9.12. Graphic 6. Study of the position of the individuals associated with the variable age. Representation in the factor space (1, 2).

TSI is transformed into 100-TSI then into TSI_1, TSI_2, TSI_3;
TCO is transformed into 100-TCO then into TCO_1, TCO_2, TCO_3.

N-D correspondence analysis is performed on the data set involving the classes of indicators ($IGS_1, \ldots, TCO_3$).

TABLE 9.27. Results of numerical computations: eigenvalues and percentages of variance from the analyses of k_{IJ} and b_{JJ}.

α	e.v. (k_{IJ})	perc. (k_{IJ})	cum. (k_{IJ})	e.v. (b_{JJ})	perc. (b_{JJ})	cum. (b_{JJ})
2	0.43	21.74	21.74	0.18	38.29	38.29
3	0.35	17.43	38.90	0.12	25.53	63.82
4	0.25	12.57	51.48	0.06	12.76	76.58
5	0.19	9.75	61.23	0.04	8.51	85.09
6	0.15	7.61	68.85	0.03	6.38	91.47

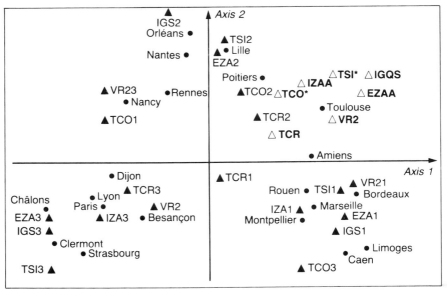

FIGURE 9.13. Quality of service analysis. Representation of the variables ($\triangle$), the forms ($\blacktriangle$), and the regions ($\bullet$) in the factor space (1, 2).

10. Case Studies

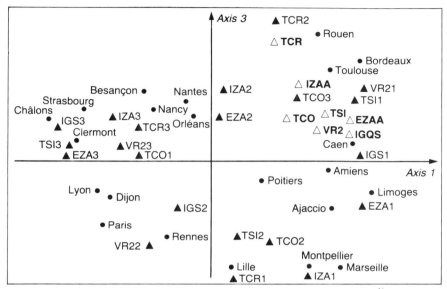

FIGURE 9.14. Quality of service analysis. Representation of the variables (△), the forms (▲), and the regions (•) in the factor space (1, 3).

10.2.3. Numerical Results

The eigenvalues and the percentage of variance associated with the analysis are given in Table 9.27.

10.2.4. Graphics

The main graphics related to the N-D correspondence analysis are given in Figs. 9.13 and 9.14.

Chapter 10
Classification of Individuals–Variables Data Sets

1. Introduction

Classification methods are well known in both numerical taxonomy and pattern recognition, and more generally in any topic related to data analysis. How are these methods associated with statistical data processing? There are three main methods: building classes, recognizing classes, and classifying unknown elements to a system of classes already built. These methods explain the role of classification in multivariate and statistical data processing.

We mention different authors who have chosen different ways for classification (also called clustering methods): MacQueen (1967), Cormack (1971), Jardine and Sibson (1971), Anderberg (1973), Sokal and Sneath (1973), Benzecri (1973), Hartigan (1975), Jambu (1978, 1983), Diday (1980), Lerman (1981).

According to Cormack, the literature on clustering has been as abundant as that on regression, and it still continues to flourish. To delve more deeply into the subject, the interested reader should study the specialized literature. In this chapter, devoted to clustering methods and applications, only major methods are presented. They were selected by the author with a view towards actual and concrete applications.

The plan of this chapter is as follows: First, the data sets used in clustering are studied in Section 2; then, the mathematical description of classifications is given in Section 3; then, the two main groups of methods are presented: partitioning in Section 4 and hierarchical clustering in Section 5; finally, some specific applications and case studies are presented in Section 6.

2. Basic Data Sets

Two types of data used in clustering can be distinguished: the individuals–variables data sets and the proximities data sets. Since they were presented in Chapter 2, only a brief description is recalled here.

2.1. Individuals–Variables Data Sets

The individuals–variables data sets are those that involve several variables observed on different statistical units, also called individuals. The following must be considered as individuals–variables data sets: contingency data sets, logical data sets, results of surveys and questionnaires, and measurements studies. Any data sets used in factor analysis and for elementary statistical descriptions can be used in clustering.

2.2. Proximities Data Sets

Proximities data sets are sets of numerical values corresponding to a distance matrix between two sets of variables or two sets of individuals. Some examples are presented in Appendix 2. Data processing of these data sets is studied in Chapter 11.

3. The Mathematical Description of Classifications

Under the name of classification, usually two different structures are understood: partitions and hierarchical classifications (more accurately, binary indexed hierarchical classifications, but more briefly, we speak of hierarchical classifications).

3.1. Partitions

Consider a set of elements, denoted by I. The set $Q(I)$ is a partition of I if $Q(I)$ satisfies the following relations:

$$\begin{cases} I = \{q_1, q_2, \ldots, q_k\}; \\ \bigcup_{i=1}^{k} q_i = I \quad \text{with } q_i \neq \emptyset; \\ q_i \cap q_j = \emptyset \quad \text{for } i \neq j. \end{cases}$$

3. Basic Data Sets

3.2. Hierarchical Classifications

Using Fig. 10.1, notations and terminology associated with classification structures and algorithms are given:

Consider a set I of five elements, denoted by 1, 2, 3, 4, 5:

$$I = \{1, 2, 3, 4, 5\}.$$

Consider now a hierarchical classification based on I, denoted by $C(I)$:

$$C(I) = \{\{1, 2, 3, 4, 5\}, \{3, 2, 1, 5\}, \{3, 2\}, \{1, 5\}, \{1\}, \{2\}, \{3\}, \{4\}, \{5\}\}.$$

$C(I)$ is a subset of $\mathcal{P}(I) - \emptyset$.

$C(I)$ is formed by subsets of I, ordered by the inclusion relation; it satisfies the following axioms:

1. $\forall a, b \in C(I), a \cap b = \{a, b, \emptyset\}$;
2. $\forall a \in C(I), \{b \in C(I); b \neq a \text{ and } b \cup a\} \in \{a, \emptyset\}$.

Consider the set of terminal classes of $C(I)$, denoted by Ter $C(I)$. This is the set of minimal elements of $C(I)$ with respect to the inclusion relation. Since $C(I)$ is a total binary hierarchical classification, Ter $C(I)$ contains only single-element classes (or singleton classes):

$$\text{Ter } C(I) = \{\{1\}, \{2\}, \{3\}, \{4\}, \{5\}\}.$$

Consider the set of nonterminal classes of $C(I)$, denoted by Nod $C(I)$. These classes are also called nodes of $C(I)$. We have

$$\text{Nod } C(I) = \{\{1, 2, 3, 4, 5\}, \{3, 2\}, \{1, 5\}, \{3, 2, 1, 5\}\}.$$

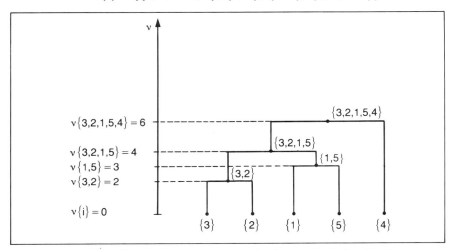

FIGURE 10.1. Classical representation of a hierarchical classification.

Since $C(I)$ is a total binary hierarchical classification, the number of elements of $\text{Nod } C(I)$ is $\text{Card } I - 1$ (where $\text{Card } I$ is the number of elements of I). Furthermore,

$$\text{Card Nod } C(I) \cup \text{Card Ter } C(I) = \text{Card } C(I) = 2 \text{ Card } I - 1.$$

Thus, the terminal classes can be numbered from 1 to $\text{Card } I$ and the nonterminal classes (or nodes) can be numbered from $\text{Card } I + 1$ to $2 \text{ Card } I - 1$. The nodes are numbered according to their construction. The first class formed has the number $\text{Card } I + 1$; the last class formed, which is the whole set I, has the number $2 \text{ Card } I - 1$. The latter is also called the summit of $C(I)$.

Consider the set of summit-classes of $C(I)$ denoted by $\text{Som } C(I)$. $\text{Som } C(I)$ is the set of maximal elements of $C(I)$ with respect to the inclusion relation. For a total binary hierarchical classification $\text{Som } C(I)$ is equal to $\{I\}$.

All these notations and explanations are summarized in Fig. 10.1.

Consider now, for each class a of $C(I)$, the level index of the class a, denoted by $v(a)$. The level index $v(a)$ is the height at which the class a is placed when a is formed (cf. Table 10.1). In most cases, the following relation holds:

$$\forall a, b \in C(I) \text{ and } a \subset b \quad \Rightarrow \quad v(a) \leq v(b).$$

This yields

$$a \subset a \cup b \quad \Rightarrow \quad v(a) \leq v(a \cup b),$$
$$b \subset a \cup b \quad \Rightarrow \quad v(b) \leq v(a \cup b).$$

TABLE 10.1. Numbering and level index associated with the hierarchical classification $C(I)$.

	class	class number	Level index
terminals Ter $C(I)$	{1}	1	0
	{2}	2	0
	{3}	3	0
	{4}	4	0
	{5}	5	0
nodes Nod $C(I)$	{3, 2}	6	2
	{1, 5}	7	3
	{3, 2, 1, 5}	8	4
	{3, 2, 1, 5, 4}	9	6

3. Basic Data Sets

As each new upper node is a disjoint union of two lower previous nodes, the union node has a level index greater than each of the two forming nodes. This corresponds to the representation given in Fig. 10.1. v is thus considered as a proximity measure between two forming nodes. Consider the set of classes immediately below each class a of Nod $C(I)$; this set, denoted by Suci $(a, c(I))$, forms a partition of the class a into only two classes, called the *eldest* and the *youngest* of a. Thus, each node of $C(I)$ possesses one eldest class and one youngest class. This is summarized in Table 10.2.

In order to present notations we will be using, we consider all the distinct levels v of $C(I)$. A partition, denoted by Som C_v, can be associated with each level. This partition is called the summit of $C(I)$ at the level v. For our example, this gives

Som $C_{v=0} = \{\{1\}, \{2\}, \{3\}, \{4\}, \{5\}\}$,
Som $C_{v=2} = \{\{1\}, \{3, 2\}, \{4\}, \{5\}\}$,
Som $C_{v=3} = \{\{1, 5\}, \{3, 2\}, \{4\}\}$,
Som $C_{v=4} = \{\{1, 5, 3, 2\}, \{4\}\}$,
Som $C_{v=6} = \{\{1, 5, 3, 2, 4\} = \{I\} =$ Som $C(I)\}$.

Therefore, a hierarchical classification can be viewed as a series of nested partitions where each term is formed by the union of two classes in the preceding term, until the final partition, I itself, is obtained.

The graphical representation of a hierarchical classification is called a tree (or dendrogram). For our example, this yields the representation in Fig. 10.2(b). Sometimes, however, dendograms produced by computer are a little different. A computer plot for our example gives the representation in Fig. 10.2(a). The reader should study them both in order to have the two plots in mind.

TABLE 10.2. Eldest and youngest classes associated with the hierarchical classification $C(I)$.

class	class number	eldest class	eldest class number	youngest class	youngest class number
$\{3, 2\}$	6	$\{3\}$	3	$\{2\}$	2
$\{1, 5\}$	7	$\{1\}$	1	$\{5\}$	5
$\{3, 2, 1, 5\}$	8	$\{3, 2\}$	6	$\{1, 5\}$	7
$\{3, 2, 1, 5, 4\}$	9	$\{4\}$	4	$\{3, 2, 1, 5\}$	8

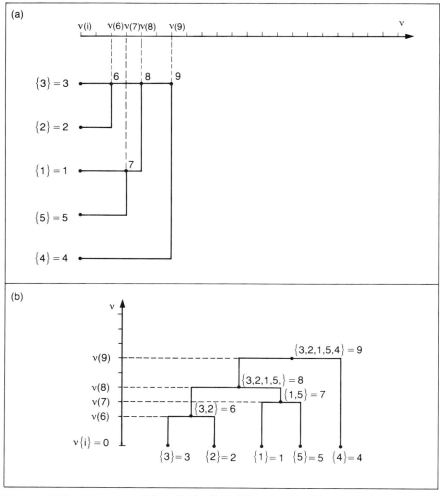

FIGURE 10.2. Computer (a) and classical (b) representations of a hierarchical classification.

4. Partitioning Methods

4.1. Introduction

To look for the best partition of a set I, all the possible partitions of the set I need to be examined. From a computer point of view, it is difficult to compute all of the partitions for large data sets. This is shown by some computational features: For a set I of cardinality n, consider the number of partitions of I into k classes, denoted by par(n, k). For $k > n$,

4. Partitioning Methods

$\text{par}(n, k) = 0$; $\text{par}(n, n) = 1$, and $\text{par}(n, 1) = 1$. The number of possible partitions of I in 2 classes, $\text{par}(n, 2) = 2^{n-1} - 1$. An analytic formula is not known for calculating the number of partitions of a set I of cardinality n into k classes. A recurrence formula is given by Volle (1985):

$$\text{par}(n, k) = k \, \text{par}(n - 1, k) + \text{par}(n - 1, k - 1).$$

This shows how difficult it is to compute par for large data sets, and consequently to find an optimal partition. Even if it is not possible to find an optimal partition, partitioning methods have a real interest in practice, because they allow the needs of users to be satisfied. Partitioning methods are based on the principle that a constraint is required in order to be able to find a solution. Two main methods (or algorithms) are examined here; the first is based on the principle of fixing the number of classes in the partition, often called "the algorithm of moving centers with k classes fixed." There are many variants (k-means of MacQueen (1967), Isodata of Ball and Hall (1965), dynamic clouds of Diday (1980), etc.); the second is based on the principle of fixing the diameter of the classes that are to be found, often called "the algorithm of moving centers with the diameter d fixed" (optimized balls of Flamenbaum, 1979). These two algorithms are studied in what follows.

4.2. Algorithm of Moving Centers with k Initially Fixed Classes

4.2.1. Principle of Construction

Consider an individuals–variables data set, denoted by X_{IJ}. We know that the elements i of I can be considered as points of a cloud set in the Euclidean space R_J and, conversely, the elements j of J can be considered as points of a cloud set in the Euclidean space R_I. The partitioning algorithm can be performed on only one set (I or J), not on both sets simultaneously. We assume that the partition is built on the set I.

> The basic principle of the algorithm is to determine a partition of I into k classes, the number of classes k being fixed initially by the user. k is thus a dependent parameter of the algorithm.

4.2.2. The algorithm [PAR.NC = k]

Step 0.   (a) k centers are arbitrarily chosen. These centers are either points in the space of variables, groups of elements of I, or centers of gravity of classes chosen by the user.
Let $C_1^{(0)}, C_2^{(0)}, \ldots, C_l^{(0)}, \ldots, C_k^{(0)}$ be these centers.

(b) Each individual i is assigned to a class and only one whose center is one of the previously built k centers according to the following rule:

$$\begin{pmatrix} i \text{ belongs to the class,} \\ \text{denoted by } P_l^{(0)}, \\ \text{with center } C_l^{(0)} \end{pmatrix} \Leftrightarrow \begin{pmatrix} \|i - C_l^{(0)}\| \text{ is its minimum} \\ \text{on the whole set of centers,} \\ C_1^{(0)}, C_2^{(0)}, \ldots, C_l^{(0)}, \ldots, C_k^{(0)} \end{pmatrix}.$$

At the end of this step, k classes are built, denoted by $P_1^{(0)}, P_2^{(0)}, \ldots, P_l^{(0)}, \ldots, P_k^{(0)}$, with centers $C_1^{(0)}, C_2^{(0)}, \ldots, C_l^{(0)}, \ldots, C_k^{(0)}$, respectively; and any element i of I belongs to one of the classes $P_l^{(0)}$.

Step 1. (a) k new centers are built as follows:
The centers of gravity denoted by $C_1^{(1)}, C_2^{(1)}, \ldots, C_l^{(1)}, \ldots, C_k^{(1)}$ are computed from the classes $P_1^{(1)}, P_2^{(1)}, \ldots, P_l^{(1)}, \ldots, P_k^{(1)}$.

(b) Each individual i of I is assigned to a class, and only one whose center is one of the k centers $C_1^{(1)}, C_2^{(1)}, \ldots, C_l^{(1)}, \ldots, C_k^{(1)}$, according to the following rule:

$$\begin{pmatrix} i \text{ belongs to the class,} \\ \text{denoted by } P_l^{(1)}, \\ \text{with center } C_l^{(1)} \end{pmatrix} \Leftrightarrow \begin{pmatrix} \|i - C_l^{(1)}\| \text{ is its minimum} \\ \text{on the whole set of centers,} \\ C_1^{(1)}, C_2^{(1)}, \ldots, C_l^{(1)}, \ldots, C_k^{(1)} \end{pmatrix}.$$

At the end of this step, k new classes built, denoted by $P_1^{(1)}, P_2^{(1)}, \ldots, P_l^{(1)}, \ldots, P_k^{(1)}$, with centers $C_1^{(1)}, C_2^{(1)}, \ldots, C_l^{(1)}, \ldots, C_k^{(1)}$, respectively, and any element i of I belongs to one of the classes $P_l^{(1)}$.

Step h. (a) k new centers are built as follows:
The centers of gravity, denoted by $C_1^{(h)}, C_2^{(h)}, \ldots, C_l^{(h)}, \ldots, C_k^{(h)}$, are computed from the classes $P_1^{(h-1)}, P_2^{(h-1)}, \ldots, P_l^{(h-1)}, \ldots, P_k^{(h-1)}$ built in the previous step $(h-1)$.

(b) Each individual i of I is assigned to a class and only one whose center is one of the k centers $C_1^{(h)}, C_2^{(h)}, \ldots, C_l^{(h)}, \ldots, C_k^{(h)}$, according to the following rule:

$$\begin{pmatrix} i \text{ belongs to the class,} \\ \text{denoted by } P_l^{(h)}, \\ \text{with center } C_l^{(h)} \end{pmatrix} \Leftrightarrow \begin{pmatrix} \|i - C_l^{(h)}\| \text{ is its minimum} \\ \text{on the whole set of centers,} \\ C_1^{(h)}, C_2^{(h)}, \ldots, C_l^{(h)}, \ldots, C_k^{(h)} \end{pmatrix}.$$

4. Partitioning Methods

|Step End.|At the end of Step h, a partition is built in k classes, $P_1^{(h)}$, $P_2^{(h)}$, ..., $P_l^{(h)}$, ..., $P_k^{(h)}$, with centers $C_1^{(h)}$, $C_2^{(h)}$, ..., $C_l^{(h)}$, ..., $C_k^{(h)}$, respectively.
Several rules can be applied for stopping the iteration:
(a) The contents of classes are not modified during two successive steps.
(b) The number of steps is given by the user.
(c) The value of the criteria does not change during two successive steps.|

4.2.3. The Criterion

4.2.3.1. Supplementary Notations. To study the criteria on which the algorithm is based, some geometric properties need to be recalled:

Let I be a finite set, and $N(I)$ be the cloud of individuals associated with I in the space R_J; any $i \in I$ is described by the coordinates $i_1, i_2, \ldots, i_j, \ldots, i_{\text{Card } J}$, to which masses m_i have been assigned.

We will use the notation

$$m_I = \sum_{i \in I} m_i.$$

The letters a, b, q, s, and t usually represent the classes (or subsets) of I. The coordinates of these classes in R_J are denoted by a_J, b_J, q_J, s_J, and t_J, respectively, and the masses assigned to classes are denoted by m_a, m_b, m_q, m_s, and m_t, respectively. The following relations occur:

$$a_J = \sum_{i \in a} \left(\frac{m_i}{m_a}\right) i_J;$$

$$g_J = \sum_{i \in I} \left(\frac{m_i}{m_I}\right) i_J$$

(g is the center of gravity of I).

The square of the distance between two individuals i and i' is denoted by $\|i_J - i'_J\|^2$ ($\|\cdot\|$ is the mathematical norm associated with the space R_J). For the usual Euclidean distance, we have the following relation:

$$\|i_J - i'_J\|^2 = \sum_{j \in J} (i_j - i'_j)^2$$

The second order central moment of class q in R_J is denoted by $M^2(q)$, and the variance of class q in R_J is denoted by $V(q)$. Let Q be a partition of I; the second order central moment of Q is denoted by $M^2(Q)$, and the variance of Q is denoted by $V(Q)$. $M^2(Q)$ is the second order central moment of the cloud containing the centers of gravity of the classes that make up the partition Q. The following relations are obtained:

$$M^2(q) = \sum_{i \in q} m_i \, \|i_J - q_J\|^2;$$

$$V(q) = \sum_{i \in q} \left(\frac{m_i}{m_q}\right) \|i_J - q_J\|^2;$$

$$M^2(Q) = \sum_{q \in Q} m_q \, \|q_J - g_J\|^2;$$

$$V(Q) = \sum_{q \in Q} \left(\frac{m_q}{m_I}\right) \|q_J - g_J\|^2.$$

From these relations, we get the following relations:

Rule 1. Let I be a Euclidean space of finite dimension; q a subset of I, q_J the coordinates of its center of gravity, and m_q its mass. Let q' be a subset of I with $q' \cap q = \varnothing$. Then,

$$m_q \, \|q_J - q'_J\| + M^2(q) = \sum_{i \in q} \|i_J - q'_J\|^2.$$

Rule 2.

$$2M^2(I) \cdot m_I = \sum_{i,i' \in I} m_i m_{i'} \, \|i_J - i'_J\|^2.$$

Rule 3. Let I be a Euclidean space of finite dimension, and let Q be a partition of I. Then,

$$M^2(I) = M^2(Q) + \sum_{q \in Q} M^2(q).$$

Rule 4. Then, let I be a Euclidean space of finite dimension and let a and b be two subsets of I. Then

$$\sum_{\substack{i \in a \\ i' \in b}} m_i m_{i'} \, \|i_J - i'_J\|^2 = m_b M^2(a) + m_a M^2(b) + m_a m_b \, \|a_J - b_J\|^2.$$

Rule 5. If $Q = \{a, b\}$ and $a \cap b = \varnothing$, then

$$\begin{aligned} M^2\{a, b\} &= M^2(a \cup b) - M^2(a) - M^2(b) \\ &= m_a \, \|a_J - g_J\rangle^2 + m_b \, \|b_J - g_J\|^2 \\ &= \frac{m_a \cdot m_b}{m_a + m_b} \, \|a_J - b_J\|^2. \end{aligned}$$

4. Partitioning Methods

Rule 6.

$$V(I) = V(Q) + \sum_{q \in Q} \left(\frac{m_q}{m_I}\right) V(q).$$

4.2.3.2. Criterion Associated with the Algorithm. The criterion based on convergence of the algorithm is given as follows: From a certain step onwards, the second order central moment of the partition does not decrease and, therefore, the partition and the centers of classes are stable. To prove convergence, the passage from the centers $C_1^{(h-1)}, C_2^{(h-1)}, \ldots, C_l^{(h-1)}, \ldots, C_k^{(h-1)}$ established at Step $(h-1)$ to the centers $C_1^{(h)}, C_2^{(h)}, \ldots, C_l^{(h)}, \ldots, C_k^{(h)}$ established at the Step h needs to show an increase in the second order central moment $M^2(Q)$. According to Huyghens' theorem, this means an increase in the sum of the second order central moment of classes.

For convergence of the algorithm, it must be proved that

$$\sum_{l=1}^{k} M^2(P_l^{(h-1)}, C_l^{(h-1)}) \geq \sum_{l=1}^{k} M^2(P_l^{(h)}, C_l^{(h)}), \tag{1}$$

where $M^2(P_l^{(h)}, C_l^{(h)})$ is the second order central moment of the class $P_l^{(h)}$ relative to its center of gravity $C_l^{(h)}$.

Based on Huyghens' theorem, the following relations hold:

$$M^2(P_l^{(h-1)}, C_l^{(h-1)}) = M^2(P_l^{(h-1)}, C_l^{(h)}) + m_{P_l^{(k-1)}} \|C_l^{(h-1)} - C_l^{(h)}\|^2$$

$$\Rightarrow M^2(P_l^{(h-1)}, C_l^{(h-1)}) \geq M^2(P_l^{(h-1)}, C_l^{(h)})$$

$$\Rightarrow \sum_{l=1}^{k} M^2(P_l^{(h-1)}, C_l^{(h-1)}) \geq \sum_{l=1}^{k} M^2(P_l^{(h-1)}, C_l^{(h)}).$$

It must be proved that this implies that

$$\sum_{l=1}^{k} M^2(P_l^{(h-1)}, C_l^{(h)}) \geq \sum_{l=1}^{k} M^2(P_l^{(h)}, C_l^{(h)}).$$

This then yields the final result.

The previous inequality may be written as follows (with μ_j = mass of j):

$$\sum_{l=1}^{k} M^2(P_l^{(h-1)}, C_l^{(h)}) \stackrel{\text{def}}{=} \sum_{l=1}^{k} \left(\sum_{j \in P^{(k-1)}} \mu_j \|j - C_l^{(h)}\|^2 \right)$$

$$\geq \sum_{l=1}^{k} M^2(P_l^{(h)}, C_l^{(h)}) \stackrel{\text{def}}{=} \sum_{l=1}^{k} \left(\sum_{j \in P^{(k)}} \mu_j \|j - C_l^{(h)}\|^2 \right).$$

In the right hand side of Eq. (1), the points j are the points nearest to $C_l^{(h)}$, whereas in the left hand side of Eq. (1), the points j of $P^{(h-1)}$ are not necessarily the points nearest to $C_l^{(h)}$ (by definition of classes). Therefore, the right hand side of Eq. (1) is less than the left hand side of Eq. (1); this proves the algorithm's convergence.

4.2.3.3. Variants of Criterion The algorithm depends on the number k of initially attributed classes, as well as on the choice of the contents of the k initial classes or centers. In the k-means method of MacQueen (1967), the initial centers are given by random sampling techniques, and then their positions are modified after each point i of I is assigned. This means that the final partition depends on the order in which the points were assigned. In the dynamic clouds method of Diday (1980), the centers are computed from subclasses of I; the assignment rule involves maximizing a distance criterion between one element and one class.

4.2.3.4. Reminder
(a) The algorithm results in a final partition that depends on the number of classes k, fixed initially before applying the algorithm to real data. The consequence is that a partition into k classes can be greatly different from a partition into $(k + 1)$ classes or $(k - 1)$ classes.

(b) The algorithm results in a final partition that depends on the choice of initial centers or classes.

(c) To avoid these two disadvantages, some researchers propose that the algorithm should be applied as many times as there are choices of initial centers. The partitions obtained are crossed between themselves in order to obtain a final partition. However, the partition that results does not necessarily have k classes (as fixed initially), and the criterion associated with the algorithm is not necessarily optimal for it.

(d) The field of application of the algorithm is that where the number of initially fixed classes k corresponds to k of real classes recognized in practice. This means that before applying the algorithm, a preliminary data analysis must be done to know if k is a good estimate of the number of classes to be recognized.

(e) The final partition is sensitive to "the chaining effect," meaning that classes may not correspond to any relation in the data.

4.3. *Algorithm of Moving Centers with an Initially Fixed Diameter of Classes d*

As in the previous case, the general framework of partitioning is related to a set I, described by a set J of variables. The corresponding data set

4. Partitioning Methods

X_{IJ} is similar to Euclidean coordinates, and masses m_i are assigned to points i of I.

> The basic principle of the algorithm is to determine a partition of I into classes whose diameters are less than a given value, denoted by $d = 2R$, and fixed initially by the user. $d = 2R$ is thus a dependent parameter of the algorithm.

4.3.1. The algorithm [PAR.DI = 2R]

Step 0.
(a) A diameter of classes, denoted by $d = 2R$, is arbitrarily fixed; the first individual i of I is the first center $C_1^{(0)}(=i_0)$. Each individual i of I is assigned to a center according to the following rule:

$\|i - C_l^{(0)}\| \leq R \Rightarrow i$ belongs to the class $P_l^{(0)}$ with the center $C_l^{(0)}$;

$\|i - C_l^{(1)}\| > R \Rightarrow i$ is the center $C_{l+1}^{(0)}$ of a new class $P_{l+1}^{(0)}$.

At the end of the substep (a), k centers of classes are determined, denoted by $C_1^{(0)}, C_2^{(0)}, \ldots, C_l^{(0)}, \ldots, C_k^{(0)}$; each individual is assigned to one class and only of the set of formed classes, denoted by $P_1^{(0)}, P_2^{(0)}, \ldots, P_l^{(0)}, \ldots, P_k^{(0)}$.

(b) The centers of gravity of the classes $P_1^{(0)}, P_2^{(0)}, \ldots, P_l^{(0)}, \ldots, P_k^{(0)}$ are computed; they are denoted by $C_1^{(0)}, C_2^{(0)}, \ldots, C_l^{(0)}, \ldots, C_k^{(0)}$, respectively.

Step 1.
(a) The individuals i of I are reassigned according to the following rule:

$\|i - C_l^{(1)}\| \leq R \Rightarrow i$ belongs to the class $P_l^{(1)}$ with the center $C_l^{(1)}$;

$\|i - C_l^{(1)}\| > R \Rightarrow i$ is the center $C_{l+1}^{(1)}$ of a new class $P_{l+1}^{(1)}$.

At the end of the substep (a), k' classes are formed, denoted by $P_1^{(1)}, P_2^{(1)}, \ldots, P_l^{(1)}, \ldots, P_{k'}^{(1)}$, with respective centers denoted by $C_1^{(1)}, C_2^{(1)}, \ldots, C_l^{(1)}, \ldots, C_{k'}^{(1)}$.

(b) The centers of gravity of the previous classes are computed; they are denoted by $C_1^{(1)}, C_2^{(1)}, \ldots, C_l^{(1)}, \ldots, C_{k'}^{(1)}$.

| Step h. | (a) The individuals i of I are reassigned according to the following rule:

$\|i - C_l^{(h)}\| \leq R \Rightarrow i$ belongs to the class $P_l^{(h)}$ with the center $C_1^{(h)}$;

$\|i - C_l^{(h)}\| > R \Rightarrow i$ is the center $C_{l+1}^{(h)}$ of a new class $P_{l+1}^{(h)}$.

(b) At the end of the substep (a), $k(h)$ classes are formed, denoted by $P_1^{(h)}, P_2^{(h)}, \ldots, P_l^{(h)}, \ldots, P_{k(h)}^{(h)}$, with respective centers denoted by $C_1^{(h)}, C_2^{(h)}, \ldots, C_l^{(h)}, \ldots, C_{k(h)}^{(h)}$. |
| Step End. | End of the procedure. The partiton in $k(h)$ classes is formed.

—When the maximum number of classes fixed by the user is reached before the end, the procedure is reiterated with a new diameter value.

—When the maximum number of classes is not fixed by the user, only the diameter value plays a role in the algorithm. Generally the diameter value is related to the total variance of the cloud of points.

—There is no optimal criterion to reach. |

4.3.2. Reminder

(a) The partition-solution depends on the initially fixed diameter, which can be modified by successive attempts or chosen by preliminary data analysis.

(b) With this algorithm, the "chaining effect" does not exist as in the algorithm of moving centers with k initially fixed classes.

(c) The algorithm does not fix a distance; this means that the distance chosen must be coherent with the type of data set (*cf.* Table 10.3).

(d) The algorithm is applied to fields related to building classes and not to recognizing classes. This algorithm can be considered as a preliminary data analysis because it reduces the number of individuals associated with a data set into a smaller number of classes.

4.4. Patterns of acceptable data sets

What type of data sets may be used with the two previous algorithms? Any of the individuals–variables data sets described in the previous chapters (Chapters 7, 8, and 9) may be classified by such partitioning

4. Partitioning Methods

TABLE 10.3. Data sets and available distances for algorithms of classification.

DATA SETS	I	J
principal components data sets	$d^2(i,i') = \sum_{j \in J}(X_{ij} - X_{i'j})^2$	
standardized principal components data sets	$d^2(i,i') = \sum_{j \in J}(X_{ij} - X_{i'j})^2/\sigma_j$	$2(1 - r_{jj'})$
2-D correspondence data sets	$d^2(i,i') = \sum_{j \in J}(f_j^i - f_j^{i'})^2/f_j$	$d(j,j') = 2(1 - r_{jj'})$
N-D correspondence data sets (disjunctive form data sets)	$d^2(i,i') = \sum_{j \in J}(f_j^i - f_j^{i'})^2/f_j$	$d^2(j,j') = \sum_{i \in I}(f_i^j - f_i^{j'})^2/f_i$
N-D correspondence data sets (N-contingency data sets)	Computations on the points i of I as supplementary elements	$d^2(j,j') = \sum_{i \in I}(f_i^j - f_i^{j'})^2/f_i$
factor coordinates data sets from principal components analysis	$d^2(i,i') = \sum_{\alpha}(u_\alpha(i) - u_\alpha(i'))^2$	$d^2(j,j') = \sum_{\alpha}(v_\alpha(j) - v_\alpha(j'))^2$
factor coordinates data sets from standardized principal components analysis	$d^2(i,i') = \sum_{\alpha}(u_\alpha(i) - u_\alpha(i'))^2$	$d^2(j,j') = \sum_{\alpha}(v_\alpha(j) - v_\alpha(j'))^2$
factor coordinates data sets from 2-D correspondence analysis	$d^2(i,i') = \sum_{\alpha}(\psi_\alpha(i) - \psi_\alpha(i'))^2$ or	$d^2(j,j') = \sum_{\alpha}(\phi_\alpha(j) - \phi_\alpha(j'))^2$ or
factor coordinates data sets from N-D correspondence data analysis of disjunctive form data sets	$d^2(i,i') = \lambda_\alpha^{-1/2}\sum_{\alpha}(\psi_\alpha(i) - \psi_\alpha(i'))^2$	$d^2(j,j') = \lambda_\alpha^{-1/2}\sum_{\alpha}(\phi_\alpha(j) - \phi_\alpha(j'))^2$
factor coordinates data sets from N-D correspondence analysis of N-contingency data sets	Computations on the points i of I as supplementary elements	

methods. The only problem is to select a distance coherent with these individuals–variables data sets. In order to simplify the problem, the individuals–variables data sets may be classified into three main groups:

- Principal components data sets. These are individuals–variables data sets, similar to Euclidean coordinates, on which the distance associated with principal components analysis may be applied.
- Correspondence data sets. These are individuals–variables data sets, similar to profiles-coordinates, on which the distributional distance between profiles may be applied.
- Factor coordinates data sets. These are data sets, similar to Euclidean coordinates, resulting from principal components analysis or correspondence analysis.

4.5. Interpretation of Partitions

As in any factor analysis of individuals–variables data sets, the final result (here, a partition) must be interpreted, meaning that the differences between classes of individuals must be explained in terms of variables, and conversely. In the following, we present the basic elements needed to interpret partitions.

4.5.1. Hierarchical Representation of a Partition
Consider a partition Q of I. $I = \{i; i \in I\}$; $Q = \{q \in Q; q' \cap q'' = \emptyset\}$. The individuals $i \in I$ are considered as *terminals* of a hierarchical classification

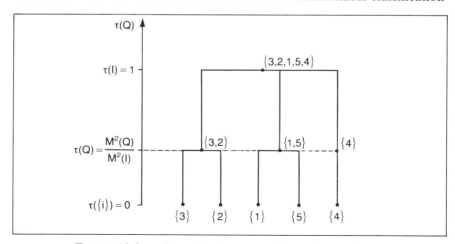

FIGURE 10.3. Hierarchical representation of a partition.

4. Partitioning Methods

with three levels (*cf.* Fig. 10.3).

- The first level, made up of the individuals $i \in I$, are the terminals of the classification; the index level associated with i, denoted by $\tau(\{i\})$, is equal to zero.
- The second level is made up of the classes $q \in Q(I)$; the index level of Q, denoted by $\tau(q)$, is equal to $M^2(Q)/M^2(I)$.
- The third level is concerned with the summit of the hierarchical classification, I itself, whose index level $\tau(I)$ is equal to 1.

The index level $\tau(Q)$ can be considered as an index indicating the quality of the partition; the smaller $\tau(Q)$ is, the more homogeneous the partition Q is.

4.5.2. Contributions

The principle of contributions is based on explaining two quantities: the eccentricity of classes q of I and the deviation between pairs of classes of I. This explanation also can be made for the elements of J, if the classification is built on J.

4.5.2.1. Explanation of Eccentricities of Classes as Related to Euclidean Axes. Consider a set I, provided with Euclidean coordinates denoted by $X_{IJ} = \{X_{i1}, X_{i2}, \ldots, X_{ij}, \ldots, X_{i\,\text{Card}\,J};\ i \in I\}$, and with masses m_i associated with i of I; consider a class of elements of I, denoted by q: $q = \{i \in I; i \in q\}$. The eccentricity of a class q of I is denoted by $\rho^2(q)$; it is the squared distance between the center of the class q and the center of gravity of I (*cf.* Fig. 10.4).

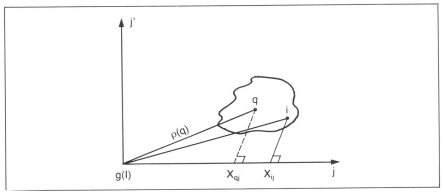

FIGURE 10.4. Geometric representation of the class q relative to the center of gravity $g(I)$.

The following relations hold:

$$X_{qj} = \frac{\sum_{i \in q} m_i X_{qj}}{\sum_{i \in q} m_i} \quad \text{(coordinates of the center of } q\text{)};$$

$$m_q = \sum_{i \in q} m_i \quad \text{(mass of } q\text{)};$$

$$\rho^2(q) = d^2(q, g(I)) \quad \text{(eccentricity of } q\text{)}$$

$$= \sum_{j \in J} X_{qj}^2$$

$$= \frac{M^2(q)}{m_q};$$

$\text{cor}_j(q)$ = relative contribution of j to the eccentricity of q
 = cosine squared of the angle formed by the vector radius of q and the axis j (the vector radius is the vector joining the center of gravity of I and the center of gravity of q);

$\text{cor}_j(q) \# 0 \Rightarrow j$ does not explain the eccentricity of q;

$\text{cor}_j(q) \# 1 \Rightarrow j$ explains the eccentricity of q;

$$\text{ctr}_j(q) = \frac{m_q X_{qj}^2}{\sigma_j^2}$$

= relative contribution of q to the variance of j;

$\text{ctr}_j(q)$ is high $\Rightarrow q$ explains the variance of j;

$\text{ctr}_j(q)$ is small $\Rightarrow q$ does not explain the variance of j;

$$\text{qlt}_s(q) = \sum_{j \leq s} \text{cor}_j(q)$$

= relative contribution of q in s variables

= 1 if $s = \text{Card } J$;

$$\text{inr}(q) = \frac{M^2(q)}{M^2(I)}$$

= rate of variance of q related to I.

The previous computations can be applied to the following data sets:

- standardized principal components data sets. Such a data set is an individuals–variables data set where the variables are normalized as follows:

$$X_{ij} \text{ is replaced by } X_{ij} = \frac{X_{ij} - \bar{X}_j}{\sigma_j} \quad \text{and}$$

the masses assigned to i are equal to 1 or $1/n$.

4. Partitioning Methods

- correspondence data sets. Such a data set gives a system of Euclidean coordinates as follows:

  X_{ij} is replaced by $f_j^i f_j^{-1/2}$ or $f_i^j f_i^{-1/2}$;

  the masses assigned are equal to 1, f_i or f_j

- factor coordinates data sets. Any factor analysis gives factor coordinates on I or J:

  X_{ij} is replaced by the values of the factor coordinates and

  the masses assigned to i (or j) are equal to 1, f_i (or f_j).

4.5.2.2. Explanation of Dipoles $dp(q, q')$ as Related to Euclidean Axes. Consider a set I provided with Euclidean coordinates, denoted by $X_{IJ} = \{X_{i1}, X_{i2}, \ldots, X_{ij}, \ldots, X_{i,\,\text{Card}\,j};\ i \in I\}$, and with the masses m_i assigned to $i \in I$. Consider a partition Q of I into q_k classes and the dipoles, denoted by $dp(q, q')$, formed by any pair of distinct classes of Q (*cf.* Fig. 10.5):

$$dp = dp(q, q') = \{(q, q');\ q \cap q' = \varnothing\}.$$

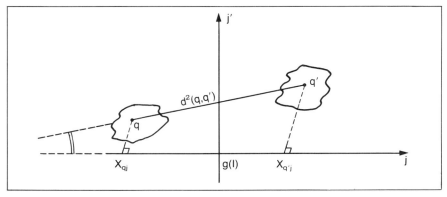

FIGURE 10.5. Geometric representation of a dipole $dp(q, q')$. $\text{cod}_j(dp(q, q'))\#1$: the dipole $dp(q, q')$ is correlated with the j-axis. $\text{cod}_j(dp(q, q'))\#0$: the dipole $dp(q, q')$ is not explained by the j-axis.

The following relations are obtained:

$$\text{inr}(\text{dp}(q, q')) = \frac{M^2\{q, q'\}}{M^2(I)};$$

$$\text{cod}_j(\text{dp}(q, q')) = \frac{(X_{qj} - X_{q'j})^2}{d^2(q, q')} \quad \text{with } d^2(q, q') = \sum_{j \in J}(X_{qj} - X_{q'j})^2$$

= relative contribution of the j-axis to the deviation between two classes q and q'.

= cosine squared of the angle formed by the vector radius of qq' and the j-axis (cf. Fig. 10.5);

FIGURE 10.6. Representation of a partition in a hierarchical classification. The level index associated with the partition, denoted by $\tau(Q)$, is equal to $M^2(Q)/M^2(I)$. The labels are explained in Appendix 2, §10.

4. Partitioning Methods

$\text{ctd}_j(\text{dp}(q, q'))$ = relative contribution of the dipole $\text{dp}(q, q')$

to the variance of the j-axis

$$= \frac{m_q \cdot m_{q'}}{(m_q + m_{q'})} \frac{(X_{qj} - X_{q'j})^2}{\sigma_j^2};$$

$\text{qld}_s(\text{dp}(q, q'))$ = quality of representation of the dipole $\text{dp}(q, q')$ in s axes

$$= \sum_{j \leq s} \text{cod}_j(\text{dp}(q, q')).$$

4.5.2.3. Case Study. In order to illustrate the previous computations, a classification was built based on the example of the family timetables (*cf.* Appendix 2, §10). Classification was performed on the set of population groups described by seven factor coordinates, themselves coming from the 2-D correspondence analysis of the raw data set. A partition into five classes was given, and was represented in a hierarchical form (*cf.* Fig. 10.6). Then, the computations were made to explain the eccentricities of classes, and dipoles (*cf.* Tables 10.4 and 10.5). A synthesis of these computations is given in Tables 10.6 and 10.7.

4.6. Rules for Selecting Significant Classes, Dipoles, and Variables

In practice, the number of classes, variables (or original axes), or dipoles can be large, so that rules of selection are needed to enable the significant information to be extracted in terms of classes, dipoles and variables. The following rules can be adapted according to the needs of the data analyst.

Rule 1. Only those classes q are retained with $\text{cor}_j(q)$ greater than a given value k, representing the mean of contributions or a value similar to a squared coefficient of correlation (in general, $k = 0.5$).

Rule 2. Only those classes q are retained with $\text{ctr}_j(q)$ greater than a given rate of variance p (in general, $p = 80\%$ of variance).

Rule 3. Only those classes q are retained with $\text{cod}_j(q)$ greater than a given value k, representing the mean of contributions or a value similar to a squared coefficient of correlation (in general, $k = 0.5$).

TABLE 10.4. Table concerning the position of five classes of the partition $Q(I)$ related to seven factor axes. Origin: Data set from activities timetables (cf. Appendix 2, §10). Factor axes are from the correspondence analysis of this data set.

class[a]	weight[b]	inr[c]	qlt[d]	1st factor ψ_1[e]	cor_1[f]	ctr_1[g]	2nd factor ψ_2	cor_2	ctr_2	3rd factor ψ_3	cor_3	ctr_3	4th factor ψ_4	cor_4	ctr_4	5th factor ψ_5	cor_5	ctr_5	6th factor ψ_6	cor_6	ctr_6	7th factor ψ_7	cor_7	ctr_7
1	71	351	998	−478	325	110	185	49	227	28	1	8	−60	5	60	659	618	134	−38	2	117	5	0	9
2	179	394	997	−554	971	369	−20	1	6	−11	0	3	34	4	48	−81	21	5	30	3	175	0	0	0
3	214	335	996	82	30	10	−11	1	2	−29	4	27	0	0	0	−464	961	198	−28	3	186	−13	1	177
4	107	216	998	15	1	0	−184	118	337	87	26	119	−57	11	81	493	843	112	−19	1	44	11	0	70
5	429	392	999	316	763	288	29	6	33	−7	0	3	10	1	11	173	228	55	13	1	78	3	0	16

[a] class (q) = serial number of the class q in the partition $Q(I)$
[b] weight $(q) = fq$ = mass associated with the class q
[c] inr(q) = rate of variance of the class q relative to the whole set $I = M^2(q)/M^2(I)$
[d] qlt$_s(q)$ = quality of representation of the class q in s axes (here, $s = 7$) = $\sum_{\alpha \leq s} \mathrm{cor}_\alpha(q)$
[e] $\psi_{\alpha q}$ = factor coordinate of the center of the class q on the α-axis
[f] $\mathrm{cor}_\alpha(q) = \psi_{\alpha q}^2/\rho^2(q)$ = relative contribution of the α-axis to the eccentricity of the class q = cosine squared of the angle formed by the vector radius of the class q and the α-axis
[g] $\mathrm{ctr}_\alpha(q) = f_q \psi_{\alpha q}^2/\lambda_\alpha$ = relative contribution of the class q to the variance of the α-axis

4. Partitioning Methods

TABLE 10.5. Table concerning the dipoles formed from a partition in five classes, related to seven factor axes. Origin: Data set from activities timetables (cf. Appendix 2, §10) Factor axes are from the correspondence analysis of this data set.

class[a]	class[a]	inr[b]	qld[c]	1st factor			2nd factor			3rd factor			4th factor			5th factor			6th factor			7th factor		
				ψ_{d_1}[d]	cod_1[e]	ctr_1[f]	ψ_{d_2}	cod_2	ctr_2	ψ_{d_3}	cod_3	ctr_3	ψ_{d_4}	cod_4	ctr_4	ψ_{d_5}	cod_5	ctr_5	ψ_{d_6}	cod_6	ctr_6	ψ_{d_7}	cod_7	ctr_7
2	1	31	992	−76	10	2	−205	69	198	−40	3	12	94	14	105	−741	897	120	68	8	364	−5	0	6
3	1	87	1000	560	194	113	−196	24	190	−58	2	26	60	2	45	−1123	778	291	11	0	7	−18	0	85
3	2	54	994	637	728	265	9	0	1	−18	1	5	−34	2	27	−383	263	61	−58	6	360	−13	0	81
4	1	18	999	493	591	70	−370	333	541	58	8	22	3	0	0	−166	67	5	19	1	18	6	0	9
4	2	47	996	569	462	146	−165	39	168	98	14	95	−91	12	129	375	471	95	−49	3	178	11	0	44
4	3	69	999	−68	5	2	−174	31	199	116	14	142	−57	3	53	957	946	282	9	0	6	24	1	211
5	1	55	997	794	700	259	−157	27	139	−35	1	11	70	6	71	−486	263	62	51	3	178	−2	0	1
5	2	104	1000	870	918	642	48	3	27	5	0	0	−24	1	16	254	78	35	−17	0	40	3	0	4
5	3	66	996	234	118	53	39	3	21	22	1	11	11	0	4	637	873	249	41	4	263	15	1	174
5	4	22	996	301	359	52	213	179	360	−94	35	111	67	18	90	−320	405	38	32	4	97	−9	0	33

[a] class (q); class (q') = dipole $d = \text{dp}(q, q')$ made of the two classes of the partition q and q'.
[b] inr(d) = rate of variance of the dipole $d = \text{dp}(q, q')$ relative to the whole set I = rate of variance of the partition $\{q, q'\}$ relative to the whole set I
[c] qld$_s(d)$ = quantity of representation of the dipole $d = \text{dp}(q, q')$ in s axes (here, $s = 7$) = $\sum_{\alpha \leqslant s} \text{cod}_\alpha(d)$
[d] $\psi_{d_\alpha}(d) = \psi_{\alpha q} - \psi_{\alpha q'}$
[e] $\text{cod}_\alpha(d)$ = relative contribution of the α-axis to the dipole $d = \text{dp}(q, q')$ = cosine squared of the angle formed by the dipole d and the α-axis
[f] $\text{ctd}_\alpha(d)$ = relative contribution of the dipole $d = \text{dp}(q, q')$ to the variance of the α-axis

TABLE 10.6. Table of relative contributions to the eccentricity ($\text{cor}_j(q) \geq 0.50$).

classes	$\text{cor}_j(q)$	
	negative coordinate of the center of class	positive coordinate of the center of class
class 1		Axis 5 (0.61)
class 2	Axis 2 (0.97)	
class 3	Axis 5 (0.96)	
class 4		Axis 4 (0.84)
class 5		Axis 1 (0.76)

Rule 4. Only those classes q are retained with $\text{ctd}_j(q)$ greater than a given rate of variance p (in general, $p = 80\%$ of variance).

The parameters p and k are fixed by the data analyst according to his needs.

4.7. Classifying Supplementary Points into Partitions

It is known that the greatest advantage of factor analysis is the possibility of classifying supplementary points onto factor graphics. This is extended to partitioning methods, using the formulas that follow.

4.7.1. The Classifying Principle

Consider a set I provided with Euclidean coordinates, a distance d on I,

TABLE 10.7. Table of relative contributions of axes to the subdivisions into classes ($\text{cod}_j(\text{dp}) \geq 0.50$).

dipoles	$\text{cod}_j(q)$
dp(2, 1)	Axis 5 (0.89)
dp(3, 1)	Axis 5 (0.77)
di(3, 2)	Axis 1 (0.72)
dp(4, 1)	Axis 1 (0.59)
dp(4, 2)	None
dp(4, 3)	Axis 5 (0.94)
dp(5, 1)	Axis 1 (0.70)
dp(5, 2)	Axis 1 (0.91)
dp(5, 3)	Axis 5 (0.87)
dp(5, 4)	None

4. Partitioning Methods

and a partition Q on I into k classes $Q = \{q_1, q_2, \ldots, q_l, \ldots, q_k\}$ Let i_s be a supplementary element of I, provided with Euclidean coordinates. The principle used to classify the supplementary element i_s is expressed as follows:

$$\begin{pmatrix} i_s \text{ is assigned to} \\ \text{the class } q_l \text{ of } Q \end{pmatrix} \Leftrightarrow \begin{pmatrix} \delta(\{i_s\}, q_l) \text{ is minimum when all} \\ \text{the classes of } Q \text{ are examined} \end{pmatrix}$$

$$\Leftrightarrow \begin{pmatrix} \delta(\{i_s\}, q_l) < \delta(\{i_s\}, q_{l'}); \\ \text{with } l' = 1, k \text{ and } l' \neq l \end{pmatrix};$$

where δ is a distance function between the class $\{i_s\}$ and a class q_l of $Q(I)$. Therefore, the formulas presented in the following can be applied.

4.7.2. Distance Formulas and Assignment Criteria
Only a few of the distance formulas are given.

(a) Minimum distance between classes

$$i_s \in q_l \quad \Leftrightarrow \quad \begin{pmatrix} \delta(\{i_s\}, q_l) = \inf_{l'=1,k} \delta(\{i_s\}, q_{l'}) \\ \text{with } \delta(\{i_s\}, q_{l'}) = \min_{i \in q_{l'}} d(i_s, i) \end{pmatrix}.$$

(b) Maximum distance between classes

$$i_s \in q_l \quad \Leftrightarrow \quad \begin{pmatrix} \delta(\{i_s\}, q_l) = \inf_{l'=1,k} \delta(\{i_s\}, q_{l'}) \\ \text{with } \delta(\{i_s\}, q_{l'}) = \max_{i \in q_{l'}} d(i_s, i) \end{pmatrix}.$$

(c) Mean of distances between elements

$$i_s q_l \quad \Leftrightarrow \quad \begin{pmatrix} \delta(\{i_s\}, q_l) = \inf_{l'=1,k} \delta(\{i_s\}, q_{l'}) \\ \text{with } \delta(\{i_s\}, q_{l'}) = \dfrac{\sum_{i \in q_{l'}} d(i_s, i)}{n_{q_{l'}}} \\ \text{and } n_{q_{l'}} \text{ the number of elements of } q_{l'} \end{pmatrix}.$$

(d) Center of gravity

$$i_s \in q_l \quad \Leftrightarrow \quad \begin{pmatrix} \delta(\{i_s\}, q_l) = \inf_{l'=1,k} \delta(\{i_s\}, q_{l'}) \\ \text{with } \delta(\{i_s\}, q_{l'}) = \|i_s - q_{l'}\|, \\ \text{where } q_{l'} \text{ is center of gravity of } q_{l'}, \\ i_s \text{ is center of gravity of } i_s, \text{ and} \\ \| \ \| \text{ is the Euclidean norm associated} \\ \text{with the set } I \end{pmatrix}$$

(e) Second order central moment of the union of two classes

$$i_s \in q_l \quad \Leftrightarrow \quad \begin{pmatrix} \delta(\{i_s\}, q_l) = \inf_{l'=1,k} \delta(\{i_s\}, q_{l'}) \\ \text{with } \delta^2(\{i_s\}, q_{l'}) = M^2(\{i_s\} \cup q_{l'}), \\ \text{where } M^2(\{i_s\}, q_{l'}) \text{ is the second order central} \\ \text{moment of } (\{i_s\} \cup q_{l'}) \end{pmatrix}.$$

since

$$M^2(\{i_s\} \cup q_{l'}) = M^2(\{i_s\}, q_{l'}) + M^2(\{i_s\}) + M^2(q_{l'})$$

and

$$M^2(\{i_s\}) = 0 \quad \text{and} \quad M^2(\{i_s\}, q_{l'}) = \frac{m_{i_s} \cdot m_{q_{l'}}}{m_{i_s} + m_{q_{l'}}} \|i_s - q_{l'}\|^2.$$

Then the criterion is expressed as follows:

$$i_s \in q_l \quad \Leftrightarrow \quad \begin{pmatrix} \delta(\{i_s\}, q_l) = \inf_{l'=1,k} \delta(\{i_s\}, q_{l'}) \\ \text{with } \delta^2(\{i_s\}, q_{l'}) = M^2(q_{l'}) + \frac{m_{i_s} \cdot m_{q_{l'}}}{m_{i_s} + m_{q_{l'}}} \|i_s - q_{l'}\|^2 \end{pmatrix}.$$

(f) Second order central moment of a partition

$$i_s \in q_l \quad \Leftrightarrow \quad \begin{pmatrix} \text{the second order central moment} \\ \text{of the partition is maximum} \end{pmatrix}$$

$$\Leftrightarrow \quad \begin{pmatrix} M^2\{q_1, q_2, \ldots, q_l \cup \{i_s\}, \ldots, q_k\} \\ \geq M^2\{q_1, q_2, \ldots, q_{l'} \cup \{i_s\}, \ldots, q_k\} \\ \text{for } l' = 1, \ldots, k, \ l' \neq l \end{pmatrix}.$$

4.7.3. The Algorithm [SUP.PAR]

This algorithm shows an example of how to look for the minimum of δ on the classes of the partition $Q(I)$.

$$
\begin{array}{|l}
\text{Begin}_1 \\
\text{Min} = \delta(\{i_s\}, q_1) \\
l = 1 \\
\quad \begin{array}{|l} \text{Begin}_2 \\ \text{From } k = 2 \text{ to } k = k\max(\text{number of classes of the partition}) \\ \text{if } \delta(\{i_s\}, q_k) \leq \text{Min} \\ \quad \text{then Min} = \delta(\{i_s\}, q_k) \\ \qquad l = k \\ \text{End}_2 \end{array} \\
\text{End}_1
\end{array}
$$

4.8. Partitioning Reminder

1. Some partitioning algorithms are adapted to recognizing natural classes, whereas some others are adapted to building classes. They should be chosen according to the user's requirements.

2. There is no absolute optimum to be attained in classification.

3. Once a classification is formed, it must be interpreted according to statistical considerations.

4. Partitioning algorithms must be used either in the spirit of exploratory data analysis or in the spirit of factor analysis.

5. Hierarchical Classification Methods

5.1. Introduction

There are two main ways to form a hierarchical classification. One way is to form small classes from elements that are similar to each other, and then, from these, to build successively less homogeneous classes until the final class obtained is the whole set. This process is called *ascending hierarchical classification*. The other procedure consists in dividing the whole set into two parts, and then each of these two parts is divided into another two parts and so on, until the individual elements are obtained. This process is called *descending hierarchical classification*. The present section is devoted to ascending classification, which is of interest to us because of its importance in practical exploratory data analysis. Note that the hierarchical clustering algorithms classify a few thousand elements in a few seconds of central unit processing time on mainframe computers (*cf.* Jambu, 1978). After a hierarchical classification is built, the results must be interpreted and assessed, as for any data analysis method. Since 1973, generalization of contributions computations in factor analysis has allowed the study and interpretation of classes as well as the deviation between classes in terms of significant variables or individuals (*cf.* Jambu, 1978). This section is divided as follows: the algorithms of hierarchical classifications (Sections 5.2–5.4); the patterns of acceptable data sets (Section 5.5); the interpretation of a hierarchical classification (Section 5.6); the graphical representations (Section 5.7); the rules for selecting significant dipoles and variables (Section 5.8), and the classifying supplementary elements (Section 5.9). The major applications to real data are presented at the end of this chapter. (*cf.* Section 7).

5.2. *Algorithms of Hierarchical Classification*

5.2.1. Classical Algorithms of Hierarchical Classification

A simplified picture of the hierarchical process is given in what follows. Consider a data set that is similar to an individuals–variables data set. From this data set, a set I or J is chosen, on which the hierarchical classification is formed. Then, a distance is computed on the chosen set, and the classification process is applied. Initially, the two most similar elements are merged together to form the first class and then, step by step, the other classes are formed according to the same principle, applied to classes instead of elements Finally, two classes remain, and upon merging them the whole set is obtained. Two main algorithms are described here:

- The algorithm [CAH.DIST], based on updating distances between classes without any updating on the basic individuals–variables data set;
- The algorithm [CAH.EUCL], based on updating both distances between classes and coordinates of classes from the individuals–variables data set.

5.2.1.1. *The Algorithm [CAH.DIST]*

Step 0. | Initialization. I is chosen to be classified.
$h = 0$; $C_h = \{\{i\}; i \in I\}$; Som $C_h = \{\{i\}; i \in I\}$; Card$(\{i\}) = 1$;
$v(\{i\}) = 0$; $m(\{i\}) =$ mass assigned to the point i of I;
Num$(\{i\}) = i$; $\delta(\{i\}, \{i'\}) = f(d(i, i'))$, where f is an in-

5. Hierarchical Classification Methods

itialization function related to the criterion used (for example, $f(d(i, i')) = d(i, i')$ or $(m_i \cdot m_{i'}/(m_i + m_{i'})) \|i - i'\|^2)$. Card I is the number of elements of I;

Comments. At Step 0, a distance is supposed to be computed between all the pairs of elements of I and a mass is assigned to each element i of I. The initialization function is the extension of the distance function (on pairs of points) to single-element classes; C_h and Som C_h are formed by a partition of I into classes reduced to single-element classes.

Step h. Recursion from Step h to Step $h + 1$
—At the beginning of Step h, Som C_{h-1} and δ^{h-1} are known.
—Consider the two classes of Som C_{h-1}, denoted by s_h and s'_h, that realize the minimum of δ^{h-1} (this means that $\delta^{h-1}(s_h, s'_h)$ is the minimum on Som C_{h-1}). s_h and s'_h are respectively the eldest class and the youngest class of the class a_h defined as follows:
—$a_h = s_h \cup s'_h$
—Som C_h = Som $C_{h-1} \cup \{a_h\} - \{s_h\} - \{s'_h\}$
—$C_h = C_{h-1} \cup \{a_h\}$
—$v(a_h) = \min\{\delta^{h-1}(s, s'); s \neq s'; s, s' \in $ Som $C_{h-1}\}$
—Num(a_h) = Card $I + h$
—Card(a_h) = Card(s_h) + Card(s'_h)
—$m(a_h) = m(s_h) + m(s'_h)$
—Eldest$(a_h) = s_h$
—Youngest$(a_h) = s'_h$
—If h = card $I - 1$ go to Step *End*
    Compute $\delta^{h-1}(t, a_h) = \delta^{h-1}(t, s_h \cup s'_h)$ for $t \in $ Som C_h with $t \neq s_h \neq s'_h$
    $h = h + 1$ (End of the *Step h*)

Step End. h = Card $I + 1$—End of the clustering procedure
At the end of the clustering process, the parameters for plotting the hierarchical clustering are determined as well as the hierarchy itself. The following data tables can be computed for n = Card $I + 1$ to 2 Card $I + 1$:
Num(n) = number of the nodes n
Eldest(n) = eldest of the nodes n
Youngest(n) = youngest of the nodes n
$m(n)$ = mass of the nodes n
Card(n) = number of elements of the nodes n
$v(n)$ = level index of the nodes n

Comments. The algorithm depends on the updating of δ^h from δ^{h-1}, which requires the distances between classes of elements of I to be recomputed. Formulas for distances and aggregation criteria must be given for classes of elements (and not only for elements). They must be used to update the distances computed during the previous step.

5.2.1.2. Distances between Classes and Recursion Formulas

(a) SINGLE LINKAGE OR NEAREST NEIGHBOR: δ_{uim}. Let $d(i, i')$ be the initial distance calculated on the set I, and let a and b be two subsets (or classes) of I; then,

$$\delta_{\text{uim}}(a, b) = \inf\{d(i, i'); i \in a; i' \in b\}.$$

The distance between classes is defined as the minimum of the (initial) distances between the elements of classes.

Recursion formula:

$$\delta^h_{\text{uim}}(t, s_h \cup s'_h) = \inf\{\delta^{h-1}_{\text{uim}}(t, s_h), \delta^{h-1}_{\text{uim}}(t, s'_h)\}$$
$$\text{for } t \neq s_h; t \neq s'_h; t, s_h, s'_h \in \text{Som } C_{h-1};$$

$$\delta^h_{\text{uim}}(t, t') = \delta^{h-1}_{\text{uim}}(t, t')$$
$$\text{for } t \neq t' \neq s_h \neq s'_h; t, t', s_h, s'_h \in \text{Som } C_{h-1}.$$

Disadvantage of the criterion: since only a single link between two elements suffices to unite two classes, this method can lead to uniting extended classes due to the chaining effect. Single linkage clustering is also known as maximal inferior ultrametric or subdominant ultrametric clustering. These names have been given to this criterion because the hierarchical classification method leads to an indexed hierarchy whose associated ultrametric δ is the greatest ultrametric less than or equal to the initial distance d. This clustering criterion is cited in various forms, and is used by many authors such as: Jardine and Sibson (1971), Sokal and Sneath (1973), Benzecri (1973), Hartigan (1975), Bock (1979), and Lerman (1981).

(b) COMPLETE LINKAGE CLUSTERING: δ_{usm}. If a and b are two subsets of I, and $d(i, i')$ is a distance computed on I, then

$$\delta_{\text{usm}}(a, b) = \sup\{d(i, i'); i \in a; i' \in b\}.$$

The distance between two classes is defined as the maximum of the distances between the elements of the two classes.

5. Hierarchical Classification Methods

Recursion formula:

$$\delta^h_{\text{usm}}(t, s_h \cup s'_h) = \sup\{\delta^{h-1}_{\text{usm}}(t, s_h), \delta^{h-1}_{\text{usm}}(t, s'_h)\}$$
$$\text{for } t \neq s_h; t \neq s'_h; t, s_h, s'_h \in \text{Som } C_{h-1};$$
$$\delta^h_{\text{usm}}(t, t') = \delta^{h-1}_{\text{usm}}(t, t')$$
$$\text{for } t \neq t' \neq s_h \neq s'_h; t, t', s_h, s'_h \in \text{Som } C_{h-1}.$$

Complete linkage clustering is also known as minimal superior ultrametric clustering. This name is given to the criterion because the classification method leads to an indexed hierarchy whose associated ultrametric δ is the smallest ultrametric that is greater or equal to the initial distance d. This criterion is cited by many authors such as: Benzecri (1973), Sokal and Sneath (1973), and Hartigan (1975).

(c) AVERAGE LINKAGE CLUSTERING: δ_{moy}. If a and b are two subsets of I, d is a distance computed on I, and Card a and Card b are the respective numbers of elements of a and b, then

$$\delta_{\text{moy}}(a, b) = \sum \{d(i, i'); i \in a; i' \in b\}/(\text{Card } a \cdot \text{Card } b).$$

The distance between two classes is defined as the average of the (initial) distances between the elements of the two classes.

Recursion formula:

$$\delta^h_{\text{moy}}(t, s_h \cup s'_h) = \frac{\text{Card } s_h \cdot \delta^{h-1}_{\text{moy}}(t, s_h) + \text{Card } s'_h \cdot \delta^{h-1}_{\text{moy}}(t, s'_h)}{\text{Card } s_h + \text{Card } s'_h}$$
$$\text{with } t \neq s_h; t \neq s'_h; t, s_h, s'_h \in \text{Som } C_{h-1}.$$
$$\delta^h_{\text{moy}}(t, t') = \delta^{h-1}_{\text{moy}}(t, t')$$
$$\text{with } t \neq t' \neq s_h \neq s'_h; t, t', s_h, s'_h \in \text{Som } C_{h-1}.$$

This criterion is also known as the minimum of the means of distances.

(d) DISTANCE BETWEEN CENTERS OF GRAVITY: δ_{bar}. Consider I as an Euclidean space provided with its usual norm $\| \; \|$; the distance between two classes is defined as the Euclidean distance (according to $\| \; \|$) between the centers of gravity of the classes. If a and b are two subsets of I, then

$$\delta_{\text{bar}}(a, b) = \|a_J - b_J\|^2.$$

Recursion formula:

$$\delta_{\text{bar}}^h(t, s_h \cup s_h') = \frac{m_t}{m_t + m_{s_h}} \delta_{\text{bar}}^{h-1}(t, s_h) + \frac{m_t}{m_t + m_{s_h'}} \delta_{\text{bar}}^{h-1}(t, s_h')$$

$$- \frac{m_{s_h} \cdot m_{s_h'}}{(m_{s_h} + m_{s_h'})^2} \delta_{\text{bar}}^{h-1}(s_h, s_h')$$

for $t \neq s_h \neq s_h'$; $t, s_h, s_h' \in \text{Som } C_{h-1}$;

$$\delta_{\text{bar}}^h(t, t') = \delta_{\text{bar}}^{h-1}(t, t')$$

for $t \neq t' \neq s_h \neq s_h'$; $t, t', s_h, s_h' \in \text{Som } C_{h-1}$.

Disadvantage of the criterion: the procedure can lead to inversions (this means that $v(a) \leq v(b)$ with $a \supset b$).

(e) MEDIAN DISTANCE: δ_{med}. If a and b are two subsets of I, then

$$\delta_{\text{med}}(a, b) = \|a_J - b_J\|^2$$

with

$$a_J = \sum_{i \in a} \frac{i_J}{\text{Card } a} \quad \text{(arithmetic mean of the coordinates).}$$

The distance between classes is defined as the distance between the points whose coordinates are the arithmetic means of the class coordinates.

Recursion formula:

$$\delta_{\text{med}}^h(t, s_h \cup s_h') = \tfrac{1}{2} \delta_{\text{med}}^{h-1}(t, s_h) + \tfrac{1}{2} \delta_{\text{med}}^{h-1}(t, s_h') + \tfrac{1}{4} \delta_{\text{med}}^{h-1}(s_h, s_h')$$

for $t \neq s_h$; $t \neq s_h'$; $t, s_h, s_h' \in \text{Som } C_{h-1}$;

$$\delta_{\text{med}}^h(t, t') = \delta_{\text{med}}^{h-1}(t, t')$$

for $t \neq t' \neq s_h \neq s_h'$; $t, t', s_h, s_h' \in \text{Som } C_{h-1}$.

This criterion is deduced from the previous one as follows:

$$m_i = 1 \quad \forall i \in I \quad \text{and} \quad m_a = 1 \quad \forall a \in C(I).$$

5. Hierarchical Classification Methods

(f) SECOND ORDER CENTRAL MOMENT OF A PARTITION: δ_{mot}. To formalize this criterion, the classical results and notations of geometry presented in Section 4.2.2 are used as follows:

Definition of the distance between classes. Consider two successive partitions in a hierarchical representation, Som C_{h-1} and Som C_h:

$$\text{Som } C_h = \text{Som } C_{h-1} \cup \{a_k\} - \{s_h\} - \{s'_h\},$$

$$C_h = C_{h-1} \cup \{a_h\}.$$

Then,

$$M^2(I) = M^2(\text{Som } C_h) + \sum \{M^2(t); t \in \text{Som } C_h\};$$

$$M^2(I) = M^2(\text{Som } C_{h-1}) + \sum \{M^2(t); t \in \text{Som } C_{h-1}\}.$$

Therefore,

$$M^2(\text{Som } C_h) = M^2(\text{Som } C_{h-1}) + M^2(s_h) + M^2(s'_h) + M^2(s_k \cup s'_h).$$

A recursion formula is thus obtained for the second order central moment of successive partitions; the clustering method consists of clustering two classes s'_h and s_h of Som C_{h-1} such that $M^2(\text{Som } C_h)$ is maximal ($M^2(\text{Som } C_h)$ is the second order central moment of the cloud that contains the centers of the classes belonging to the partition in the reference space R_J). The distance between two classes is defined as being the difference between the second order central moment of two successive partitions:

$$\delta_{\text{mot}}(s_h, s'_h) = M^2(\text{Som } C_{h-1}) - M^2(\text{Som } C_h) \quad \text{for } s_h, s'_h \in \text{Som } C_{h-1}.$$

(The second order moment decreases when the classes are clustered.)

This distance can be expressed in different ways:

$$\delta_{\text{mot}}(s_h, s'_h) = M^2(s_h \cup s'_h) - M^2(s_h) - M^2(s'_h)$$

$$= M^2\{s_h, s'_h\}$$

$$= \frac{m_{s_h} \cdot m_{s'_h}}{m_{s_h} + m_{s'_h}} \|s_h - s'_h\|^2;$$

here, the second line is the second order central moment of the partition defined by two classes only.

Recursion formula:

$$\delta_{\text{mot}}^h(t, s_h \cup s_h')$$
$$= \frac{(m_{s_h} + m_t)\,\delta_{\text{mot}}^{h-1}(t, s_h) + (m_{s_h'} + m_t)\,\delta_{\text{mot}}^{h-1}(t, s_h') - m_t\,\delta_{\text{mot}}^{h-1}(s_h, s_h')}{m_t + m_{s_h} + m_{s_h'}}$$

$$\text{for } t \neq s_h;\ t \neq s_h';\ t, s_h, s_h' \in \text{Som } C_{h-1};$$

$$\delta_{\text{mot}}^h(t, t') = \delta_{\text{mot}}^{h-1}(t, t')$$

$$\text{for } t \neq t' \neq s_h \neq s_h';\ t, t', s_h, s_h' \in \text{Som } C_{h-1}.$$

The first node formed will have the value $\inf\{[(m_i m_{i'})/(m_i + m_{i'})]d^2(i, i');\ i, i' \in I\}$ for its level measure; other properties can be updated:

$$\sum_{n \in \text{Nod}(C(I))} v(n) = M^2(N_J(I)) = M^2(I).$$

The rate of variance corresponding to the node n is denoted by

$$\tau(n) = v(n)/M^2(N_J(I)).$$

(g) VARIANCE OF A PARTITION: δ_{var}. This criterion is obtained from the previous one; if a and b are two subsets of I, the distance between classes is defined by

$$\delta_{\text{var}}(a, b) = \frac{m_a \cdot m_b}{m_I(m_a + m_b)} \|a - b\|^2 = \frac{\delta_{\text{mot}}(a, b)}{m_I}.$$

Recursion formula:

$$\delta_{\text{var}}^h(t, s_h \cup s_h')$$
$$= \frac{(m_{s_h} + m_t)\,\delta_{\text{var}}^{h-1}(t, s_h) + (m_{s_h'} + m_t)\,\delta_{\text{var}}^{h-1}(t, s_h') - m_t\,\delta_{\text{var}}^{h-1}(s_h, s_h')}{m_t + m_{s_h} + m_{s_h'}}$$

$$\text{for } t \neq s_h \neq s_h';\ t, s_h, s_h' \in \text{Som } C_{h-1};$$

$$\delta_{\text{var}}^h(t, t') = \delta_{\text{var}}^{h-1}(t, t')$$

$$\text{for } t \neq t' \neq s_h \neq s_h';\ t, t', s_h, s_h' \in \text{Som } C_{h-1}.$$

The first node formed has the value $\inf\{[m_i m_{i'}/m_I(m_i + m_{i'})]d^2(i, i');\ i, i' \in I\}$.

5. Hierarchical Classification Methods

(h) SECOND ORDER CENTRAL MOMENT OF THE UNION OF TWO CLASSES: δ_{M^2}. If a and b are subsets of I, then $\delta_{M^2} = M^2(a \cup b)$.

This criterion involves step-by-step clustering (i.e., partition by partition) of two classes. The second order moment of their union has a minimum value.

Recursion formula:

$$\delta_{M^2}^h(t, s_h \cup s_h') = [(m_{s_h} + m_t) \delta_{M^2}^{h-1}(t, s_h) + (m_{s_h'} + m_t) \delta_{M^2}^{h-1}(t, s_h')$$
$$+ (m_{s_h} + m_{s_h'}) \delta_{M^2}^{h-1}(s_h, s_h') - m_{s_h} v(s_h) - m_{s_h'} v(s_h')$$
$$- m_t v(t)]/(m_{s_h} + m_{s_h'} + m_t)$$
$$\text{for } t \neq s_h \neq s_h'; t, s_h, s_h' \in \text{Som } C_{h-1};$$

$$\delta_{M^2}^h(t, t') = \delta_{M^2}^{h-1}(t, t')$$
$$\text{for } t \neq t' \neq s_h \neq s_h'; t, t', s_h, s_h' \in \text{Som } C_{h-1}.$$

Note that, for the first time, the recursion formula uses the level measures of classes; moreover, the first node formed has not $\inf\{d(i, i'); i, i' \in I\}$ as its level measure, but $\inf\{d^2(i, i') \cdot (m_i \cdot m_{i'})\}$, and the last node formed has $M^2(I)$ for its level measure.

(i) VARIANCE OF THE UNION OF TWO CLASSES: δ_{V^2}. If a and b are two subsets of I, the distance between classes a and b has following formula:

$$\delta_{V^2} = V^2(a \cup b).$$

Recursion formula: it is obtained from the previous one by the formula $M^2(a) = m_a V(a)$:

$$\delta_{V^2}^h(t, s_h \cup s_h') = [(m_{s_h} + m_t)^2 \delta_{V^2}^{h-1}(t, s_h) + (m_{s_h'} + m_t)^2 \delta_{V^2}^{h-1}(t, s_h')$$
$$+ (m_{s_h} + m_{s_h'})^2 \delta_{V^2}^{h-1}(s_h, s_h') - m_{s_h}^2 v(s_h) - m_{s_h'}^2 v(s_h')$$
$$- m_t^2 v(t)]/(m_{s_h} + m_{s_h'} + m_t)^2$$
$$\text{for } t \neq s_h \neq s_h'; t, s_h, s_h' \in \text{Som } C_{h-1};$$

$$\delta_{V^2}^h(t, t') = \delta_{V^2}^{h-1}(t, t')$$
$$\text{for } t \neq t' \neq s_h \neq s_h'; t, t', s_h, s_h' \in \text{Som } C_{h-1}.$$

As with the previous criterion, the first node formed has not the value $\inf\{d(i, i'); i, i' \in I\}$ for its level measure, but $\inf\{d^2(i, i') \cdot m_i m_{i'}/(m_i + m_{i'})^2\}$. The last node formed has $V(I)$ for level measure.

(j) ANGULAR DISTANCE BETWEEN CENTERS OF CLASSES: δ_{ang}. In a Euclidean space, a vector $\mathbf{v}(a)$ associated with the class a of I is defined by

$$\mathbf{v}(a) = \sum_{i \in a} \mathbf{v}(i).$$

Then,

$$\delta_{\text{ang}}(a, b) = \frac{1 - \langle \mathbf{v}(a), \mathbf{v}(b) \rangle}{\|\mathbf{v}(a)\| \, \|\mathbf{v}(b)\|}$$

$$= 1 - \cos(\mathbf{v}(a), \mathbf{v}(b)).$$

Recursion formula:

$$\delta_{\text{ang}}^h(t, s_h \cup s_h')$$
$$= 1 - \frac{\rho(s_h)(1 - \delta_{\text{ang}}^{h-1}(t, s_h)) + \rho(s_h')(1 - \delta_{\text{ang}}^{h-1}(t, s_h'))}{(\rho^2(s_h) + \rho^2(s_h') + 2\rho(s_h)\rho(s_h')(1 - \delta_{\text{ang}}^{h-1}(s_h, s_h')))^{1/2}}$$
$$\text{for } t \neq s_h \neq s_h'; \, t, s_h, s_h' \in \text{Som } C_{h-1}, \text{ with } \rho(a) = \|a\|;$$
$$\delta_{\text{ang}}^h(t, t') = \delta_{\text{ang}}^{h-1}(t, t')$$
$$\text{for } t \neq t' \neq s_h \neq s_h'; \, t, t', s_h, s_h' \in \text{Som } C_{h-1}.$$

This recursion formula could lead to inversions. The formula can be applied to any initial distance whose values are between -1 and $+1$.

5.2.1.3. General Recursion Formula. The principle of a recursion formula is to allow updating of a table of distances between subsets of a set, on a partition that is the summit of the hierarchy. Most of the recursion formulas presented here can be reduced to a single expression (the exception is the criterion for minimum angular distances)

Recursion formula:

$$\delta^h(t, s_h \cup s_h') = \alpha(s_h, t) \, \delta^{h-1}(t, s_h) + \alpha(s_h', t) \, \delta^{h-1}(t, s_h')$$
$$+ \alpha(s_h, s_h') \, \delta^{h-1}(s_h, s_h')$$
$$+ \beta(t)v(t) + \beta(s_h)v(s_h)$$
$$+ \beta(s_h')v(s_h') + \gamma \, |\delta^{h-1}(t, s_h) - \delta^{h-1}(t, s_h')|$$
$$\text{for } t \neq s_h \neq s_h'; \, t, s_h, s_h' \in \text{Som } C_{h-1};$$
$$\delta^h(t, t') = \delta^{h-1}(t, t')$$
$$\text{for } t \neq t' \neq s_h \neq s_h'; \, t, t', s_h, s_h' \in \text{Som } C_{h-1}.$$

5. Hierarchical Classification Methods

This formula generalizes the one given by Lance and Williams. In Table 10.8, a summary of all the criteria in terms of the general recursion formula is given. In conclusion, a hierarchical clustering method is expressed by:

- the algorithm [ ];
- the aggregation criterion δ;
- the initial distance d on the set I to be classified;
- (optionally) the masses assigned to the elements i of I;

$$\mathcal{M} = ([\text{algorithm}], [\text{criterion } \delta], [\text{distance d}], [\text{masses m}]).$$

5.2.1.4. The Algorithm [CAH.EUCL]

Step 0. Initialization. I is chosen to be classified
$h = 0$; $C_h\{\{i\}; i \in I\}$; Som $C_h = \{\{i\}; i \in I\}$; Card$(\{i\}) = 1$;
$v(\{i\}) = 0$; $m(\{i\}) =$ mass assigned to the point i of I;
Num$(\{i\})$;
$\delta^0(\{i\}, \{i'\}) = f(d(i, i'))$, where f is an initialization function related to the criterion used. Card I is the number of elements of I. $X_{IJ} = \{X_{ij}; i \in I; j \in J\}$ is the individuals-variables data set.

Comments. At Step 0, a distance is computed on the set I, and masses are assigned to each element i of I. The function $f(d(i, i'))$ is an extension criterion used at Step 0.

Step h. Recursion from Step h to step $h + 1$.
—At the beginning of Step h, Som C_{h-1} and δ^{h-1} are known.
—Consider the two classes of Som C_{h-1}, denoted by s_h and s'_h, that realize the minimum of δ^{h-1} (i.e., $\delta^{h-1}(s_h, s'_h)$ is the minimum on Som C_{h-1}). s_h and s'_h are respectively the eldest class and the youngest class of the class a_h defined as follows:
—$a_h = s_h \cup s'_h$
—Som $C_h = $ Som $C_{h-1} \cup \{a_h\} - \{s_h\} - \{s'_h\}$
—$C_h = C_{h-1} \cup \{a_h\}$
—$v(a_h) = \min\{\delta^{h-1}(s, s'); s \neq s'; s, s' \in $ Som $C_{h-1}\}$
—Num$(a_h) = $ Card $I + h$
—Card$(a_h) = $ Card$(s_h) + $ Card(s'_h)

TABLE 10.8. Table of parameters associated with the general recursion formula.

criteria	$\alpha(s, t)$	$\alpha(s', t)$	$\alpha(s, s')$	$\beta(s)$	$\beta(s')$	$\beta(t)$	γ
δ_{uim}	$\dfrac{1}{2}$	$\dfrac{1}{2}$	0	0	0	0	$-\dfrac{1}{2}$
δ_{usm}	$\dfrac{1}{2}$	$\dfrac{1}{2}$	0	0	0	0	$+\dfrac{1}{2}$
δ_{moy}	$\dfrac{\text{Card } s}{\text{Card } s + \text{Card } t}$	$\dfrac{\text{Card } s'}{\text{Card } s' + \text{Card } t}$	0	0	0	0	0
δ_{bar}	$\dfrac{m_s}{m_s + m_t}$	$\dfrac{m_{s'}}{m_{s'} + m_t}$	$-\dfrac{m_s m_{s'}}{(m_s + m_{s'})^2}$	0	0	0	0
δ_{med}	$\dfrac{1}{2}$	$\dfrac{1}{2}$	$-\dfrac{1}{4}$	0	0	0	0
δ_{M^2}	$\dfrac{m_t + m_s}{m_t + m_s + m_{s'}}$	$\dfrac{m_t + m_{s'}}{m_t + m_s + m_{s'}}$	$\dfrac{m_s + m_{s'}}{m_t + m_s + m_{s'}}$	$\dfrac{-m_s}{m_t + m_s + m_{s'}}$	$\dfrac{-m_{s'}}{m_t + m_s + m_{s'}}$	$\dfrac{-m_t}{m_t + m_s + m_{s'}}$	0
δ_{V^2}	$\dfrac{(m_t + m_s)^2}{(m_t + m_s + m_{s'})^2}$	$\dfrac{(m_t + m_{s'})^2}{(m_t + m_s + m_{s'})^2}$	$\dfrac{(m_s + m_{s'})^2}{(m_t + m_s + m_{s'})^2}$	$\dfrac{-m_s^2}{(m_t + m_s + m_{s'})^2}$	$\dfrac{-m_{s'}^2}{(m_t + m_s + m_{s'})^2}$	$\dfrac{-m_t^2}{(m_t + m_s + m_{s'})^2}$	0
δ_{mot}	$\dfrac{m_t + m_s}{m_t + m_s + m_{s'}}$	$\dfrac{m_t + m_{s'}}{m_t + m_s + m_{s'}}$	$\dfrac{-m_t}{m_t + m_s + m_{s'}}$	0	0	0	0
δ_{var}	$\dfrac{m_t + m_s}{m_t + m_s + m_{s'}}$	$\dfrac{m_t + m_{s'}}{m_t + m_s + m_{s'}}$	$\dfrac{-m_t}{m_t + m_s + m_{s'}}$	0	0	0	0

5. Hierarchical Classification Methods

$\quad$ — $m(a_h) = m(s_h) + m(s'_h)$
$\quad$ — Eldest$(a_h) = s_h$
$\quad$ — Youngest$(a_h) = s'_h$

$$X_{a_h j} = \frac{m_{s_h} X_{s_h j} + m_{s'_h} X_{s'_h j}}{m(a_h)} \quad \text{for each } j \text{ of } J$$

If $h = \text{Card } I - 1$ go to *Step End*
Computation of $\delta(t, a_h)$ for $t \neq s_h \neq s'_h \in \text{Som } C_{h-1}$
$h = h + 1$ (end of the *Step h*)

Step End. End of the clustering procedure ($h = \text{Card } I - 1$)
At the end of the clustering process, the parameters of the hierarchical clustering are determined. The following data tables can be computed for $n = \text{Card } I + 1$ to $2 \text{Card } I - 1$.
Num(n) = number of the nodes n
Eldest(n) = eldest of the nodes n
Youngest(n) = youngest of the nodes n
$m(n)$ = mass of the nodes
Card(n) = number of elements of the nodes n
$v(n)$ = level index of the nodes n

Comments. The algorithm [CAH.EUCL] differs from the algorithm [CAH.DIST] on the following points: The updating is made on both the initial data set (computation of the coordinates of the class centers) and on the table of distances δ, not only on the table of distances δ as in [CAH.DIST]. As a result, we need distance formulas that are adapted to the criteria used in [CAH.DIST]; these are proposed in the next section.

5.2.1.5. Distance Formulas. Only Euclidean distances may be used when the reference space associated with X_{IJ} is the standard space R^J. The recurrence formulas are expressed as follows:

$$\delta_{\text{bar}}(t, a_h) = \|t - a_h\|^2$$

$$= \sum_{j \in J} (x_{tj} - x_{a_h j})^2;$$

$$\delta_{M^2}(t, a_h) = \frac{m_{s_h} \cdot m_{s'_h}}{m_{s_h} + m_{s'_h}} \|t - a_h\|^2 + v(s_h) + v(s'_h)$$

$$= \frac{m_{s_h} \cdot m_{s'_h}}{m_{s_h} + m_{s'_h}} \sum \{(x_{tj} - x_{a_h j}); j \in J\} + v(s_h) + v(s'_h);$$

$$\delta_{\text{mot}}(t, a_h) = \frac{m_{s_h} \cdot m_{s'_h}}{m_{s_h} + m_{s'_h}} \|t - a_h\|^2;$$

$$\delta_{V^2}(t, a_h) = \frac{m_{s_h} \cdot m_{s'_h}}{m_{s_h} + m_{s'_h}} \sum \{(x_{tj} - a_{a_h j})^2 ; j \in J\}$$

$$= \frac{m_{s_h} \cdot m_{s'_h}}{(m_{s_h} + m_{s'_h})^2} \|t - a_h\|^2 + \frac{m_{s_h}}{(m_{s_h} + m_{s'_h})} v(s_h)$$

$$+ \frac{m_{s'_h}}{(m_{s_h} + m_{s'_h})} v(s'_h)$$

$$= \frac{m_{s_h} \cdot m_{s'_h}}{(m_{s_h} + m_{s'_h})^2} \sum \{(x_{tj} - x_{a_h j})^2 ; j \in J\}$$

$$+ \frac{m_{s_h}}{(m_{s_h} + m_{s'_h})} v(s_h) + \frac{m_{s'_h}}{(m_{s_h} + m_{s'_h})} v(s'_h).$$

The initialization functions $f(d(i, i'))$ are the extensions of the recursion formulas to single-element classes. ($\| \; \|$ is the usual Euclidean norm.)

5.3. Accelerated Algorithms of Hierarchical Classification

5.3.1. Introduction

Certain remarks can be made concerning classical algorithms of hierarchical classification:

- The procedures are based on consultation of a table of distances between elements or classes. Sometimes the initial data table on which modifications are continually carried out is also consulted. From a strictly computational viewpoint, neither the table of distances nor the data table needs to be put into central memory, depending on the algorithm adopted. Programming, however, is made easier if the table of distances and the initial data table are in fact found in the central memory (this is a disadvantage, however, when large data sets have to be classified).

- In the general algorithm, all distances on the partition Som C_{h-1} need to be computed. The minimum of these distances for the classes s_h and s'_h is required; to save space in central memory as well as to reduce the number of comparisons required, it would be better to eliminate the distances that have no influence on the algorithm's execution. This is precisely the aim of accelerated algorithms for hierarchical ascending classification. They are based on a particular property of distances between subsets called the reducibility property, proposed by Bruynooghe (1978).

5. Hierarchical Classification Methods

- Thus, the accelerated algorithms allow a set of two or three thousand elements to be classified in a few seconds on a mainframe computer. They can also be used on any microcomputers. Information related to these programs is given in Chapter 12.

5.3.2. Reducibility Property and Consequences

5.3.2.1. Notation and the Reducibility Property. Consider a partition Som C_{h-1}; after two classes s_h and s'_h are chosen for clustering, the general recursion formula computes the distance $\delta^h(t, s_h \cup s'_h)$ on Som C_h derived from Som C_{h-1} using the formula

$$\text{Som } C_h = \text{Som } C_{h-1} \cup \{a_h\} - \{s_h\} - \{s'_h\}.$$

Given a fixed value ρ_h the following specific sets are defined:

$$U_h(\rho_h) = \{(s, s'); s \neq s'; s, s' \in \text{Som } C_{h-1}; \delta(s, s') \leq (\rho_h)\};$$

$U_h(\rho_h)$ is the set of unordered pairs of classes in the partition Som C_{h-1}, such that the distance between these classes is less than ρ_h;

$$X_h(\rho_h) = \{s; \text{Som } C_{h-1}; \exists s' \in \text{Som } C_{h-1}; (s, s') \in U_h(\rho_h)\}$$
$$= \{s; \text{Som } C_{h-1}; \exists s' \in \text{Som } C_{h-1}; \delta(s, s') \leq (\rho_h)\}.$$

These definitions are illustrated in Table 10.9, which describes successive states of the table of distances for a hierarchical ascending classification (based on the average distance). Arbitrary values are given to ρ_h.

We next define

$$G_h(\rho_h) = (X_h(\rho_h), U_h(\rho_h));$$

$G_h(\rho_h)$ is the graph whose vertices are the classes in $X_h(\rho_h)$ and whose edges are the pairs of classes in $X_h(\rho_h)$ such that their lengths are less than ρ_h.

Consider the set denoted by $B_h(s, \rho_h)$:

$$B_h(s, \rho_h) = \{t; \text{Som } C_{h-1}; (s, t) \in U_h(\rho_h)\}$$
$$= \{t; \text{Som } C_{h-1}; \delta^{h-1}(s, t) \leq \rho_h\};$$

$B_h(s, \rho_h)$ is called the ball of center s and radius ρ_h.

TABLE 10.9. The different states of the accelerated algorithm compared to the classical one.

$h = 0$

i' \ i	1	2	3	4	5
1	0	40	10	40	36
2	40	0	48	36	20
3	10	48	0	42	40
4	40	36	42	0	18
5	36	20	40	18	0

Table of $\delta(i, i')$.

i' \ i	1	2	3	4	5
1	0		10		
2		0			20
3	10		0		
4				0	18
5		20		18	0

$\rho = 21$; table of $\delta(i, i')$;
$\delta(i, i') \leq 21$;
$U_0(\rho) = \{(1, 3); (2, 5); (4, 5)\}$;
Card $U_0(\rho) = 3$;
$X_0(\rho) = \{1, 2, 3, 4, 5\}$; Card $X_0(\rho) = 5$.

$h = 1$

s' \ s	2	6	4	5
2	0	44	36	20
6	44	0	41	38
4	36	41	0	18
5	20	38	18	0

Table of $\delta(s, s')$ after aggregation;
$(6) = \{1\} \cup \{3\}$.

s' \ s	2	6	4	5
2	0			20
6		0		
4			0	18
5	20		18	0

$\rho = 21$; table of $\delta(s, s')$;
$\delta(s, s') \leq 21$;
$U_1(\rho) = \{(2, 5); (4, 5)\}$; Card $U_1(\rho) = 2$;
$X_1(\rho) = \{2, 4, 5\}$; Card $X_1(\rho) = 3$.

5. Hierarchical Classification Methods

TABLE 10.9 (cont'd)

$h = 2$

s' \ s	2	6	7
2	0	44	28
6	44	0	39.5
7	28	39.5	0

Table of $\delta(s, s')$ after aggregation;
$(6) = \{1\} \cup \{3\}$; $(7) = \{4\} \cup \{5\}$.

s' \ s	2	6	7
2	0		28
6		0	
7	28		0

$\rho = 30$; table of $\delta(s, s')$;
$\delta(s, s') \leq 30$;
$U_2(\rho) = \{(2, 7)\}$; Card $U_2(\rho) = 1$;
$X_2(\rho) = \{2, 7\}$; Card $X_2(\rho) = 2$.

$h = 3$

s' \ s	6	8
6	0	41
8	41	0

Table of $\delta(s, s')$ after aggregation;
$(6) = \{1\} \cup \{3\}$; $(8) = \{4\} \cup \{5\} \cup \{2\}$.

s' \ s	6	8
6	0	41
8	41	0

$\rho = 45$; table of $\delta(s, s')$;
$\delta(s, s') \leq 45$;
$U_3(\rho) = \{(6, 8)\}$; Card $U_3(\rho) = 1$;
$X_3(\rho) = \{6, 8\}$; Card $X_3(\rho) = 2$.

With these notations, the reducibility property can be expressed as follows:

$$B_h(s \cup s', \rho_h) \subset B_h(s', \rho_h) \cup B_h(s, \rho_h) \quad \forall s, s' \in U_h(\rho_h);$$

or

$$\forall t, s, s' \in \text{Som } C_{h-1}, \quad \left. \begin{array}{l} \delta(t, s') \geq \rho_h, \\ \delta(t, s) \geq \rho_h, \\ \delta(s, s') < \rho_h \end{array} \right\} \Rightarrow \delta(t, s \cup s') \geq \rho_h;$$

or again

$$\left. \begin{array}{l} \delta(t, s \cup s') \leq \rho_h, \\ \delta(s, s') < \rho_h \end{array} \right\} \Rightarrow \delta(t, s') \leq \rho_h \quad \text{or} \quad \delta(t, s) \leq \rho_h.$$

5.3.2.2. Reducibility Property and Clustering Criteria.
The reducibility property is satisfied by the following criteria: single linkage, complete linkage, average linkage, maximum second order central moment of a partition, minimum variance of a partition, minimum second order central moment and variance of classes.

(a)
$$\delta_{\text{uim}}(t, s' \cup s) = \inf\{\delta_{\text{uim}}(t, s), \delta_{\text{uim}}(t, s')\} \geq \rho_h ;$$

(b)
$$\delta_{\text{usm}}(t, s' \cup s) = \sup\{\delta_{\text{usm}}(t, s), \delta_{\text{usm}}(t, s')\} \geq \rho_h ;$$

(c)
$$\delta_{\text{moy}}(t, s' \cup s) = \frac{\text{Card } s \, \delta_{\text{moy}}(t, s) + \text{Card } s' \, \delta_{\text{moy}}(t, s')}{\text{Card } s + \text{Card } s'}$$
$$\geq \frac{\text{Card } s \, \rho_h + \text{Card } s' \, \rho_h}{\text{Card } s + \text{Card } s'} \geq \rho_h ;$$

(d)
$$\delta_{\text{mot}}(t, s' \cup s) \geq \frac{((m_t + m_s)\rho_h + (m_t + m_{s'})\rho_h - m_t\rho_h)}{(m_t + m_s + m_{s'})} \geq \rho_h ;$$

(e)
$$\delta_{\text{var}}(t, s' \cup s) \geq \frac{((m_t + m_s)\rho_h + (m_t + m_{s'})\rho_h - m_t\rho_h)}{(m_t + m_s - m_{s'})} \geq \rho_h ;$$

(f)
$$\delta_{M^2}(t, s' \cup s) = \alpha(s, t) \delta_{M^2}(t, s) + \alpha(s', t) \delta_{M^2}(t, s')$$
$$+ \alpha(s', s) \delta_{M^2}(s', s)$$
$$\beta(s)v(s) - \beta(s')v(s') - \beta(t)v(t)$$

with

$$\delta_{M^2}(t, s) = M^2(t \cup s'); t, s, s' \in \text{Som } C_{h-1}; t \neq s \neq s';$$
$$\alpha(s', s) \delta_{M^2}(s', s) + \beta(s)v(s) - \beta(s')v(s')$$
$$= (m_s + m_{s'})M^2(s \cup s') - m_s M^2(s) - m_{s'} M^2(s').$$

5. Hierarchical Classification Methods

Then,

$$M^2(s \cup s') > M^2(s) + M^2(s')$$
$$\Rightarrow (m_s + m_{s'})M^2(s \cup s') \geq (m_s + m_{s'})(M^2(s) + M^2(s'))$$
$$\Rightarrow (m_s + m_{s'})M^2(s \cup s') \geq m_s M^2(s) + m_{s'} M^2(s')$$
$$+ m_{s'} M^2(s) + m_s M^2(s')$$
$$\Rightarrow (m_s + m_{s'})M^2(s \cup s') \geq m_s M^2(s) + m_{s'} M^2(s')$$
$$\Rightarrow \alpha(s', s) \, \delta_{M^2}(s', s) + \beta(s)v(s) - \beta(s')v(s') \geq 0.$$

Moreover, $v(t) = M^2(t) \leq \rho_k$; if $M^2(t) > \rho_h$. Therefore, $t \notin \text{Som } C_{h-1}$, which finally implies that

$$\delta_{M^2}(t, s' \cup s) \geq \frac{((m_t + m_s)\rho_h + (m_t + m_{s'})\rho_h - m_t \rho_h)}{(m_t + m_s + m_{s'})} \geq \rho_h$$

Therefore, $\delta_{M^2}(t, s' \cup s) \geq \rho_h$.

It is proved (Jambu, 1978) that the distances that do not satisfy the reducibility property lead to inversion, and conversely. Due to this property, a new algorithm can be proposed that is faster than the classical one. The same solution is obtained, but the solution is reached in less time and requires less updating.

5.3.3. Accelerated Algorithms of Hierarchical Clustering

The two algorithms proposed, denoted by [CAHA.DIST] and [CAHA.EUCL] are derived from the classical ones [CAH.DIST] and [CAH.EUCL] as presented in Table 10.10.

5.3.3.1. The Algorithm [CAHA.DIST]

Step 0. Initialization. I is chosen to be classified.
$h = 0$; $C_h = \text{Som } C_h = \{\{i\}; i \in I\}$; $v(\{i\}) = 0$; $\text{Card}(\{i\}) = 1$;
$m(\{i\}) = $ mass assigned to the point i of I; $\text{Num}(\{i\}) = i$;
$\delta(\{i\}, \{i'\}) = f(d(i, i'))$, where f is an initialization function.

Comments. At Step 0 the initialization is made as in [CAH.DIST]
Step h. Recursion from *Step h* to *Step h + 1*
  Step A(h) Update the graph $G_{h-1}(X_{h-1}, U_{h-1})$ again
             Update the threshold ρ_{h-1} again
             Eliminating the edges whose lengths are less than ρ_{h-1}

TABLE 10.10. Relations between classical and accelerated algorithms.

classical algorithms	accelerated algorithms
[CAH.DIST] $\delta_{uim}, \delta_{usm}, \delta_{moy}, \delta_{med}$ $\delta_{mot}, \delta_{var}, \delta_{M^2}, \delta_{V^2},$ $\delta_{bar}, \delta_{ang}$ without using the basic data set X_{IJ}	[CAHA.DIST] $\delta_{uim}, \delta_{usm}, \delta_{moy}, \delta_{med}$ $\delta_{mot}, \delta_{var}, \delta_{M^2}, \delta_{V^2}$ without using the basic data set X_{IJ}
[CAH.DIST] $\delta_{mot}, \delta_{var}, \delta_{M^2}, \delta_{V^2},$ δ_{bar} with use of the basic data set X_{IJ} in central memory	[CAHA.EUCL *] $\delta_{mot}, \delta_{var}, \delta_{M^2}, \delta_{V^2}$ $*=1$: with use of the basic data set X_{IJ} in central memory; $*=2$: without using the basic data set in central memory.

Comments. The threshold ρ_{h-1} is determined by the size of the place available in central memory for the computations. A histogram of distances can be made to approximate the number of distances used in *Step h*, and to estimate ρ_h, which is the same during *Step B(h)*.

Step B(h)	Building the hierarchical classification according to the procedure [CAH.DIST], starting with the graph $G_{h-1}(X_{h-1}, U_{h-1})$

    —C_{h-1} and Som C_{h-1} as well as all the parameters established at *Step h* − 1 are known at the beginning of *Step B(h)*.

    Let s_h, s'_h be two classes of G_{h-1} that realize the minimum of δ^{h-1} on G_{h-1}

    —$\rho_h = \rho_{h-1}$
    —$a_h = s_h \cup s'_h$
    —Eldest$(a_h) = s_h$
    —Youngest$(a_h) = s'_h$
    —Som C_h = Som $C_{h-1} \cup \{a_h\} - \{s_h\} - \{s'_h\}$
    —$C_h = C_{h-1} \cup \{a_h\}$
    —$v(a_h) = \min\{\delta^{h-1}(s, s'); s \neq s'; s, s' \in$ Som $C_{h-1}\}$
    —Card(a_h) = Card(s_h) + Card(s'_h)

5. Hierarchical Classification Methods 351

—Num$(a_h) = h + $ Card I
—$m(a_h) = m(s_h) + m(s'_h)$

If $h = $ Card $I - 1$: the classification is built; go to *Step End*.

If $h \neq $ Card $I - 1$

—Updating δ^h as function of δ^{h-1} according to the recursion formula associated with [CAH.DIST]
—Updating $G_h = (X_h, U_h)$ as a function of $G_{h-1}(X_{h-1}, U_{h-1})$ as follows:
—Computation of $B(s_h, \rho_h) = \{t; t \in X_{h-1}; (t, s_h) \in U_{h-1}\}$
—Computation of $B(s'_h, \rho_h) = \{t; t \in X_{h-1}; (t, s'_h) \in U_{h-1}\}$
—Computation of $\delta^h(t, a_h) = \delta^h(t, s_h \cup s'_h)$ on the set $B(s_h, \rho_h) \cup B(s'_h, \rho_h) - \{s_h, s'_h\}$ according to the recursion formula given in [CAH.DIST], only for distances satisfying the reducibility property.
—Computation of $B(a_k, \rho_h)$
—Updating U_h as a function of U_{h-1} as follows:
$U_h = U_{h-1} - \{(t, s_h); t \in B(s_h, \rho_h)\}$

$\qquad - \{(t, s'_h); t \in B(s'_h, \rho_h)\}$
$\qquad + \{(t, a_h); t \in B(a_h, \rho_h)\}$

If $U_h = \emptyset$
| go to *Step A(h)* for updating the graph G_{h-1}

If $U_h \neq \emptyset$

| continue the building.
| go to *Step B(h)*

Step End. End of the procedure ($h = $ Card $I - 1$)
The following data tables are computed for $n = $ Card $I + 1$ to $n = 2$ Card $I - 1$.
Num$(n) = $ serial number of the node n
Card$(n) = $ number of elements of the node n
Eldest$(n) = $ number of the eldest of n
Youngest$(n) = $ number of the youngest of n
$v(n) = $ level index of the node n
$m(n) = $ mass assigned to the node n

Comments. The algorithm is similar to [CAH.DIST], though it is based on a smaller number of distances, selected by the successive values of ρ. However, recall that the final solution obtained is independent of ρ. Therefore, how is ρ chosen? (*Cf.* (b).) This introduces the following comments:

(a) At a certain *step*, the set $\{\delta^{h-1}(s, s'); s, s' \in \text{Som } C_{h-1}\}$ must be recomputed from the set $\{d(i, i'); i, i' \in I\}$. For each class s of Som C_{h-1}, the elements of I belonging to s are known. It is then easy to compute $\delta(s, s')$ as a function of $d(i, i')$ by the distance formula used for the aggregation criterion. At another *Step k*, δ^{k-1} must be recomputed from a previous value of the set δ^{h-1}. The same process is thus applied.

(b) Once δ^{h-1} is computed, the value ρ_h needs to be computed. If ρ_h is too small, the number of times that the graph G_h will be built, will be too large; if ρ_h is too large, the number of distances selected will be too large. However, the maximum number of distances can be fixed by the amount of space in central memory. With experience, it has been shown that Card I is a reasonable number of distances for *Step B(h)* (instead of Card I (Card $I - 1$)/2 at the first step, and Card Som C_h (Som $C_h - 1$)/2 at *Step h* in the classical algorithm). The best procedure is to fix a maximum number of distances, then to approximate this number by a histogram of distances, and then to determine the value ρ_h as the maximum value of the set of distances selected.

(c) Prototypes of such a program are described in Jambu (1983).

5.3.3.2. The Algorithm [CAHA.EUCL1]

Step 0. Initialization. I is chosen to be classified.

$h = 0$; $C_h = \{\{i\}; i \in I\}$; $v(\{i\}) = 0$; Num$(\{i\}) = I$; $m(\{i\})$ = mass assigned to the point i of I; X_{IJ} basic data set, similar to an individuals–variables data set; Card$(\{i\}) = 1$; $\rho_0 = f(M^2(I))$

Comments. The initialization is identical to that in [CAHA.EUCL]. The value of the threshold ρ_0 is determined as a function of the variance of the set I, to avoid computing all the distances.

Step h. Recursion from *Step h* to *Step h* + 1

Step A(h) Update $G_{h-1}(X_{h-1}, U_{h-1})$ again
Update $\rho_{h-1} = f(M^2(\text{Som } C_{h-1}))$
If $U_h = \varnothing$ do $\rho_{h-1} = 2 \rho_{h-1}$
Eliminate the distances whose lengths are smaller than ρ_{h-1}

5. Hierarchical Classification Methods

Comments. As in [CAHA.EUCL], the threshold is determined by the space available in central memory. This value will be the same during all *Step B(h)*.

Step B(h)	Build the hierarchical classification according to the procedure [CAHA.EUCL] on the set $G_{h-1}(X_{h-1}, U_{h-1})$ C_{h-1} and Som C_{h-1}, as well as the parameters formed at the *Step h − 1* are known at the beginning of *Step B(h)*

Let s_h and s'_h be two classes of G_{h-1} that realize the minimum of δ^{h-1} or G_{h-1}
$\rho_h = \rho_{h-1}$
$a_h = s_h \cup s'_h$
Eldest$(a_h) = s_h$
Youngest$(a_h) = s'_h$
Som C_h = Som $C_{h-1} \cup \{a_h\} - \{s_h\} - \{s'_h\}$
$C_h = C_{h-1} \cup \{a_h\}$
$v(a_h) = \min\{\delta^{h-1}(s, s'); s \neq s'; s, s' \in \text{Som } C_{h-1}\}$
Card(a_h) = Card(s_h) + Card(s'_h)
Num$(a_h) = h +$ Card I
$m(a_h) = m(s_h) + m(s'_h)$
Computation of X_{a_hj} for $j \in J$ with

$$X_{a_hj} = \frac{m(s_h)X_{s_hj} + m(s'_h)X_{s'_hj}}{m(a_h)}$$

If $h =$ Card $I - 1$:
 The classification is built
 go to *Step End*
If $h \neq$ Card $I - 1$:
 Update δ^h as a function of δ^{h-1}, according to the recursion formula associated with [CAH.EUCL]
 —Compute $B(s_h, \rho_h)$
 —Compute $B(s'_h, \rho_h)$
 —Compute of $\delta^h(t, a_h)$ for $t \in B(s'_h, \rho_h) \cup B(s_h, \rho_h) - \{s_h, s'_h\}$
 —Compute $B(a_h, \rho_h)$
 —Compute $U_h = U_{h-1}$
  $- \{(t, s_h); t \in B(s_h, \rho_h)\}$
  $- \{(t, s'_h); t \in B(s'_h, \rho_h)\}$
  $+ \{(t, a_h); t \in B(a_h, \rho_h)\}$
 If $U_h = \emptyset$ go to *Step A(h)* for updating G_{h-1}
 If $U_h \neq \emptyset$ go to *Step B(h)*

Step End. End of the procedure ($h = \text{Card } I - 1$)

> The following data tables are computed for $n = \text{Card } I + 1$ to $n = 2\,\text{Card } I - 1$
> Num(n) = serial number of the node n
> Card(n) = number of elements of the node n
> Eldest(n) = number of the eldest of n
> Youngest(n) = number of the youngest of n
> $v(n)$ = level index of the node n
> $m(n)$ = mass assigned to the node n

Comments. As in the algorithm [CAH.EUCL], the only problem is related to the choice of the successive values of ρ_h.

(a) Since the coordinates of the class centers are updated *step-by-step* it is not necessary to compute δ^{h-1} as a function of $d(i, i')$; then $\delta^h(s, s')$ is computed directly from the coordinates of the classes.

(b) To avoid computing the distances at *Step* 0, a specific value is given to ρ_0: $\rho_0 = 2M^2(I)/n_I(n_I - 1)$. At *Step* $A(h)$, the value ρ_h is $2M^2(\text{Som } C_h)/n_{s_h}(n_{s_h} - 1)$, where n_{s_h} is equal to the number of elements of Som C_h. An example of clustering real data is given in Table 10.11.

(c) Prototypes of programs are described in Jambu (1983).

(d) Based on the computer constraints, another algorithm, denoted by [CAHA.EUCL2], can be proposed that mixes the two algorithms [CAHA.DIST] and [CAHA.EUCL] in order to avoid keeping the basic individuals–variables data set in central memory (*cf.* Jambu, 1983).

5.4. Correspondence Hierarchical Clustering

5.4.1. Criterion

Consider the case of a correspondence data set, denoted by k_{IJ}, on which the following criterion can be built:

"to find a partition $Q(I)$ so that $V(Q)$ is minimal, where

$$V(Q) = \{\|f_{IJ} - f_I f_J\|^2_{f_I f_J} - \|f_{QJ} - f_Q f_J\|^2_{f_Q f_J}\}."$$

$V(Q)$ is formed from two quantities: the χ^2 associated with the frequency data set f_{IJ}, and the χ^2 associated with the frequency data set f_{QJ}. The χ^2 associated with the frequency data set is also the total variance of the cloud $N_J(I)$ used in correspondence analysis (*cf.* Chapter 8). At each *step*, the clustering criterion, known as the correspondence clustering criterion, maximizes the between-classes variance of the upper partition examined, the lower partition being fixed in the hierarchical clustering

5. Hierarchical Classification Methods

TABLE 10.11. Statistics associated with the behavior of the algorithm [CAHA.EUCL]. Number of elements to classify = 5000. Number of variables involved = 4. Maximum number of distances available in central memory = 10,000. Number of distances from 5000 elements to classify = 12,497,500. Software used: DACL (*cf.* Chapter 12).

h	Partition at the level N	Number of distances kept
1	5001	28 216
.		then 14 103
.		then 8 547
.	6803	9 277
.	9178	899
.	9495	525
.	9679	279
.	9790	176
.	9855	121
.	9905	76
.	9936	47
.	9955	32
.	9970	22
.	9981	16
.	9987	4
.	9995	2
.	9996	1
.	9997	2
.	9998	3
4999	9999	1

process. The method associated with the correspondence clustering criterion is known as *correspondence hierarchical clustering*.

5.4.2. Algorithm.

In terms of hierarchical clustering, the correspondence clustering criterion is expressed as follows: To find, *step-by-step* (partition-by-partition) a partition $Q(I)$, deduced from $Q'(I)$ so that $Q = Q' - \{s, s'\} \cup \{s \cup s'\}$ and $V(Q)$ is minimal.

It can be shown that:

$$V(Q) = \frac{f_s \cdot f_{s'}}{f_s + f_{s'}} \|f_J^s - f_J^{s'}\|_{f_J}^2$$

where the correspondence analysis notation f_J^s is the conditional profile of the class s on J. Thus, the recursion formula is expressed as follows:

$$\delta(t, s \cup s') = \frac{f_t(f_s + f_{s'})}{f_t + f_s + f_{s'}} \|f_J^t - f_J^{s \cup s'}\|_{f_J}^2.$$

Then,

$$\delta(t, s \cup s') = \frac{1}{(f_t + f_s + f_{s'})}((f_t + f_s)\,\delta(t, s) + (f_t + f_{s'})\,\delta(t, s') - f_t\,\delta(s, s')).$$

This is the recursion formula of [CAH.DIST] associated with the criterion δ_{mot}, maximization of the second order central moment of a partition. Then, the method $\mathcal{M}$ is expressed as follows:

> Correspondence hierarchical clustering;
> $\mathcal{M} = ([\text{CAH.DIST}], \delta_{\text{mot}}, \|f_J^i - f_J^{i'}\|_{f_J}, f_i)$

Thus, the algorithms [CAHA.DIST], [CAH.EUCL], and [CAHA.EUCL1] can also be applied with the same parameters as above.

5.5. Patterns of Acceptable Data Sets

All the data sets similar to individuals–variables data sets can be used for any hierarchical clustering procedure (a contingency data set can be viewed as an individuals–variables data set). Thus, all the data sets presented for principal components analysis, and 2-D and N-D correspondence analyses are available for classification. (*cf.* Chapters 7, 8, and 9); the problem of using classification procedures is not in the data sets but in the distances chosen. To simplify matters, three main types of data sets have been distinguished:

(a) Principal components data sets.

for I: $N(I) = \left\{ i \in I; X_{ij} = \dfrac{x_{ij} - \bar{x}_j}{\sigma_j} \right\};$

$d^2(i, i') = \sum_{j \in J}(x_{ij} - x_{i'j})^2; \qquad m_i = \dfrac{1}{\text{Card } I};$

for J: $d^2(j, j') = 2(1 - r(j, j'))$, where $r(j, j')$ is the linear coefficient of correlation between j and j'.

5. Hierarchical Classification Methods

(b) Correspondence data sets. Because of the symmetry of the sets I and J, only the characteristics for one set are given (the set I):

for I: $N(I) = \{i \in I; f_J^i, m_i = f_i\}$;

$$d^2(i, i') = \|i - i'\|^2 = \|f_J^i - f_J^{i'}\|_{f_J}^2 = \sum_{j \in J} \frac{(f_J^i - f_J^{i'})^2}{f_j};$$

$m_i = f_i$;

for J: I is replaced by J, and conversely.

(c) Factor coordinates data sets.

for I: $N(I) = \{i \in I; \psi_\alpha(i)\}$;

$$d^2(i, i') = \sum_\alpha (\psi_\alpha(i) - \psi_\alpha(i'))^2;$$

$m_i = f_i$ (for factor coordinates from correspondence analysis);

$m_i = \dfrac{1}{\text{Card } I}$ (for factor coordinates from principal components analysis);

for J: $N(J) = \{j \in J; \phi_\alpha(i)\}$;

$$d^2(j, j') = \sum_\alpha (\phi_\alpha(i) - \phi_\alpha(i'))^2 \text{ or}$$

$$d^2(j, j') = \sum_\alpha \frac{(\phi_\alpha(i) - \phi_\alpha(i'))^2}{\lambda_\alpha};$$

$m_j = f_j$ (for factor coordinates from correspondence analysis);

$m_j = \dfrac{1}{\text{Card } J}$ (for factor coordinates from principal components analysis).

Remarks. In general, the classification built on J does not have the same meaning as that built on I; the classification on J is made in view of studying redundancy of variables, whereas the classification on I is made in view of recognizing or identifying elements.

5.6. *Interpretation of Hierarchical Classifications*

5.6.1. Aim and Scope

One of the important advantages of factor analysis is that the two sets I and J are represented on the same graph, thus not only showing the structure of I and that of J, but also the relation of these structures to each other. In clustering, however, it is not the same; the classification is built on one of the sets, and the relations with the other set do not appear on the graph mathematically as in factor analysis. It has to be guessed if such a class is different from another class and which variable causes the difference. To avoid this guesswork, we introduce the computations of contributions in hierarchical classification which generalize the same computations in factor analysis, so that we have formal procedures in the interpretation of the forming of classes or the deviation between classes in terms of elements of the set not involved in the classifying process. Moreover, these computations have been extended to interpret hierarchical classifications according to factor analysis results. There, computations give an elegant solution to the explanation of hierarchical classifications. As in factor analysis, rules for selecting significant classes, dipoles, and variables are formalized to achieve the hierarchical clustering process.

5.6.2. Reference Data Set

To illustrate the computation of contributions, the contingency data set associated with the semantic field related to colors has been chosen (*cf.* Appendix 2, §11). It concerns 89 adjectives (I) and 11 colors (J); $k_{IJ}(i, j)$ represents the number of times that the adjective i of I is associated with j of J. The set I was chosen to be classified.

5.6.3. Graphical Representation of a Hierarchical Classification

After performing the classification procedure, the parameters for building and plotting the hierarchical classification are ready to be used. They are given below, printed in the reverse order in which the classification was built (i.e., from the node 2 Card $I - 1$ to the node Card $I + 1$); (*cf.* Tables 10.12, 10.13). The index of a node is denoted by J.

$I(J)$: level index of class J; $I(J)$ is denoted by $v(n)$ in the usual description of classifications.

$A(J)$: eldest of class J; $A(J)$ is denoted by $a(n)$ or Eldest(n) in the usual description of classifications

$B(J)$: youngest of class J; $B(J)$ is denoted by $b(n)$ or Youngest(n) in the usual description of classifications.

$T(J)$: rate of variance of class J associated with the level index $I(J)$; $T(J)$ is denoted by $\tau(n)$ in the usual description of classifications.

5. Hierarchical Classification Methods

$T(Q)$: rate of variance of the partition at level J; $T(Q)$ is denoted by $\tau(Q_n)$ in the usual description of classifications.

$P(J)$: number of the elements of class J; $P(J)$ is denoted by $\text{Card}(n)$ in the usual description of classifications.

From these parameters, the hierarchical representation is plotted automatically. (*cf.* Fig. 10.7). The corresponding name is assigned to each element i of I. For the upper classes, it is possible to assign to each node its number in the hierarchical clustering, in order to facilitate a better understanding of the classification.

5.6.4. Computations of Contributions

As in factor analysis, graphical representations are not sufficient to understand how classes are formed and what their differences are. The aim of these contributions is to *explain* and to *quantify* that *explanation*. That is the only way to obtain an *explained* classification.

5.6.4.1. Interpretation of Level Indices. Consider the set I of elements i provided with Euclidean coordinates

$$X_{iJ} = \{X_{i1}, X_{i2}, \ldots, X_{ij}, \ldots, X_{i\,\text{Card}\,J}\}$$

and associated masses m_i. Consider a hierarchical classification $C(I)$ built on I, provided with classes q of I. Let $\|i - i'\|^2$ be the squared distance between two elements i and i' of I. Then the following relations hold:

- Coordinates of the center of gravity of q denoted by X_{qj}:

$$X_{qj} = \frac{\sum\limits_{i \in q} m_i X_{ij}}{\sum\limits_{i \in q} m_i};$$

- Second order central moment of I, denoted by $M^2(I)$ or $M^2(N(I))$:

$$M^2(I) = \sum_{i \in I} m_i \|i - g\|^2;$$

- Second order central moment of q, denoted by $M^2(q)$:

$$M^2(q) = \sum_{i \in q} m_i \|i - q\|^2;$$

TABLE 10.12. Analysis of the semantic field associated with colors. J is the number of the node; $A(J)$ is the eldest of J; $B(J)$ is the youngest of J; $I(J)$ is the level multiplied by 1000; $T(J)$ it the rate of variance of J multiplied by 1000. $T(Q)$ is the cumulative proportion of variance associated with each partition of the hierarchical classification multiplied by 1000.

J	$I(J)$	$A(J)$	$B(J)$	$T(J)$	$T(Q)$
177	623	176	173	139	139
176	490	174	175	109	248
175	477	172	156	106	354
174	417	153	170	93	447
173	369	158	118	82	529
172	353	159	171	79	608
171	230	166	163	51	659
170	220	168	169	49	708
169	174	161	151	39	747
168	150	165	167	34	780
167	137	126	164	31	811
166	78	160	143	17	828
165	73	128	141	16	844
164	61	162	144	14	858
163	41	134	56	9	867
162	38	152	149	8	875
161	33	155	117	7	883
160	33	140	157	7	890
159	30	105	125	7	897
158	26	136	147	6	902
157	22	15	146	5	907
156	22	150	154	5	912
155	20	120	111	4	917
154	19	138	59	4	921
153	19	121	142	4	925
152	19	5	132	4	929
151	19	130	85	4	934
150	18	53	148	4	938
149	17	14	115	4	941
148	13	122	135	3	944
147	13	131	145	3	947
146	13	62	36	3	950
145	11	127	106	3	952
144	11	21	84	2	955
143	11	29	139	2	957
142	11	133	137	2	960
141	10	108	79	2	962
140	10	11	69	2	964
139	10	129	95	2	966
138	9	7	113	2	968

5. Hierarchical Classification Methods

TABLE 10.12 (cont'd)

J	I(J)	A(J)	B(J)	T(J)	T(Q)
137	9	7	113	2	970
136	8	104	3	2	972
135	8	109	58	2	974
134	8	112	35	2	975
133	7	25	49	2	977
132	7	114	124	2	979
131	6	103	101	1	980
130	6	119	68	1	981
129	6	119	68	1	983
128	6	12	30	1	984
127	5	96	26	1	985
126	4	39	91	1	986
125	4	65	46	1	987
124	4	17	42	1	988
123	4	10	47	1	898
122	4	71	45	1	989
121	3	97	81	1	990
120	3	54	55	1	991
119	3	20	24	1	992
118	3	107	72	1	992
117	3	77	76	1	993
116	3	40	119	1	994
115	3	22	44	1	994
114	3	16	19	1	993
113	3	27	84	1	996
112	3	13	102	1	996
111	3	99	67	1	997
110	2	89	73	1	997
109	2	82	33	0	998
108	2	78	83	0	998
107	2	100	98	0	999
106	1	60	74	0	999
105	1	94	66	0	999
104	1	2	87	0	999
103	1	18	50	0	1000
102	1	60	88	0	1000
101	0	82	86	0	1000
100	0	20	37	0	1000
99	0	64	75	0	1000
98	0	32	93	0	1000
97	0	4	28	0	1000
96	0	8	70	0	1000
95	0	48	61	0	1000
94	0	41	57	0	1000
93	0	43	92	0	1000
92	0	9	31	0	1000
91	0	51	90	0	1000
90	0	6	38	0	1000

TABLE 10.13. (first part). Analysis of the semantic field associated with colors. Parameters of the indexed hierarchical classification. The column J contains node numbers. The column $I(J)$ contains level measures of nodes ($v \times 1000$). The column $P(I)$ contains numbers of elements of nodes. The column Description of Classes in the Hierarchy contains the names of the adjectives associated with a class J, labelled by their first four letters.

J	$I(J)$	$A(J)$	$B(J)$	$P(J)$	DESCRIPTION OF CLASSES IN THE HIERARCHY
177	622	176	173	89	
176	489	174	175	70	ACID SONO DAZZ LUMI ASIA SUNN SLY SOLF EMPT ATTR COLO DECO COMI JOYF HOT HARD UGLY HARM MASC AUTU ICY SOBE SAD DARK MIST SMOO WINT UNST DIRT SINI SILE DEEP SEVE REFI DEAD MYST DISC ECCL RELI OLD FINI SUGA FLOW PALE ROMA LIGH FOOL CHIL FRAG PERF BALA CELE EXTE FARA PATR CALM REST CLEA LOST COLD RUST VERN ALIV FRAN NATU LIMP CLEN IMMA SNOW PURE
175	477	156	172	30	FINI SUGA FLOW PALE ROMA LIGH FOOL CHIL FRAG PERF BALA CELE EXTE FARA PATR CALM REST CLEA LOST COLD RUST VERN ALIV FRAN NATU LIMP CLEN IMMA SNOW PURE
174	416	153	170	40	ACID SONO DAZZ LUMI ASIA SUNN SLY SOFT EMPT ATTR COLO DECO COMI JOYF HOT HARD UGLY HARM MASC AUTU ICY SOBE SAD DARK MIST SMOO WINT UNST DIRT SINI SILE DEEP SEVE REFI DEAD MYST DISC ECCL RELI OLD
173	369	158	118	19	ACRE VIOL ANXI RAW ILL FLAM KEEN BURN REVO NEU PASS BLOO DYNA FRUI MECH JUIC BURT FEMI ROUN
172	352	171	159	20	BALA CELE EXTE FARA PATR CALM REST CLEA LOST COLD RUST VERN ALIV FRAN NATU LIMP CLEN IMMA SNOW PURE
171	230	166	163	15	BALA CELE EXTE FARA PATR CALM REST CLEA LOST COLD RUST VERN ALIV FRAN NATU

170	219	163	169	33	SOFT EMPT ATTR COLO DECO COMI JOYF HOT HARD UGLY HARM MASC AUTU ICY SOBE SAD DARK MIST SMOO WINT UNST DIRT SINI SILE DEEP SEVE REFI DEAD MYST DISC ECCL RELI OLD
169	173	161	151	11	SINI SILE DEEP SEVE REFI DEAD MYST DISC ECCL RELI OLD
168	150	167	165	22	SOFT EMPT ATTR COLO DECO COMI JOYF HOT HARD UGLY HARM MASC AUTU ICY SOBE SAD DARK MIST SMOO WINT
167	136	164	126	14	SOFT EMPT ATTR COLO DECO COMI JOYF HOT HARD UGLY HARM MASC AUTU ICY
166	78	143	160	10	BALA CELE EXTE FARA PATR CALM REST CLEA LOST COLD
165	72	141	128	8	SOBE SAD DARK MIST SMOO WINT DIRT
164	60	144	162	10	SOFT EMPT ATTR COLO DECO COMI JOYF HOT HARD UGLY
163	40	134	56	5	RUST VERN ALIV FRAN NATU
162	37	152	149	8	ATTR COLO DECO COMI JOYF HOT HARD UGLY
161	32	117	155	7	SINI SILE DEEP SEVE REFI DEAD MYST
160	32	140	157	5	CALM REST CLEA LOST COLD
159	30	125	105	5	LIMP CLEN IMMA SNOW PURE
158	25	136	147	12	ACRE VIOL ANXI RAW ILL FLAM KEEN BURN REVO IRRI PASS BLOO
157	22	15	146	3	CLEA LOST COLD
156	22	150	154	10	FINI SUGA FLOW PALE ROMA LICH FOOL CHIL FRAG PERF
155	19	111	120	5	DEEP SEVE REFI DEAD MYST
154	19	138	59	4	FOOL CHIL FRAG PERF
153	19	142	121	7	ACID SONO DAZZ LUMI ASIA SUNN SLY

TABLE 10.13. (second part). Parameters of the indexed hierarchical classification.

152	18	5	132	5	ATTR COLO DECO COMI JOYF
151	18	130	85	4	DISC ECCL RELI OLD
150	17	53	148	6	FINI SUGA FLOW PALE ROMA LIGH
149	17	14	115	3	HOT HARD UGLY
148	13	135	122	5	SUGA FLOW PALE ROMA LIGH
147	12	131	145	9	RAW ILL FLAM KEEN BURN REVO IRRI PASS BLOO
146	11	62	36	2	LOST COLD
145	11	127	106	5	BURN REVO IRRI PASS BLOO
144	11	21	84	21	SOFT EMPT
143	10	29	189	5	BALA CELE EXTE FARA PATR
142	10	137	133	4	ACID SONO DAZZ LUMI
141	10	108	79	3	SOBE SAD DARK
140	9	11	69	2	CALM REST
139	9	129	95	4	CELE EXTE FARA PATR
138	8	7	113	3	FOOL CHIL FRAG
137	8	1	80	2	ACID SONO
136	8	104	3	3	ACRE VIOL ANXI
135	7	109	58	3	SUGA FLOW PALE
134	7	112	35	4	RUST VERN ALIV FRAN
133	7	25	49	2	DAZZ LUMI
132	6	114	124	4	COLO DECO COMI JOYF
131	6	103	101	4	RAW ILL FLAM KEEN
130	6	119	68	3	DISC ECCL RELI
129	5	12	30	2	CELE EXTE
128	5	123	116	5	MIST SMOO WINT UNST DIRT
127	4	96	26	3	BURN REVO IRRI
126	3	39	91	4	HARM MASC AUTU ICY
125	3	46	65	2	LIMP CLEN
124	3	17	42	2	COMI JOYF
123	3	10	47	2	MIST SMOO
122	3	71	45	2	ROMA LIGH
121	3	97	81	3	ASIA SUNN SLY
	3	51	55	2	DEAD MYST

TABLE 10.13. (third part). Parameters of the indexed hierarchical classification.

119	3	20	24	2	DISC ECCL
118	2	107	72	7	DYNA FRUI MECH JUIC BURT FEMI ROUN
117	2	77	76	2	SINI SILE
116	2	40	110	3	WINT UNST DIRT
115	2	22	44	2	HARD UGLY
114	2	16	19	2	COLO DECO
113	2	27	34	2	CUIL FRAG
112	2	13	102	3	RUST VERN ALIV
111	2	99	67	3	DEEP SEVE REFI
110	2	89	73	2	UNST DIRT
109	2	82	33	2	SUCA FLOW
108	2	78	83	2	SOBE SAD
107	1	100	98	6	DYNA FRUI MECH JUIC BURT FEMI
106	1	60	74	2	PASS BLOO
105	1	94	66	3	IMMA SNOW PURE
104	1	2	87	2	ACRE VIOL
103	0	18	50	2	RAW ILL
102	0	63	88	21	VERN ALIV
101	0	32	86	2	FLAM KEEN
100	0	23	37	2	DYNA FRUI
99	0	64	75	2	DEEP SEVE
98	0	52	93	4	MECH JUIC BURT FEMI
97	0	4	28	2	ASIA SUNN
96	0	8	70	2	BURN REVO
95	0	48	61	2	FARA PATR
94	0	41	57	2	IMMA SNOW
93	0	43	92	3	JUIC BURT FEMI
92	0	9	31	2	BURT FEMI
91	0	51	90	3	MASC AUTU ICY
90	0	6	38	2	AUTU ICY

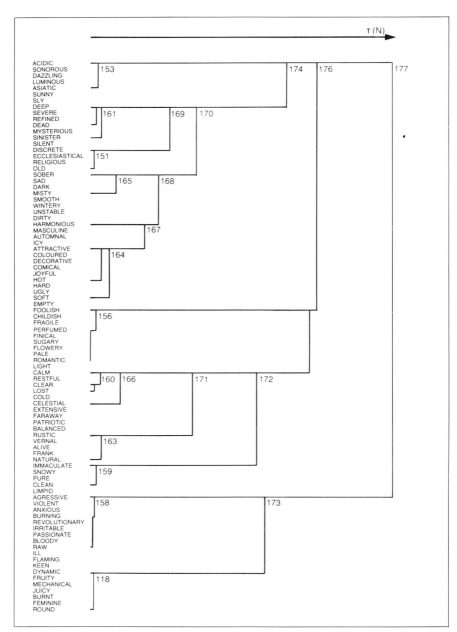

FIGURE 10.7. Analysis of the semantic field associated with colors. Hierarchical classification of adjectives; criterion: second order central moment of a partition; correspondence data set.

5. Hierarchical Classification Methods

- Level index of class n, denoted by $v(n)$:

$$v(n) = M^2(n) - M^2(a(n)) + M^2(b(n)) \quad \text{(by definition)}$$

with $a(n) = \text{Eldest }(n)$,

$b(n) = \text{Youngest }(n)$;

$$v(n) = \frac{m_{a(n)} \cdot m_{b(n)}}{m_{a(n)} + m_{b(n)}} \|a - b\|^2$$

$$= M^2(Q), \quad \text{where } Q\{a(n), b(n)\}.$$

The total variance of a class n can be divided into a sum of three terms: $M^2(a(n))$, $M^2(b(n))$, and $v(n)$, proportional to the squared distance between the centers of classes $a(n)$ and $b(n)$ (Huyghens' theorem). This yields the following relations:

$$M^2(I) = v(I) + M^2(a(I)) + M^2(b(I));$$
$$M^2(a(I)) = v(a(I)) + M^2(a(a(I))) + M^2(b(a(I)));$$
$$M^2(b(I)) = v(b(I)) + M^2(a(b(I))) + M^2(b(b(I)));$$

thus, $\quad M^2(I) = v(I) + v(a(I)) + v(b(I)) + \cdots;$

thus,
$$M^2(I) = \sum \{v(n); n \in \text{Nod } C(I)\}.$$

Note. The computation of $v(n)$ is independent of the aggregation criteria chosen for the hierarchical process; whatever criterion is used for the hierarchical classification, the following relations hold:

$v(n) = $ absolute contribution of node n to the total variance $M^2(I)$;

$$\tau(n) = \frac{v(n)}{M^2(I)} = \text{relative contribution of node } n \text{ to the total variance } M^2(I);$$

$$\tau(Q_n) = \frac{M^2(Q_n)}{M^2(I)} = \text{quality of the partition } Q_n \text{ at the level } n$$

$$= \sum \{\tau(n'); n' = 2 \text{ Card } I + 1, n\}.$$

For the different types of data sets involved in hierarchical classification, the computations are as follows:

(a) Principal components data sets.

$$v(n) = \frac{m_{a(n)} \cdot m_{b(n)}}{m_{a(n)} + m_{b(n)}} \cdot \sum_{j \in J} (X_{aj} - X_{bj})^2$$

with
$$X_{aj} = \frac{\sum_{i \in a(n)} m_i X_{ij}}{\sum_{i \in a(n)} m_i}$$

and
$$m_i = \frac{1}{\text{Card } I}, \quad a = a(n), \quad b = b(n).$$

(b) Correspondence data sets.
$$v(n) = \frac{f_{a(n)} \cdot f_{b(n)}}{f_{a(n)} + f_{b(n)}} \sum_{j \in J} \frac{(f_j^a - f_j^b)^2}{f_j}$$

with
$$f_q = \sum_{i \in q} f_i, \quad f_j^a = \frac{f_{aj}}{f_a}, \quad a = a(n), \quad b = b(n),$$

and
$$f_{aj} = \sum_{i \in a(n)} f_{ij} \quad \text{and} \quad f_{a(n)} = \sum_{i \in a(n)} f_i.$$

The distance used is the distributional distance between profiles associated with the Eldest(n) and Youngest(n) classes.

(c) Factor coordinates data sets.
$$v(n) = \frac{f_{a(n)} \cdot f_{b(n)}}{f_{a(n)} + f_{b(n)}} \sum_{\alpha=1}^{s} (F_\alpha(a) - F_\alpha(b))^2$$

with s = number of selected factors.

with
$$F_\alpha(q) = \frac{\sum_{i \in q} f_i F_\alpha(q)}{\sum_{i \in a} f_i},$$

and
$$f_q = \sum_{i \in q} f_i,$$

The formula can be different according to the type of data set used in factor analysis (principal components analysis or correspondence analysis), but the principle is the same.

Comment. In interpreting results, $v(n)$ is interpreted as the variance of the dipole n formed by $a(n)$ and $b(n)$. Thus, the hierarchical classification must be interpreted from the highest node ($n = 2\,\text{Card}\,I - 1$) to

5. Hierarchical Classification Methods

the lowest node ($n = \text{Card } I + 1$); note the analogy between the interpretation process in factor analysis and that in hierarchical clustering.

5.6.4.2. Interpretation of Class Eccentricities. Consider any class q of a hierarchical classification $C(I)$. It is necessary to study and interpret the position of class q, according to the basic variables set (J) or the factor axes from any factor analysis.

The eccentricity $\rho^2(q)$ of a class q is related to the center of gravity of I:

$$\rho^2(q) = d^2(q, g(I)) = \sum_{j \in J} X_{qj}^2.$$

The interpretation of the eccentricity is based on its decomposition relative to the variables (or axes) (*cf.* Fig. 10.8).

The formulas associated with the interpretation of the eccentricity $\rho^2(q)$ are the following:

- $\text{rho}(q) = \rho^2(q) = $ eccentricity of class q relative to the center of gravity of I;
- $\text{inr}(q) = \dfrac{M^2(q)}{\rho^2(q)} = \dfrac{m_q \rho^2(q)}{M^2(I)}$;
- $\text{cor}_j(q) = \dfrac{X_{qj}^2}{M^2(I)} = $ relative contribution of the variable j to the eccentricity.

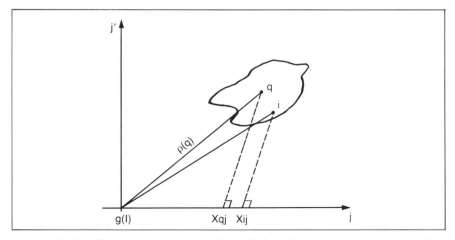

FIGURE 10.8. Geometric representation of the class q relative to variables (identified as axes).

$cor_j(q)$ represents the squared cosine of the angle formed by the vector radius of q and the axis j.

$cor_j(q) \# 0 \Rightarrow j$ does not explain the deviation of q from the center of gravity.

$cor_j(q) \# 1 \Rightarrow j$ explains the deviation of q from the center of gravity.

Generally, a sign ($+$ or $-$) is assigned to $cor_j(q)$ according to whether the relative contribution is high ($+$) or low ($-$), because a value is higher or lower the average ($\bar{x}_j$).

- $ctr_j(q) = \dfrac{m_q X_{qj}^2}{\sigma^2(j)} =$ relative contribution of the jth variance of q related to the total variance of j.

$ctr_j(q) \# 0 \Rightarrow q$ does not explain the variance of j.
$ctr_j(q) \# 1 \Rightarrow q$ explains the variance of j.

- $qlt_s(q) = \sum\limits_{j \leq s} cor_j(q) =$ quality of representation of class q in s axes

(if $s = $ Card J, $qlt_s = 1$).

(a) Principal components data sets.

$$\left. \begin{array}{l} X_{ij} = \dfrac{x_{ij} - \bar{x}_j}{\sigma_j} \\ \\ m_i = \dfrac{1}{\text{Card } I} \end{array} \right\}$$ no change relative to the previous formulas.

(b) Correspondence data sets.

$$f_{IJ} = \{f_{ij}, i \in I; j \in J\};$$
$$g_j(I) = f_j;$$
$$M^2(I) = \sum_{i \in I} f_i \rho^2(i) = \sum_{i \in I, j \in J} \frac{f_i(f_i^j - f_j)^2}{f_j};$$
$$M^2(Q) = \sum_{q \in Q} f_q \rho^2(q);$$
$$M^2(q) = \sum_{i \in q} f_i \rho^2(i);$$
$$inr(q) = \frac{M^2(q)}{M^2(I)};$$
$$rho(q) = \rho^2(q) = \sum_{j \in J} \frac{(f_j^q - f_j)^2}{f_j};$$
$$cor_j(q) = \frac{(f_j^q - f_j)^2 / f_j}{\rho^2(q)}.$$

5. Hierarchical Classification Methods

The sign (+ or −) is assigned to $\text{cor}_j(q)$ according to the value of $(f_j^q - f_j)$.

$$\text{ctr}_j(q) = \frac{(f_j^q - f_j)^2 / f_j}{M_j^2(q)}$$

= relative contribution of the variance of q to the projected variance onto the axes j.

(c) Factor coordinates data sets.

$$\text{ctr}_\alpha(q) = \frac{m_q \phi_\alpha^2(q)}{\lambda_\alpha};$$

$$\text{qlt}_s(q) = \sum_{j \leq s} \text{cor}_\alpha(q);$$

$$\text{inr}(q) = \frac{M^2(q)}{M^2(I)};$$

$$\text{rho}(q) = \sum_\alpha \phi_\alpha^2(q);$$

$$\text{cor}_\alpha(q) = \frac{\phi_\alpha^2(q)}{\rho^2(q)};$$

with

$$m_i = \frac{1}{\text{Card } I} \quad \text{in the case of factor coordinates from principal components analysis;}$$

$$m_i = f_i \quad \text{in the case of factor coordinates from correspondence analysis.}$$

5.6.4.3. *Case Study.* Consider the hierarchical classification built on adjectives associated with colors (*cf.* Fig. 10.9). Only the 11 upper classes are studied (i.e., classes whose level indices are the highest). The results are given in Table 10.14(a)(b).

Comment. The same classification $C(I)$ can be studied according to factor coordinates even if the classification is not built from factor coordinates data sets, because the formulas are independent of the criterion chosen.

5.6.4.4. *Interpretation of Dipoles.* Interpretation of the dipoles, denoted by $\text{dp}(q, q')$, is based on the double projection of the second order central moment $M^2(I)$ onto the system of Euclidean axes (variables

TABLE 10.14 (a). Table concerning the position of 11 classes of the hierarchical classification related to 11 colors (part (a) = blue, red, yellow, white, gray, pink). Origin: analysis of semantic field data set (cf. Appendix 2, §11).

class[a]	a[b]	b[c]	weight[d]	inr[e]	qlt$_s$[f]	rho2[g]	BLUE pr$_1$[h]	cor$_1$[i]	ctr$_1$[j]	RED pr$_2$	cor$_2$	ctr$_2$	YELLOW pr$_3$	cor$_3$	ctr$_3$	WHITE pr$_4$	cor$_4$	ctr$_4$	GRAY pr$_5$	cor$_5$	ctr$_5$	PINK pr$_6$	cor$_6$	ctr$_6$
177	176	173	1000	0	0	0	85	0	0	91	0	0	91	0	0	91	0	0	91	0	0	88	0	0
176	174	175	832	23	1000	125	102	27	7	14	−517	108	99	5	1	110	30	7	110	30	11	104	25	5
175	172	156	355	81	1000	1025	219	206	181	5	−79	57	26	−46	39	162	54	45	33	−36	45	211	169	125
174	153	170	478	51	1000	483	15	−118	66	21	−112	51	153	86	47	71	−10	5	166	127	102	25	−93	44
173	158	118	168	115	1000	3089	0	−27	34	473	517	534	54	−5	6	0	−30	35	0	−30	53	5	−25	26
172	159	171	251	89	1000	1591	295	326	314	7	−49	39	18	−37	35	211	99	91	25	−30	41	22	−31	25
171	166	163	201	90	1000	2016	332	357	348	9	−37	30	23	−25	24	55	−7	7	32	−19	27	23	−24	20
170	168	169	394	63	1000	723	12	−87	60	19	−80	45	39	−41	27	76	−3	2	197	169	167	30	−52	30
169	161	151	125	70	1000	2503	0	−34	26	15	−26	16	0	−36	27	109	1	1	58	−5	5	7	−29	19
168	165	167	269	43	1000	713	17	−76	35	20	−77	29	58	−17	8	61	−14	6	261	444	295	41	−35	14
167	126	164	130	11	1000	389	21	−123	15	42	−68	7	91	0	0	105	5	1	77	−6	1	63	−18	2

[a] class(q) = serial number of q
[b] a(q) = eldest of q
[c] b(q) = youngest of q
[d] weight (q) = weight associated with q = fq
[e] inr(q) = relative variance of q with respect to the total variance
[f] qlt$_s$(q) = quality of representation of q in s original axes (or variables); (here, s = Card J)
[g] rho2(q) = $\rho^2(q) = d^2(g, q)$ = eccentricity of q
[h] pr$_j$(q) = profile of q on the j-axis = f_j^q
[i] cor$_j$(q) = relative contribution of j to the eccentricity of q assigned by + (if > average) or − (if < average)
[j] ctr$_j$(q) = relative contribution of the variance of class q to the variance of N(I) projected on the j-axis

5. Hierarchical Classification Methods

TABLE 10.14 (b). Table concerning the position of 11 classes of the hierarchical classification related to 11 colors (part (b) = brown, purple, black, orange, green). Origin: analysis of semantic field data set (cf. Appendix 2, §11)

class[a]	a[b]	b[c]	weight[d]	inr[e]	qlt$_s$[f]	rho2[g]	BROWN pr_7[h]	BROWN cor_7[i]	BROWN ctr_7[j]	PURPLE pr_8	PURPLE cor_8	PURPLE ctr_8	BLACK pr_9	BLACK cor_9	BLACK ctr_9	ORANGE pr_{10}	ORANGE cor_{10}	ORANGE ctr_{10}	GREEN pr_{11}	GREEN cor_{11}	GREEN ctr_{11}
177	176	173	1000	0	0	0	91	0	0	91	0	0	90	0	0	100	0	0	90	0	0
176	174	175	832	23	1000	125	108	26	9	99	5	1	106	23	7	41	−284	60	108	29	7
175	172	156	355	81	1000	1025	82	−1	1	18	−57	59	8	−74	76	10	−79	58	226	200	175
174	153	170	478	51	1000	483	128	31	22	158	103	67	179	182	120	63	29	13	21	−110	61
173	158	118	168	115	1000	3089	5	−26	43	54	−5	7	11	−23	33	397	284	295	0	−29	36
172	159	171	251	89	1000	1591	95	0	0	11	−44	50	11	−44	50	11	−50	40	295	291	279
171	166	163	201	90	1000	2016	118	4	5	14	33	37	14	−32	37	14	−37	30	368	425	414
170	168	169	394	63	1000	723	150	53	48	190	148	118	218	249	202	56	−28	16	14	−89	61
169	161	151	125	70	1000	2503	36	−13	13	416	463	407	358	317	282	0	−40	25	0	−36	27
168	165	167	269	43	1000	713	203	194	117	85	−1	0	153	60	33	81	−5	2	20	−76	35
167	126	164	130	11	1000	389	231	549	88	105	5	1	63	−21	3	168	117	12	35	−87	11

[a] class(q) = serial number of q
[b] $a(q)$ = eldest of q
[c] $b(q)$ = youngest of q
[d] weight (q) = weight associated with $q = fq$
[e] inr(q) = relative variance of q with respect to the total variance
[f] qlt$_s$(q) = quality of representation of q in s original axes (or variables); (here, s = Card J)
[g] rho2(q) = $\rho^2(q) = d^2(g, q)$ = eccentricity of q
[h] pr$_j^q$(q) = profile of q on the j-axis = f_i^q
[i] cor$_j$(q) = relative contribution of j to the eccentricity of q assigned by + (if > average) or − (if < average)
[j] ctr$_j$(q) = relative contribution of the variance of class q to the variance of $N(I)$ projected on the j-axis

identified as Euclidean axes, factor axes) and onto the set of nodes of the hierarchical classification, denoted by Nod $C(I)$.

The following formulas hold:

$$M^2(N(I)) = \sum_{n \in \text{Nod } C(I)} v(n),$$

$$v(n) = \frac{m_{a(n)} \cdot m_{b(n)}}{m_{a(n)} + m_{b(n)}} \sum_{j \in J} (X_{aj} - X_{bj})^2.$$

This yields

$$M^2(N(I)) = \sum_{n \in \text{Nod } C(I)} \sum_{j \in J} \frac{m_a \cdot m_b}{m_a + m_b} (X_{aj} - X_{bj})^2.$$

Defining $v(n, j)$ by

$$v(n, j) = \frac{m_a \cdot m_b}{m_a + m_b} (X_{aj} - X_{bj})^2,$$

we get

$$M^2(N(I)) = \sum \{v(n, j); n \in \text{Nod } C(I); j \in J\},$$

where $v(n, j)$ = absolute contribution of the dipole $\text{dp}(a(n), b(n))$ and the variable j to the total variance of I, $M^2(N(I))$.

Comment. The double projection is independent of the criterion used for building the hierarchical classification; any hierarchical classification may be therefore interpreted in terms of dipoles and significant variables.

Let

$$\text{In}(j) = \sum_{n \in \text{Nod } C(I)} v(n, j) = j\text{th component of } M^2(N(I)).$$

Consider the quantities $v(n, j)/v(n)$ and $v(n, j)/\text{In}(j)$:

$$\text{cod}_j(n) = \frac{v(n, j)}{v(n)}$$

= relative contribution of the axis j to the variance of the dipole $\text{dp}(a(n), b(n))$ formed from the node n
= squared cosine of the angle formed by the axis j and the straight line joining the centers of classes $a(n)$ and $b(n)$ (cf. Fig. 10.9);

$\text{cod}_j(n) \# 1 \Rightarrow j$ explains the deviation between $a(n)$ and $b(n)$;

5. Hierarchical Classification Methods

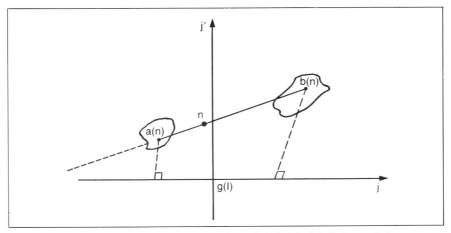

FIGURE 10.9. Geometric representation of $dp(a(n), b(n))$ formed from n.

$cod_j(n) \# 0 \Rightarrow j$ does not explain the deviation between $a(n)$ and $b(n)$;

$ctd_j(n) = \dfrac{v(n, j)}{In(j)}$ = relative contribution of the dipole $dp(a(n), b(n))$ to the variance of j;

$ctd_j(n) \# 1 \Rightarrow$ the dispersion of the cloud on the axis j is given exclusively by the elements of $a(n)$ and $b(n)$;

$qlt_s(n) = \sum\limits_{j \leq s} cod_j(n)$ = quality of the representation of the dipole $dp(a(n), b(n))$ in s axes;

$ind(n) = \dfrac{M^2\{a(n), b(n)\}}{M^2(I)}$ = variance of the dipole $dp(a(n), b(n))$ related to the total variance;

$d2ab(n) = d^2(a(n), b(n))$ = squared distance between the centers of the classes $a(n)$ and $b(n)$.

These different formulas are expressed for the following types of data sets:

(a) Principal components data sets:

$$\left. \begin{array}{l} X_{ij} = \dfrac{x_{ij} - \bar{x}_j}{\sigma_j}, \\ m_i = \dfrac{1}{\text{Card } I}; \end{array} \right\}$$ no change related to the previous formulas.

(b) Correspondence data sets:

$$f_{IJ} = \{f_{ij}; i \in I; j \in J\},$$

$$v(n, j) = \frac{f_a f_b}{f_n} \frac{(f_j^a - f_j^b)^2}{f_j},$$

$$\frac{v(n, j)}{v(n)} = \frac{(f_j^a - f_j^b)^2/f_j}{\|f_j^a - f_j^b\|^2}.$$

(c) Factor coordinates data sets:

$$\{m_i; \psi_\alpha(i); i \in I\} \quad \text{or} \quad \{m_j; \phi_\alpha(j); j \in J\},$$

For I, we have:

$$v(n, \alpha) = \frac{m_a \cdot m_b}{m_a + m_b} (\psi_\alpha(a) - \psi_\alpha(b))^2,$$

$$\frac{v(n, \alpha)}{v(n)} = \frac{(\psi_\alpha(a) - \psi_\alpha(b))^2}{\sum_\alpha (\psi_\alpha(a) - \psi_\alpha(b))^2}.$$

for J, we have:

$$v(n, \alpha) = \frac{m_a \cdot m_b}{m_a + m_b} (\phi_\alpha(a) - \phi_\alpha(b))^2$$

$$v(n, \alpha) = \frac{(\phi_\alpha(a) - \phi_\alpha(b))^2}{\sum_\alpha (\phi_\alpha(a) - \phi_\alpha(b))}^2$$

5.6.4.5. Case Study. Consider the hierarchical classification or the adjectives from the semantic field data set. Only 11 dipoles are studied, selected according to their variances. The results are given in Table 10.15(a) and (b), and are interpreted in order to explain the separation between classes.

For example:

dp(176, 173) is mainly explained by *orange* and *red*;

dp(174, 175) is explained by several colors;

dp(172, 156) is mainly explained by *red*;

dp(153, 170) is mainly explained by *yellow*.

Graphical results are summarized in Fig. 10.10.

5. Hierarchical Classification Methods

TABLE 10.15 (a). Table concerning the upper dipoles of a hierarchical classification (11 dipoles), related to 11 colors (part (a) = blue, red, yellow, white, gray, pink). Origin: analysis of semantic field data set (cf. Appendix 2, §11).

							BLUE				RED			YELLOW			WHITE			GRAY			PINK		
$n = \text{dp}^a$	a^b	b^c	weight^d	ind^e	qld$_s{}^f$	d2ab^g	pd$_1{}^h$	cod$_1{}^i$	ctr$_1{}^j$	pd$_2$	cod$_2$	ctr$_2$	pd$_3$	cod$_3$	ctr$_3$	pd$_4$	cod$_4$	ctr$_4$	pd$_5$	cod$_5$	ctr$_5$	pd$_6$	cod$_6$	ctr$_6$
177	176	173	1000	139	1000	446	102	27	41	−459	517	642	44	5	7	110	30	42	110	30	64	99	25	32
176	174	175	832	109	1000	241	−203	202	239	16	1	1	127	73	85	−91	38	43	133	80	136	−186	164	164
175	172	156	355	106	1000	650	259	122	141	7	0	0	−26	1	1	167	47	52	−27	1	2	−645	731	710
174	153	170	478	93	1000	603	21	1	1	14	0	0	645	758	747	−33	2	2	−175	56	81	−30	2	1
173	158	118	168	82	1000	1050	0	0	0	623	405	298	78	6	5	0	0	0	0	0	0	8	0	0
172	159	171	251	79	1000	879	−186	47	40	9	0	0	−23	1	1	782	762	620	−32	1	2	−5	0	0
171	166	163	201	51	1000	501	435	445	247	−26	1	1	35	3	1	65	9	5	49	5	4	−5	0	0
170	168	169	394	49	1000	258	17	1	1	6	0	0	58	14	7	−48	10	5	203	175	133	33	5	2
169	161	151	125	39	1000	655	0	0	0	21	1	0	0	0	0	55	5	2	−53	5	3	11	0	0
168	165	167	269	34	1000	224	−8	0	0	−42	9	3	−65	20	7	−85	36	12	357	625	327	−43	10	3
167	126	164	130	31	1000	873	−28	1	0	9	0	0	−106	14	5	−122	19	6	−89	10	5	−73	7	7

^{a} dp = n = serial number of dipole n
^{b} a(dp) = a(n) = eldest of n
^{c} b(dp) = b(n) = youngest of n
^{d} weight (dp) = relative frequency of $n = f_a(n) + f_b(n)$
^{e} ind (dp) = variance of dipole dp related to the total variance = $\tau(n)$ = rate of variance of $n = a(n) \cup b(n)$
^{f} qld$_s$(dp) = quality of representation of dipole dp in s axes or variables (here, s = Card J)
^{g} d2ab(dp) = $d^2(a(n), b(n))$ = squared distance between the centers of classes a(n) and b(n)
^{h} pd$_j$(dp) = $(f_j^{a(n)} - f_j^{b(n)})$ = projection of the dipole dp on the j-axis
^{i} cod$_j$(dp) = cod$_j$(dp(a(n), b(n))) = relative contribution of j-axis to the variance of dipole dp = cosine squared of the angle formed by the j-axis and the straight line joining the centers of classes a(n), b(n)
^{j} ctd$_j$(dp) = ctd$_j$(dp(a(n), b(n))) = relative contribution of dipole dp with respect to the variance of j-axis

TABLE 10.15 (b). Table concerning the upper dipoles of a hierarchical classification (11 dipoles), related to 11 colors (part (b) = brown, purple, black, orange, green). Origin: analysis of semantic field data set (*cf.* Appendix 2, §11)

$n = dp^a$	a^b	b^c	weight^d	ind^e	qld$_s{}^f$	d2ab^g	BROWN pd$_7{}^h$	cod$_7{}^i$	ctd$_7{}^j$	PURPLE pd$_8$	cod$_8$	ctd$_8$	BLACK pd$_9$	cod$_9$	ctd$_9$	ORANGE pd$_{10}$	cod$_{10}$	ctd$_{10}$	GREEN pd$_{11}$	cod$_{11}$	ctd$_{11}$
177	176	173	1000	139	1000	446	103	26	51	44	5	8	95	23	40	−356	284	355	108	29	44
176	174	175	832	109	1000	241	46	9	15	140	90	124	172	136	189	53	12	11	−205	194	229
175	172	156	355	106	1000	650	42	3	4	−24	1	1	11	0	0	2	0	0	233	93	107
174	153	170	478	93	1000	603	−129	30	40	−179	58	68	−218	87	103	42	3	2	40	3	3
173	158	118	168	82	1000	1050	−18	0	0	78	6	7	16	0	0	−783	582	431	0	0	0
172	159	171	251	79	1000	879	−118	17	19	−14	0	0	−14	0	0	−14	0	0	−368	171	145
171	166	163	201	51	1000	501	−95	20	14	21	1	1	21	1	1	−19	1	0	−482	514	285
170	168	169	394	49	1000	258	167	119	82	−331	468	289	−205	181	113	81	26	11	20	2	1
169	161	151	125	39	1000	655	18	1	0	−567	539	264	516	450	223	0	0	0	0	0	0
168	165	167	269	34	1000	224	−53	14	7	−39	7	3	174	150	64	−168	125	38	−28	4	1
167	126	164	130	31	1000	873	836	878	379	−122	19	7	−73	7	3	−195	43	12	−41	2	1

^{a} dp = n = serial number of dipole n
^{b} a(dp) = $a(n)$ = eldest of n
^{c} b(dp) = $b(n)$ = youngest of n
^{d} weight (dp) = relative frequency of $n = f_a(n) + f_b(n)$
^{e} ind (dp) = variance of dipole dp related to the total variance = $\tau(n)$ = rate of variance of $n = a(n) \cup b(n)$
^{f} qld$_s$(dp) = quality of representation of dipole dp in s axes or variables (here, s = Card J)
^{g} d2ab(dp) = $d^2(a(n), b(n))$ = squared distance between the centers of classes $a(n)$ and $b(n)$
^{h} pd$_j$(dp) = $(f_j^{a(n)} - f_j^{b(n)})$ = projection of the dipole dp on the j-axis
^{i} cod$_j$(dp) = cod$_j$(dp($a(n)$, $b(n)$)) = relative contribution of j-axis to the variance of dipole dp = cosine squared of the angle formed by the j-axis and the straight line joining the centers of classes $a(n)$, $b(n)$
^{j} ctd$_j$(dp) = ctd$_j$(dp($a(n)$, $b(n)$)) = relative contribution of dipole dp with respect to the variance of j-axis

5. Hierarchical Classification Methods

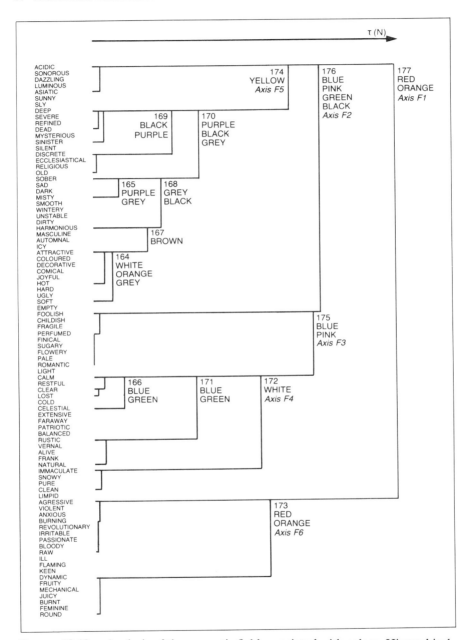

FIGURE 10.10. Analysis of the semantic field associated with colors. Hierarchical classification of adjectives. The most significant variables and factor axes are plotted near the nodes.

5.7. Graphical Representations

Using contribution values, new graphics can be easily produced that are more explicative than the usual representations of hierarchical classification. Below two ones are proposed: the first corresponds to a dipole-by-dipole data processing procedure; the second corresponds to a synthesis procedure using only significant classes, dipoles, and explaining variables.

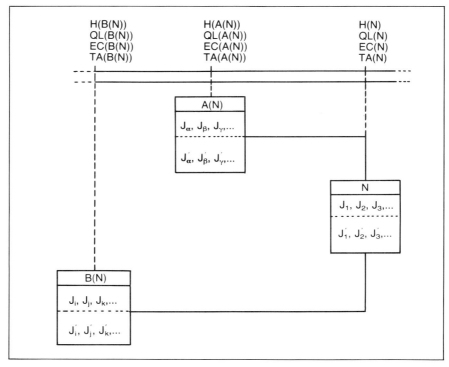

FIGURE 10.11. Elements of representation of a dipole $dp(A(N), B(N))$. N represents the number of the node; $H(N)$ represents the homogeneity of node $N = (M^2(N)/M^2(I))$; $QL(N)$ represents the quality of the partition at level $N = (M^2(Q)/M^2(I))$; $EC(N)$ represents the relative proportion of ρ^2 to the maximum of ρ^2; $\tau(N)$ represents the proportion of variance of the subdivision N; J_1, J_2, and J_3 represent the explaining variables of the variance of the subdivision of N into $A(N)$ and $B(N)$; J'_1, J'_2, and J'_3 represent the explaining variables of the deviation from the center. At each J_α, we can assign the value of the contribution, or a specific code corresponding to an interval of contribution values.

5. Hierarchical Classification Methods

5.7.1. Dipole-by-Dipole Graphic

A graph is given in Fig. 10.11 that summarizes the main features involved in explaining a hierarchical classification dipole-by-dipole.

5.7.2. Explained Hierarchical Classification Graphic

The explained hierarchical classification is summarized in Fig. 10.12. This graphic highlights the real interest of contributions in interpreting hierarchical classification. Each subdivision of the hierarchical classification can be explained by one or several variables, considered as the significant variables (or the explicative variables) of the classification. On the graphical representation, the explicative variables are plotted near the node corresponding to the dipole formed from it.

5.8. Rules for Selecting Significant Classes, Dipoles, and Variables

5.8.1. Aim and Scope

In practice, the number of individuals and variables involved in hierarchical clustering can be high, so that data processing of contributions must be made automatically. Therefore, some rules are presented allowing us to select *automatically* the significant information. These rules can be easily introduced into any computer programs for hierarchical classification. How have these rules been established? Cases studies from 1976 to 1989 were re-explored and reanalyzed, and from this re-exploration a procedure for using contributions was developed. It is a practical procedure and not a theoretical one. The rules are expressed as follows:

5.8.2. Selection of Nodes: First Class Significant Subdivisions

Rule 1. N is chosen such that

$$\sum_{n=2\,\text{Card}\,I-1}^{N} \tau(n) \geq p.$$

The given value p corresponds to a percentage of variance of I (in general, $p = 0.8$).

Rule 2. N is chosen such that

$$\frac{v(N)}{v(N+1)} \quad \text{is maximal.}$$

N is chosen such that the difference of two successive levels is higher than the others.

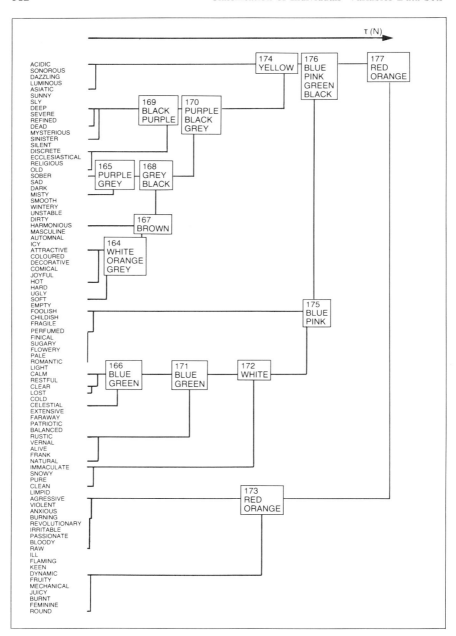

FIGURE 10.12. Analysis of the semantic field associated with colors. Explained hierarchical classification with selection of significant classes and explaining variables.

5. Hierarchical Classification Methods

Rule 3. N is chosen such that

$$\tau(N) \geq \frac{1}{\text{Card } I},$$

representing the average value of $\tau(n)$ on the whole set of dipoles.

Rule 4. Consider $v(N)$ and the sets $A(N)$ and $B(N)$. The values $v_s(N)$ can be computed:

$$v_s(N) = \frac{m_a \cdot m_b}{m_a + m_b} \|X_{aj} - X_{bj}\|^2,$$

where the elements j of J are randomly permuted. If $v_s(N) \leq v(N)$, N is chosen as a significant subdivision.

Comments. These rules can be applied to eliminate the lower branches of the hierarchical classification. Sometimes, however, it can be interesting to study medium or lower branches, because local relationships can be discovered in hierarchical clustering.

5.8.3. Selection of Nodes: Second Class Significant Subdivisions
Local relationships can be obtained at the low level of the hierarchical classification. To select (or to detect) the nodes involved with local relations, we use the following rules:

Rule 5. $N' \neq N$ is chosen such that

$$\sum_{n \geq N'} \tau(n) \geq p.$$

p is a given value corresponding to a percentage of variance of I.

Rule 6. The values $v(n, j)/M^2(I)$ are ordered in decreasing order. N' is chosen such that

$$\frac{v(n, j)}{M^2(I)} \geq \text{the average value of the total variance} \left(\frac{1}{\text{Card } I \cdot \text{Card } J}\right)$$

$$\sum_{n,j} \frac{v(n, j)}{M^2(I)} \geq p,$$

where p is a given value corresponding to a percentage of variance of I.

Particular case. $\rho^2(N') = 0$ and $v(N') = 0$ and $M^2(N') = 0$ with $N' \neq I$. In this case, N' represents a class whose profile is the same as that of I.

5.8.4. Rules for Selecting Significant Explaining Variables

To explain a classification means that class subdivisions can be expressed in terms of percentages of explaining variables j of J. Variables are selected such that their contributions to the deviation from the center and/or the deviation between two classes are high. For this, it is necessary to determine a threshold. This explanation is made only for significant nodes or subdivisions N or N' determined by the previous rules.

Rule 1. A variable j of J is an explaining variable if

$$\text{cor}_j(N) \geq k$$

or

$$\text{cor}_j(N) \geq \text{the average value of contributions } \text{cor}_j(N) \text{ on } J.$$

This rule gives the explaining variables of the subdivisions into classes (k is a given value similar to a squared cosine).

Rule 2. A variable j of J is an explaining variable if

$$\text{cod}_j(N) \geq k$$

or

$$\text{cod}_j(N) \geq \text{the average value of contributions } \text{cod}_j(N) \text{ on } J.$$

This rule gives the explaining variables of the subdivisions into classes (k is a given value similar to a cosine squared).

5.8.5. Case Study

The previous rules have been applied to the hierarchical classification of the semantic field of colors, giving the following results:

First class nodes
Rule 1: With 80% of the total variance $M^2(I)$

$$N = 167.$$

Rule 2: Maximum of $v(N)/v(N+1)$

$$N = 165.$$

Rule 3:

$$N = 164.$$

Second class nodes

5. Hierarchical Classification Methods

Rule 6:

$$N = 150.$$

Explaining variables. The explaining variables are computed for selected first class nodes only (from $N = 177$ to $N = 164$). The results are given in Tables 10.16 and 10.17.

5.9. Classifying Supplementary Points into a Hierarchical Classification

5.9.1. Basic Principle

Classifying supplementary points into a hierarchical classification is based on the same principle as that used in factor analysis:

- Suppressing particular points and reintroducing them into the graphics.
- Studying the stability and validity of the hierarchical classification.
- Classifying unknown points.
- Classifying points whose coordinates are missing.

Two methods can be adapted for this classifying process. The first method is to seek the terminal i of I nearest to the supplementary part i_s.

TABLE 10.16. Explaining variables of the eccentricities of classes from $N = 164$ to $N = 176$ given in terms of $\text{cor}_j(N)$.

Nodes	Explaining variables
$N = 176$	−RED (50%), −ORANGE (28%)
$N = 175$	RED (+20%), PINK (16%), GREEN (20%)
$N = 174$	−BLUE (11%), −RED (11%), GRAY (12%), BLACK (12%), PURPLE (12%)
$N = 173$	RED (50%), ORANGE (28%)
$N = 172$	BLUE (32%), GREEN (29%)
$N = 171$	BLUE (16%), GREEN (42%)
$N = 170$	GRAY (16%), BROWN (18%), PURPLE (24%)
$N = 169$	PURPLE (46%), BLACK (31%)
$N = 168$	GRAY (44%), BROWN (19%)
$N = 167$	−BLUE (12%), BROWN (45%)
$N = 166$	BLUE (71%)
$N = 165$	GRAY (61%)
$N = 164$	ORANGE (44%), −BLUE (19%), −GREEN (13%)

TABLE 10.17. Explaining variables of the subdivisions into classes from $N = 177$ to $N = 164$ in terms of $cod_j(N)$.

Subdivisions	Explaining variables
$N = 177$, $A(N) = 176$, $B(N) = 173$	RED (50%), ORANGE (28%)
$N = 176$, $A(N) = 174$, $B(N) = 175$	BLUE (20%), GREEN (19%) PINK (16%), BLACK (12%)
$N = 175$, $A(N) = 156$, $B(N) = 172$	BLUE (12%), PINK (73%)
$N = 174$, $A(N) = 153$, $B(N) = 170$	YELLOW (75%)
$N = 173$, $A(N) = 158$, $B(N) = 118$	RED (40%), ORANGE (50%)
$N = 172$, $A(N) = 171$, $B(N) = 159$	WHITE (72%)
$N = 171$, $A(N) = 166$, $B(N) = 163$	BLUE (44%), GREEN (51%)
$N = 170$, $A(N) = 168$, $B(N) = 169$	PURPLE (46%), BLACK (15%) GRAY (17%)
$N = 169$, $A(N) = 161$, $B(N) = 151$	BLACK (45%), PURPLE (53%)
$N = 168$, $A(N) = 167$, $B(N) = 165$	GRAY (62%), BLACK (15%), ORANGE (12%)
$N = 167$, $A(N) = 164$, $B(N) = 120$	BROWN (87%)
$N = 166$, $A(N) = 143$, $B(N) = 160$	BLUE (60%), GREEN (34%)
$N = 165$, $A(N) = 141$, $B(N) = 128$	PURPLE (62%), GRAY (17%)
$N = 164$, $A(N) = 144$, $B(N) = 162$	WHITE (32%), ORANGE (33%) GRAY (15%).

The second method is to assign the supplementary element i_s to one terminal of the tree starting class-by-class from the top. Different representations of hierarchical classifications with supplementary elements are given in Fig. 10.13.

5.9.2. Algorithms for Classifying Supplementary Elements

5.9.2.1. The Algorithm [SUP.CAH.TER]. This algorithm involves assigning any supplementary point i_s to the nearest point i of the hierarchical classification. It can be expressed as follows:

Step A   Do inf $= \|i_s - i\|$
Step B   From $i = i_2$ to Card I
         If $\|i_s - i\| \leq$ inf; inf $= \|i_s - i\|$
               $i_k = i$
         If not go to *Step B*
Step C   Inf $= \|i_s - i_k\|$; i_s is assigned to i_k

5. Hierarchical Classification Methods

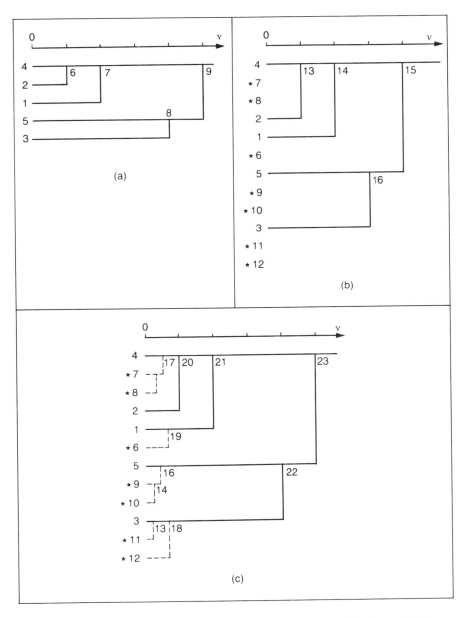

FIGURE 10.13. Representation of different hierarchical classifications. (a) Basic hierarchical classification. (b) Hierarchical classification with supplementary elements. (c) Hierarchical classification with supplementary elements with renumbering.

5.9.2.2. The Algorithm [SUP.CAH.DIST]. This algorithm involves assigning any supplementary point i_s subdivision-by-subdivision to a point i, according to the criterion of distances between classes.

(a) THE ALGORITHM
Initialization: $n = 2 \operatorname{Card} I - 1$ (number of the top of the hierarchical classification)

Begin$_1$ If $\delta(\{i_s\}, a(n)) \geq \delta(\{i_s\}, b(n))$
$\quad\quad\begin{vmatrix} i_s \in a(n) \\ \text{if } a(n) \leq \operatorname{Card} I; \text{ go to } End_1 \\ \text{if not } n = a(n); \text{ go to } Begin_1 \end{vmatrix}$
$\quad$If not
$\quad\quad\begin{vmatrix} i_s \in b(n) \\ \text{if } b(n) \leq \operatorname{Card} I; \text{ go to } End_1 \\ \text{if not } n = b(n); \text{ go to } Begin_1 \end{vmatrix}$
End$_1$

Comments. The element i_s is assigned to a series of classes until the terminal class is reached. δ is a distance between sets.

(b) DISTANCES USED FOR [SUP.CAH.DIST]: $\delta(\{i_s\}, q)$. Consider a class q of $C(I)$.

$$\delta_{\text{uim}}(\{i_s\}, q) = \inf_{i \in q} \{\|i_s - i\|\};$$

$$\delta_{\text{usm}}(\{i_s\}, q) = \sup_{i \in q} \{\|i_s - i\|\};$$

$$\delta_{\text{moy}}(\{i_s\}, q) = \frac{\sum_{i \in q} \|i_s - i\|}{\operatorname{Card} q};$$

$$\delta_{M^2}(\{i_s\}, q) = \frac{m_{i_s} \cdot m_q}{m_{i_s} + m_q} \|i_s - q\| + M^2(q);$$

$$\delta_{\text{bar}}(\{i_s\}, q) = \|i_s - q\|.$$

After developing δ_{mot}, we see that the following relation holds:

$$\delta_{\text{mot}}(\{i_s\}, q) = \frac{m_q}{m_{i_s} + m_q} \|q - i_s\|^2.$$

5.9.2.3. The Algorithm [SUP.CAH.HYPERPLAN]. This algorithm represents a process based on discrimination methods applied to two

5. Hierarchical Classification Methods

classes; in what follows, the principle is applied to any couple $(a(n), b(n))$ from the top of the hierarchical classification.

(a) HYPERPLANE CRITERION (FISHER). Consider the coordinates of the classes q, q', and s (q and q' are classes identified by their centers; s is a supplementary element):

$$q_J = \{q_1, q_2, \ldots, q_j, \ldots, q_{\text{Card }J}\},$$
$$q'_J = \{q'_1, q'_2, \ldots, q'_j, \ldots, q'_{\text{Card }J}\},$$
$$s_J = \{s_1, s_2, \ldots, s_j, \ldots, s_{\text{Card }j}\}.$$

This process is based on the determination of Fisher's hyperplane, for which the two classes q and q' are separated, and then on the assignment of s to one side of the hyperplane, given by the sign of the scalar product defined as follows:

$$\langle \overrightarrow{2s - (q + q')}, \overrightarrow{(q' - q)} \rangle_{M = \sigma^{-1}(I)}$$
$$= (q_J - q'_J)\sigma^{-1}(I)^{JJ}(2s_J - q'_J - q_J)$$
$$= \sum_{j,j' \in J} \sigma_{jj'}^{-1}(q_j - q'_j)(2s_{j'} - q_{j'} - q'_{j'}),$$

where $M = \sigma^{-1}$ is the matrix of variances–covariances of I.
This gives the following algorithm:

(b) THE ALGORITHM

Initialization
    Computation of coordinates of class centers.
    Computation of $\sigma(I)$, then $\sigma^{-1}(I)$
    Do $n = 2 \operatorname{Card} I - 1, \operatorname{Card} I + 1$
$Begin_1$
    If $\sum_{j,j' \in J} \sigma_{jj'}^{-1}(X_{aj} - X_{bj})(2X_{sj} - X_{aj'} - X_{bj'}) \geq 0$

        $i_s \in a(n)$
        if $a(n) \leq \operatorname{Card} I$; go to End_1
        if not; $n = a(n)$; go to $Begin_1$
    If not
        $i_s \in b(n)$
        if $b(n) \leq \operatorname{Card} I$; go to End_1
        if not; $n = b(n)$; go to $Begin_1$
End_1

Comments. $s = i_s$ = supplementary element; $a = a(n) = \text{Eldest}(n)$; $b = b(n) = \text{Youngest}(n)$; $X_{aj} = j^{\text{th}}$ coordinate of the class center of $a(n)$.

(c) ALTERNATIVE ALGORITHM. The previous algorithm can be modified as follows: instead of taking the Fisher's hyperplane for each pair of classes, the hyperplane that passes through the center of each class n (union of $a(n)$ and $b(n)$) can be used. This gives the following algorithm:

Initialization
   Computation of coordinates of class centers.
   Computation of $\sigma(I)$, then $\sigma^{-1}(I)$
   Do $n = 2 \text{ Card } I - 1$, $\text{Card } I + 1$
Begin$_1$

If $\sum_{j,j' \in J} \sigma_{jj'}^{-1}(X_{aj} - X_{bj'})(X_{aj} - X_{bj'}) \geq 0$

$\quad \begin{vmatrix} i_s \in a(n) \\ \text{if } a(n) \leq \text{Card } I; \text{ go to } End_1 \\ \text{if not}; n = a(n); \text{ go to } Begin_1 \end{vmatrix}$

If not

$\quad \begin{vmatrix} i_s \in b(n) \\ \text{if } b(n) \leq \text{Card } I; \text{ go to } End_1 \\ \text{if not}; n = b(n); \text{ go to } Begin_1 \end{vmatrix}$

End$_1$

5.9.2.4. The Algorithm [SUP.CAH.QUADR].

This algorithm involves a process based on the variance metric associated with a class; the process proposed in the following is due to Benzecri (1977), and it is expressed as follows:

(a) CRITERION. The distance between a supplementary element s and a class q is given as follows:

$$d^2(s, q) = \|s_J - q_J\|^2_{\sigma^{-1}(q)} = (s_J - q_J)(\sigma^{-1}(q))'(s_J - q_J).$$

To decide if s must be assigned to $a(q)$ or $b(q)$, the following quantity is computed:

$$R(s, a(q), b(q)) = \frac{s \cdot a(q)}{s \cdot b(q)} = \frac{\|s - a(q)\|^2_{\sigma^{-1}(a(q))}}{\|s - b(q)\|^2_{\sigma^{-1}(b(q))}}.$$

R is compared with a threshold $S = S(a(q), b(q))$ chosen such that the

5. Hierarchical Classification Methods

number of errors $F(s)$ described in the following equation is minimal:

$$F(S) = \text{Card}(\{i \in a(q); S < R(i, a(q), b(q))\}$$
$$\cup \{i' \in b(q); R(i, a(q), b(q)) < S\}).$$

$F(S)$ is the number of elements i of I belonging to $a(q)$ whose ratio $R(i, a(q), b(q))$ of distances to $a(q)$ and $b(q)$ is greater than S, plus the number of elements i' of I belonging to $b(q)$ whose ratio $R(i, a(q), b(q)) < S$. Thus, the surface of separation between the classes $a(q)$ and $b(q)$ is a quadric whose equation is

$$\sum_{j,j' \in J} a_{jj'}(x_{s_j} - x_{a_j})(x_{s_{j'}} - x_{a_{j'}}) - Sb_{jj'}(x_{s_j} - x_{b_j})(x_{s_{j'}} - x_{b_{j'}}) = 0.$$

This yields the following algorithm:

(b) THE ALGORITHM

Initialization

    Computation of coordinates of class centers.

Step 1

    Computation of $S(a(n), b(n))$

    $Begin_1$ | Do $n = \text{Card } I + 1$ to $2 \text{ Card } I - 1$
               Computation of $S(a(n), b(n))$
               Computation of $F(S)$
               Computation of $\text{rate}(n)$

    End_1

Step 2

    Assignment of supplementary elements.
    Do $s = 1$ to Card $S(=$ number of supplementary elements$)$
    $Begin_2$ | Do $n = 2 \text{ Card } I - 1$ to $\text{Card } I + 1$
               Computation of $R(s, a(n), b(n))$
               If $R(s, a(n), b(n)) < S(n)$
                   $s \in a(n)$
                   if $a(n) \leq \text{Card } I$; go to End_2
                   if not; $n = a(n)$; go to $Begin_2$
               If not
                   $s \in b(n)$
                   if $b(n) \leq \text{Card } I$; go to End_2
                   if not; $n = b(n)$; go to $Begin_2$

    End_2

This algorithm is available in DACL (*cf.* Chapter 12).

5.9.3. Case Study: Identification of an Unknown Skull

The example used is the biometric study of wolf and dog skulls defined by six principal characteristics (data in Appendix 2, §7). The procedure is as follows: a hierarchical classification is built on the population of wolf and dog skulls without the unknown skull; the unknown skull is then assigned to that classification by the previous algorithm. The results are given in Fig. 10.14.

5.9.4. Validity of a Hierarchical Classification

Validation of a hierarchical classification is based on the following principle: Each element involved in the hierarchical classification is also taken as a supplementary element. All the elements are then classified as supplementary elements in the basic hierarchical classification, by the algorithm [SUP.CAH.QUADR]. A class whose supplementary elements are not the same as its basic elements is considered as invalid. A certain proportion of errors may, however, be accepted. This procedure allows us to eliminate the insignificant branches of a hierarchical classification.

5.10. Hierarchical classification reminder

(a) Accelerated algorithms allow a hierarchical classification to be built on large data sets such as more than 10,000 individuals on a mainframe computer. The partitioning methods are not necessary useful for reducing the number of elements into a small number of classes.

(b) Two sets are involved in individuals–variables data sets: I (individuals) and J (variables). The algorithm can be performed on both of these sets, individually.

(c) The roles of the two sets I and J are not symmetric. A hierarchical classification on J is generally done to study the redundancy of variables, so that representative subsets of variables are selected (as in discriminant analysis). A hierarchical classification on I is generally done in order to discover typologies of elements and to study identification of elements.

(d) Classification methods are well adapted to building classes and to recognizing classes or elements.

(e) Computer programs that handle any classification method, classify supplementary elements, and interpret classifications are available in DACL (*cf.* Chapter 12).

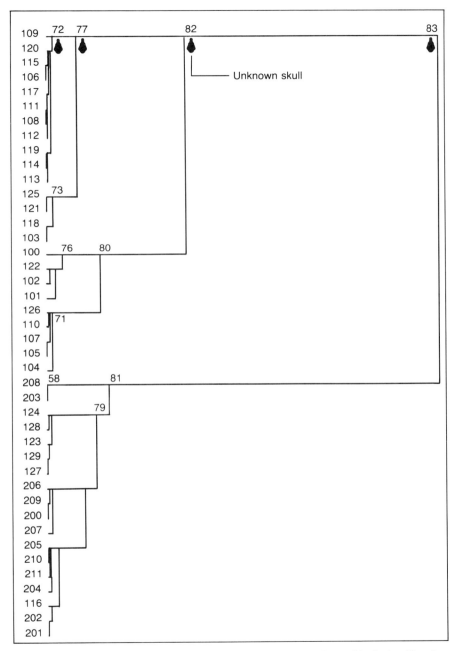

FIGURE 10.14. Classifying an unknown skull into the hierarchical classification of dog and wolf skulls. The assignment procedure was stopped at the level $N = 72$ (data in Appendix 2, §7).

6. Specific Applications

6.1. Cartography

One of the specific applications of classification methods is in cartography, because they allow a systems of classes (partitions, or nested partitions) to be built. Consider a set of individuals identified by areas on a map; a certain number of variables are associated with each area to give a basic data set, as presented in Fig. 10.15. There are different ways to handle the application in cartography. One method is to take only one variable and then divide the variable into classes (or categories). A hachure (or a color) is then assigned to each category. Then all the hachures (or colors) are transferred onto the map in the corresponding areas. The second way is to take into account all the variables simultaneously, not only one-by-one. For this, a factor analysis or a classification can be performed, as these are methods synthesizing information involved on a set of variables. From factor analysis we get factor axes, which are now considered as variables (taken one-by-one), and the previous cartography process is applied for each factor axis. Thus, a map synthesizing relationships between variables (given by factor analysis) is obtained. In a similar way, classification methods also produce partitions, which may be identified with qualitative variables. A hachure is assigned to each class of a partition, so that the cartography process previously described can then be applied. The applications of cartography in data analysis are numerous, covering various fields (taxonomy, economics, demography, geology, archeology, zoology, ecology, politics, history, geography, biology, etc.).

6.2. Taxonomy

Recall that taxonomy is the science, laws, or principles of classification; it is also the theory, principle, and process of classifying organisms into established categories such as phylum, order, family, genus, or species, described by common characteristics in varying degrees of distinction. Practical taxonomy involves determining hierarchical or nonhierarchical classifications from organisms described by a set of variables, followed by the possibility of classifying any unknown organism into the resulting classification. This is the purpose of the partitioning process of classifying supplementary elements. For more details, the reader is requested to study Jardine and Sibson (1971), Sokal and Sneath (1973), or Benzecri

6. Specific Applications

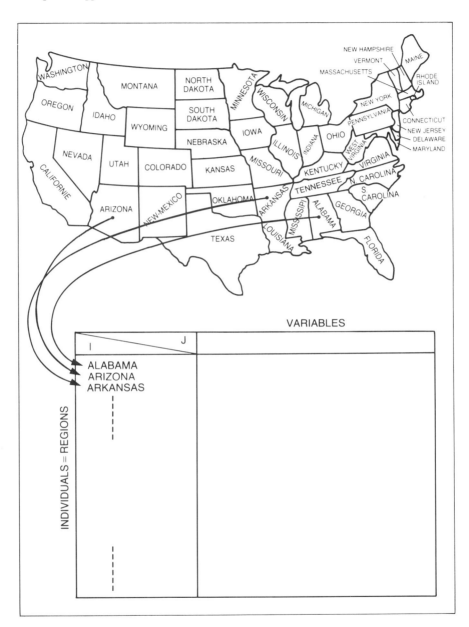

FIGURE 10.15. Data set associated with maps. Example of crimes data set (cf. Appendix 2, §15).

(1973), which are the basic references in taxonomic methods.

6.3. Pattern Recognition

Pattern recognition is related closely to data analysis and taxonomy in the following sense: any pattern must be described by a certain number of parameters or relationships between parameters; recognizing a pattern means identifying a pattern from its parameters only. The process of pattern recognition involves three steps:

Step (a). Determining, coding, and extracting parameters from the patterns to be recognized; it must be possible to process them automatically.

Step (b). Building the learning data sets. The learning data sets are those that are representative of all the patterns; they must be classified into partitions in order to obtain a dictionary of patterns; patterns must not overlap, because of the next *step*.

Step (c). The pattern recognition process itself.
Given a dictionary of patterns, any unknown pattern is first coded (as in *Step (a)*), and then compared to each member in the dictionary using the minimum distance criterion, which allows the unknown pattern to be identified. The previous process is exactly the same as that described for data analysis.

An example based on speech recognition by word (*cf.* Fig. 10.16(a), (b)) is now given.

Step (a). Determining parameters associated with a word.
A word is identified as a speech signal, which is transformed into a spectral coding as presented in Fig. 10.16(a). After spectral analysis of the signal, each word is represented by a rectangular matrix whose dimensions are as follows: the dimension M depends on the word itself; the dimension n is the number of channels; the value of each cell in the matrix is the value of the signal's energy in each channel. The final representation of each word is similar to that given in Fig. 10.16(a) for the French word "PARENTHESE." The term Δt is the basic sampling time associated with words.

7. Case Studies   397

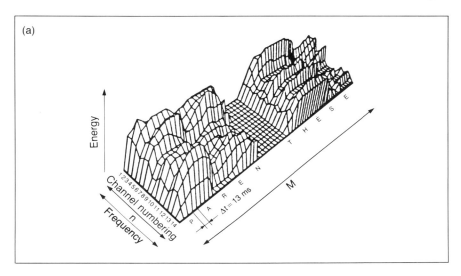

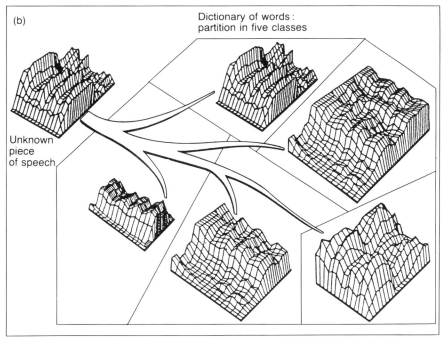

FIGURE 10.16. (a) Spectral representation of the word PARENTHESE according to 14 channels. (b) Representation of the pattern recognition process. The unknown word is compared with each of the five classes of words in the pattern dictionary.

Step (b). Learning data sets and dictionary of patterns.
All the basic words are represented by matrices as given in *Step (a)*, and then are classified in order to obtain a partitioned dictionary of words as represented in Figure 10.16 (b).

Step (c). Given the dictionary of words, the process for identifying any unknown word is the following: the unknown word is transformed into a matrix such as given in *Step (a)*. Then, data from the unknown word are compared with each class of words found in the dictionary of patterns as represented in Fig. 10.16(b). So, the unknown word is identified as the word with which it has been compared.

6.4. Statistical Information Systems

The study of information systems such as time samples, surveys, census, opinion polls, management data files, marketing questionnaires, and databases can be made by exploratory and multivariate data analysis methods such as those presented in this book, and thus by classification methods. Several case studies are given in Section 7 to show the reader what type of studies can be done with classification methods on his own data.

7. Case Studies

7.1. Quality of Service Data Set

The quality of service in the telephone network is based on performance indicators (*cf.* Appendix 2, §14). Twenty-one regions of France are involved as statistical units, and seven indicators are retained to represent the quality of service. A hierarchical classification of regions was built from the previous data set for the year 1984. The contributions for interpreting classes were computed and transferred into graphics. The results are shown in Fig. 10.17. The significant indicators are plotted near each class concerned, in order to express the significance of the subdivision into classes. For example, at the level $N = 40$, the significant indicators are IZAA, TCR, and TCOM; these indicators thus explain the separation between the class $N = 38$ and the class $N = 36$, which form the class $N = 40$ by union. From that classification, it is useful to build a series of maps since each statistical unit represents a region of France (*cf.* Fig. 10.18). The first map corresponds to a partition of France into two

7. Case Studies

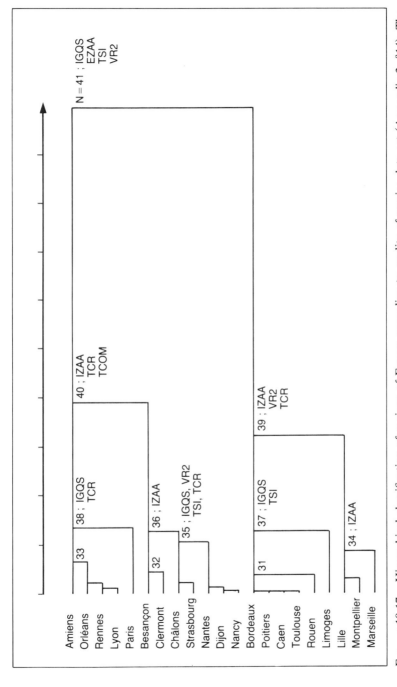

FIGURE 10.17. Hierarchical classification of regions of France according to quality of service data set (Appendix 2, §14). The uppermost significant indicators are assigned to the nodes.

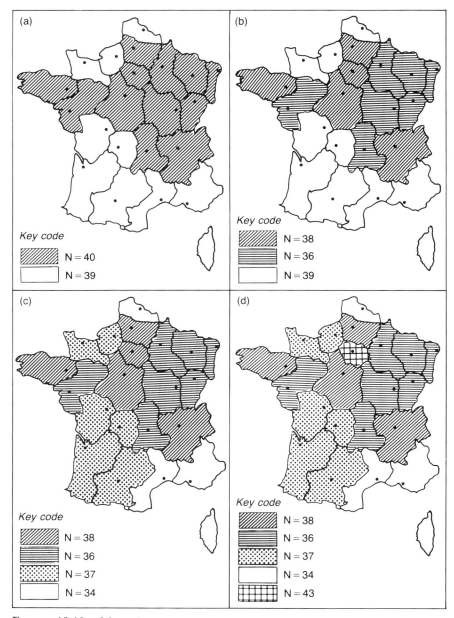

FIGURE 10.18. Maps from the hierarchical classification of regions of France built from the quality of service data set: (a) partition into two classes; (b) partition into three classes; (c) partition into four classes; (d) partition into five classes. The data are explained in Appendix 2, §14.

7. Case Studies

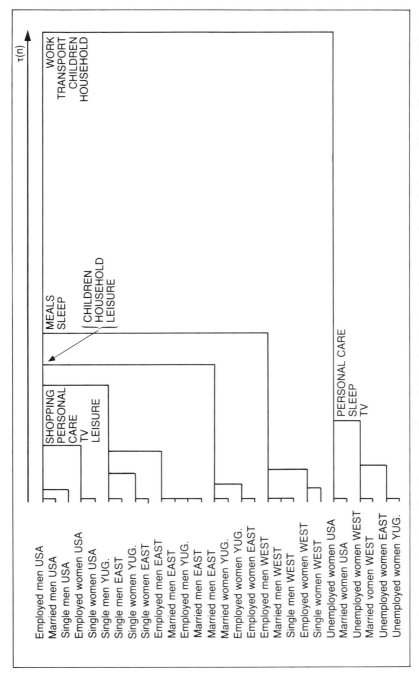

FIGURE 10.19. Family timetables data set. Hierarchical classification of groups of populations, with significant activities in terms of contributions.

402   Classification of Individuals–Variables Data Sets

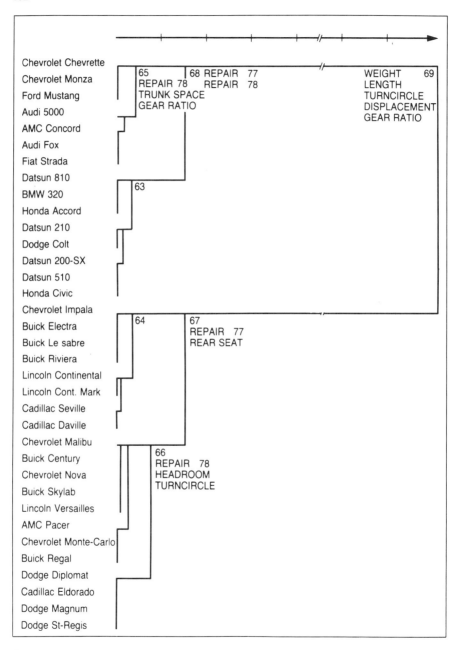

FIGURE 10.20. Car models data set. Hierarchical classification of cars, with significant variables.

7. Case Studies

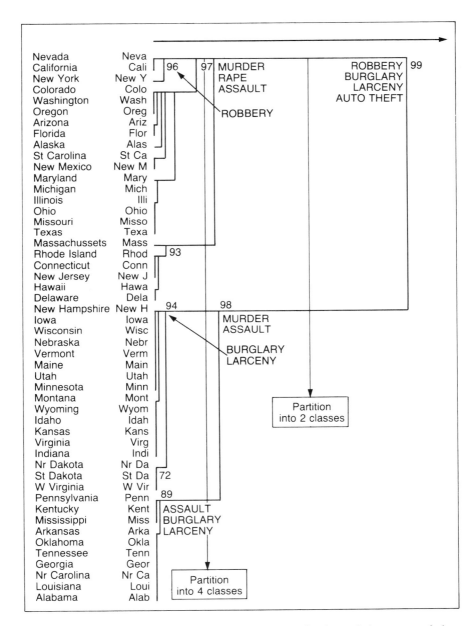

FIGURE 10.21. Crimes data set. Hierarchical classification of the states of the USA, and the most significant crimes variables associated with them.

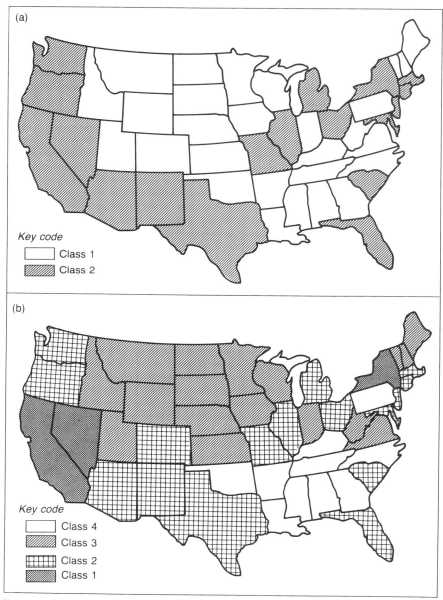

FIGURE 10.22. Crimes data set. Maps associated with the hierarchical classification of the states of the USA: (a) map with a two-class partition; (b) map with a four-class partition.

7. Case Studies

classes given by a hierarchical classification procedure; the second map corresponds to a partition of France into three classes; the third map corresponds to a partition into four classes; the fourth cartogram corresponds to a partition into five classes. This series of maps highlights the hierarchical process and allows the analyst to study the data in more detail.

7.2. Family Timetables Data Set

The family timetables data set is based on 10 daily activities observed on 28 groups of populations (cf. Appendix 2, §10). The hierarchical classification was done on the groups of populations, and the most significant variables are plotted near each node concerned (cf. Fig. 10.19). The graphic is self-explanatory.

7.3. Car Models Data Set

The car models data set involved technical variables observed on a series of cars (cf. Appendix 2, §1). The hierarchical classification was done on types of cars, and then the most significant variables were then plotted near each node concerned (cf. Fig. 10.20).

7.4. Crimes Data Set

The crimes data set contains data on the occurence of seven types of crimes in the United States of America in 1977 (cf. Appendix 2, §15). The units of observation are the individual states. The hierarchical classification was done on the states; then the most significant crimes were plotted near each class of states involved in the hierarchical classification (cf. Fig. 10.21). Then, two maps were drawn, the first (cf. Fig. 10.22(a)), shows a map using the two classes from the first stage of the classification; the second (cf. Fig. 10.22(b)) shows a map using the four classes from the third stage of the classification.

Chapter 11
Classification and Analysis of Proximities Data Sets

1. Introduction

Two types of data sets are distinguished: individuals–variables data sets with which a cloud of points can be associated, and proximities data sets representing generally distances or similarities. The latter ones are often used in psychometrics, taxonomy, humanities, natural sciences, or biology. This means that proximities data sets have a more restricted domain of application than individuals–variables data sets. However, data must be explored whatever their format. In this chapter, proximities data sets are described in Section 2, followed by an examination of proximities computations coming from individuals–variables data sets in Section 3, and then proximities data sets are studied according to three points of view: elementary description (Section 4), factor analysis (Section 5), and classification (Section 6).

2. Proximities Data Sets

Proximities data sets are sets of numerical values corresponding to a distance or a similarity matrix between two sets of variables or individuals. In Appendix 2, some examples of such data sets are given (§7, §16). Their general format is presented in Table 11.1.

It is interesting to note that a distance (or a similarity) matrix involves only one set (I or J), whereas an individuals–variables data set involves two sets (I and J). There are different ways to obtain such a distance matrix, but only two are indicated in what follows. In the first method, the distance matrix is obtained directly from an experiment or as a result of an observation procedure; in the second method, the distance matrix is

TABLE 11.1. Model of a proximities data set.

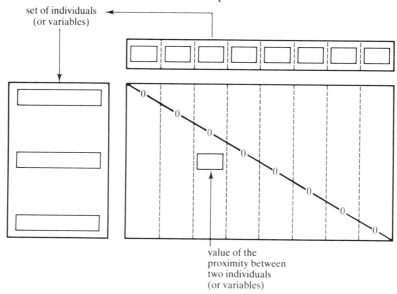

computed from individuals–variables data sets. Whatever method is used to obtain the distance matrix, its analysis and classification will be the same. Before studying this, some examples of distance computations are given in the next section.

3. Proximities Data Sets from Individuals–Variables Data Sets

3.1. Computations from Quantitative Variables Data Sets

Consider the quantitative variables data set x_{IJ}, where I is the set of individuals and J is the set of quantitative variables. x_{ij} is the value of the variable j on the individual i. A standard distance between two points of I, denoted by i and i' can be given as follows:

$$d^2(i, i') = \sum_{j,j' \in J} q^{jj'}(x_{ij} - x_{i'j})(x_{ij'} - x_{i'j'}), \tag{1}$$

where $q^{jj'}$ is the quadratic form associated with the Euclidean reference space.

3. Proximities Data Sets from Individuals–Variables

According to the different values of $\{q^{jj'}; j, j' \in J\}$, different distance formulas are obtained:

for $\begin{cases} q^{jj'} = 0 & \text{for } j \neq j', \\ q^{jj'} = 1 & \text{for } j = j', \end{cases}$ then $d^2(i, i') = \sum_{j \in J} (x_{ij} - x_{i'j})^2;$ (2)

for $\begin{cases} q^{jj'} = 0 & \text{for } j \neq j', \\ q^{jj} = \dfrac{1}{\sigma_j^2} & \text{for } j = j', \end{cases}$ then $d^2(i, i') = \dfrac{\sum_{j \in J}(x_{ij} - x_{i'j})^2}{\sigma^2}$ (3)

If
$$q^{jj'} = (\sigma^{jj'})^{-1},$$

where $\sigma^{jj'}$ is the variance–covariance matrix of the variables, then

$$d^2(i, i') = \sum_{j, j' \in J} (\sigma^{jj'})^{-1}(x_{ij} - x_{i'j})(x_{ij'} - x_{i'j'}); \qquad (4)$$

$d^2(i, i')$ is called the Mahalanobis distance.

The three previous distances are Euclidean distances. Consider now the following, based on the linear coefficient of correlation:

$$r(j, j') = \dfrac{\sum_{i \in I}(x_{ij} - \bar{x}_j)(x_{ij'} - \bar{x}_{j'})}{\sigma_j \sigma_{j'}};$$

then,
$$d(j, j') = 2(1 - r(j, j')). \qquad (5)$$

This "distance" does not satisfy the triangular inequality, so it is not a distance (from a mathematical point of view), but only a proximity. But the "distance" data processing does not need a mathematical distance.

3.2. Computations from Qualitative Variables Data Sets

From qualitative variables, a contingency data set k_{IJ} can be formed. From k_{IJ}, two distances can be computed as follows:

$$d^2(i, i') = \sum_{j \in J} (f_j^i - f_j^{i'})^2 / f_j; \qquad (6)$$

$$d^2(j, j') = \sum_{i \in I} (f_i^j - f_i^{j'})^2 / f_i. \qquad (7)$$

These two distances are distances between profiles, and they are associated with 2-D correspondence analysis (*cf.* Benzecri, 1973).

3.3. Computations from Rank Variables Data Sets

Consider a set of objects, denoted by J. These objects are ranked in increasing order by a set of individuals, denoted by I. This gives the data set denoted by r_{IJ}, where r_{ij} is the rank given by the individual i to the object j of J. The distances commonly computed are the following:

$$d(i, i') = 1 - r_s(i, i'), \qquad (8)$$

with

$$r_s(i, i') = \frac{6}{((n_J - 1)n_J)} \sum_{j \in J} (r_{ij} - x_{i'j});$$

where n_J is the number of elements of J. $d(i, i')$ is called the Spearman distance.

$$d^2(i, i) = 1 - r_T(i, i'), \qquad (9)$$

with

$$r_T(i, i') = \frac{2(n_+ - n_-)}{(n_J(n_J - 1))},$$

where n_+ is the number of times that $(r_{ij} - r_{ij'})$ has the same sign as $(r_{i'j} - r_{i'j'})$, and where n_- is the number of times that $(r_{ij} - r_{ij'})$ has the sign opposite to $(r_{i'j} - r_{i'j'})$. This distance is called the Kendall distance.

$$d^2(i, i') = n(G(i) \, \Delta G(i')), \qquad (10)$$

where n is the number of elements belonging to the set $G(i) \, \Delta G(i')$. The set $(G(i) \, \Delta G(i'))$ is determined as follows: The graphs $G(i)$ and $G(i')$ can be associated with the ranks $\{r_{ij}; j \notin J\}$ and $\{r_{i'j}; j \in J\}$, where

$$G(i) = \{(j, j'): j \neq j'; j > j'; r_{ij} < r_{ij'}\},$$

$$G(i') = \{(j, j'): j \neq j'; j > j'; r_{i'j} < r_{i'j'}\};$$

Δ is the symmetric difference between sets.

3.4. Computation from Logical Variables Data Sets

Consider a logical data set k_{IJ}, where k_{ij} is equal to 1 or 0. From k_{IJ}, different distance matrices can be computed, which are generally used in numerical taxonomy (cf. Sokal and Sneath, 1973). Consider the following

3. Proximities Data Sets from Individuals–Variables

notations:

$00_{ii'}$ = number of times that i and i' lack an attribute j of J simultaneously;

$11_{ii'}$ = number of times that i and i' possess an attribute j of J simultaneously;

$01_{ii'}$ and $10_{ii'}$ = number of times that one element possesses an attribute j and the other does not possess it.

Then,

$$00_{ii'} + 11_{ii'} + 01_{ii'} + 10_{ii'} = n_J \quad \forall i, i' \in I.$$

From this, the following distance formulas are formed:

$$d(i, i') = 1 - s_{ii'}, \tag{11}$$

with:

$$s_{ii'} = \frac{11_{ii'}}{11_{ii'} + 01_{ii'} + 10_{ii'}} ; \text{ (Jaccard)}$$

$$s_{ii'} = \frac{2 \cdot 11_{ii'}}{2 \cdot 11_{ii'} + 01_{ii'} + 10_{ii'}} ; \text{ (Czekanovski)}$$

$$s_{ii'} = \frac{11_{ii'} + 00_{ii'}}{n_J} ; \text{ (Sokal and Michener)}$$

$$s_{ii'} = \frac{11_{ii'}}{n_J} ; \text{ (Russel and Rao)}$$

$$s_{ii'} = \left(\frac{11_{ii'}}{2}\right)\left(\frac{1}{n_i} - \frac{1}{n_{i'}}\right); \text{ (Kulczinski)}$$

$$s_{ii'} = \frac{11_{ii'}}{11_{ii'} + 2(01_{ii'} + 10_{ii'})} ; \text{ (Sokal and Sneath)}$$

$$s_{ii'} = \frac{11_{ii'} + 00_{ii'}}{11_{ii'} + 00_{ii'} + 2(01_{ii'} + 10_{ii'})} . \text{ (Rogers and Tanimoto)}$$

where n_i (resp. $n_{i'}$) are the frequency of i (resp. i'). Other formulas are given in Sokal and Sneath (1973).

4. Elementary Description of Proximities Data Sets

4.1. Numerical Characteristics

A distance can be viewed as a standard quantitative variable, in which case the set of individuals is the set of pairs of elements (i, i'). The statistical characteristics such as the average, the standard deviation, the median, the interval quartile range, etc., can be computed for this quantitative variable. For details, the reader is referred to Chapter 3.

4.2. Graphics

Since the distance can be viewed as a quantitative variable, graphics such as dispersion boxes, histograms, hachured tables, or distance matrices can be plotted. For details, the reader is referred to Chapter 3.

5. Factor Analysis of Proximities Data Sets

5.1. Introduction

The question is: Although a "distance" matrix does not represent a Euclidean distance, can a factorial representation of such a distance be plotted so that it is as interpretable as a typical factor analysis of a individuals–variables data set? The method presented in the following section is from Benzecri (1973).

5.2. Notations and Formulas

Consider a finite set I, and a symmetric function d on $I \times I$, i.e., $d_{ii'} = d_{i'i}$. d is supposed to be positive, and $d_{ii} = 0$. Although d does not satisfy the triangle inequality, $d_{ii'}$ is considered as a squared distance between i and i' of I. Consider also a set of masses $\mu_J = \{\mu_i ; i \in I\}$ whose total mass is μ. f_I is given by $f_I = \{f_i = \mu_i/\mu ; i \in I\}$. The main principles of the factor analysis procedure are as follows: The cloud of points associated with the distance d, and provided with the quadratic form m, is the set of points, denoted by M_i, where

$$\forall i, i' \in I, \quad m(\overrightarrow{M_i M_{i'}}, \overrightarrow{M_i M_{i'}}) = d_{ii'}.$$

The principle is to consider the points M_i as points in Euclidean space where the distance between two points is $d_{ii'}$. Computation of the

5. Factor Analysis of Proximities Data Sets

quadratic form is based on this assumption. To compute m, the following preliminary computations are required:

- computation of $d_{\cdot i}$ for each i of I:

$$d_{\cdot i} = \sum_{i' \in I} f_{i'} d_{ii'};$$

- computation of $d_{\cdot\cdot}$:

$$d_{\cdot\cdot} = \sum_{i,i' \in I} f_i f_{i'} d_{ii'};$$

and then

$$m_{ii'} = m(\overrightarrow{GM_i}, \overrightarrow{GM_{i'}}) \quad \text{for each pair } i, i' \in I$$
$$= (-\tfrac{1}{2})(d_{ii'} - d_{\cdot i} - d_{\cdot i'} + d_{\cdot\cdot}),$$

where m is the matrix of scalar products of the vectors joining the center of gravity G and the points M_i of I, which have been chosen such that $\|M_i M_{i'}\|^2 = d_{ii'}$.

Finally, the factor axes are computed; the factors $\phi(I)$ are functions solving the following system of equations:

$$\sum_{i' \in I} \mu_i \phi_{i'} m(\overrightarrow{GM_i}, \overrightarrow{GM_{i'}}) = \lambda \phi_i \quad \text{for } i \in I.$$

These factors must be normalized as follows:

$$\sum_{i \in I} \mu_i \phi_i \phi_{i'} = \delta(\phi, \phi')$$

where

$$\delta(\phi, \phi') = 1 \text{ if } \phi \neq \phi',$$
$$= 0 \text{ if not.}$$

An approximate representation is obtained with the first p factors. The representation is given by points N_i computed as follows:

$$\overrightarrow{GN_i} = \{\sqrt{|\lambda(\phi)|} \cdot \phi_i; \phi \in \{p \text{ retained factors}\}.$$

The initial distance can be reconstructed from the factor coordinates as in a standard factor analysis.

In factor analysis proximities data sets, supplementary elements can be also classified, as follows: If $\mu_i = 0$, the distances $d_{ii'}$ have no influence on the factor computations, and so the analysis is done without the elements with zero masses. The factors are computed according to the following formula:

$$\phi_{i_s} = \frac{1}{\lambda(\phi)} \sum_{i \in I} \mu_i \phi_i m(\overrightarrow{GM_{i_s}}, \overrightarrow{GM_i}).$$

Then, the points i_s are classified as supplementary elements onto the graphics containing the basic points.

Note. If $\lambda_\alpha \geq 0$ for any value α, the Euclidean representation is exact; if there is at least one negative eigenvalue, then the representation is only approximate.

5.3. Case Study: The Family Timetables

From the family timetables data set (*cf.* Appendix 2, §10), a distance data set was computed between the population groups (*cf.* Appendix 2, §12). The formula used for the proximities factor analysis was the one used in correspondence analysis, and gave the following results:
  (a) Eigenvalues and associated parameters (Table 11.2).
  (b) Graphical representations. Only the representation in the factor space (1, 2) is given (*cf.* Fig. 11.1).
• The first axis ($\tau_1 = 25.7\%$) divides the types of activities of men and women. At the extreme right of the first axis are the women who have no

TABLE 11.2. Eigenvalues and parameters.

Num	Iter	Eigenvalues	Percentage	Cumul
1	0	0.53551	25.741	25.741
2	0	0.29166	14.020	39.760
3	1	0.22297	10.718	50.478
4	2	0.20329	9.772	60.250
5	1	0.12332	5.928	66.178
6	1	0.09415	4.525	70.703
7	1	0.7664	3.684	74.387
8	1	0.6508	3.128	77.515
9	1	0.05461	2.625	80.140
10	1	0.04768	2.292	82.432
11	1	0.04152	1.996	84.428

5. Factor Analysis of Proximities Data Sets

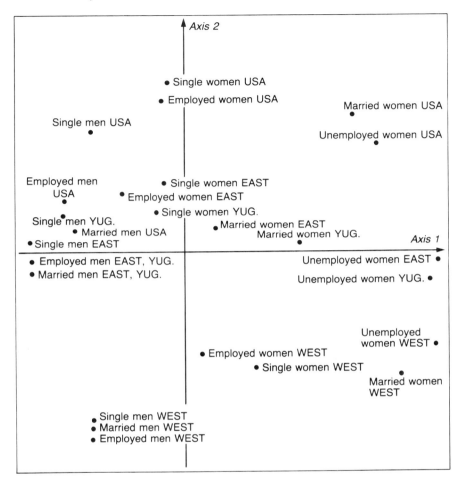

FIGURE 11.1. Factorial representation of a proximities data set, from family timetables data set, in the factor space (1, 2).

professional activity. The activities of women are more diversified than the activities of men, which are more concentrated over a few activities.

• The second axis ($\tau_2 = 14.02\%$) divides the population groups from the western countries of Europe and the USA. This corresponds to a different way of life. In the countries of western Europe, the employed men generally returned home for lunch (in 1965), while in the USA, the employed men and women had their lunch at/near their place of work.

The reader can explore the graphic himself.

6. Classification of Proximities Data Sets

6.1. Partitioning Methods

Partitioning methods have been less frequently applied to proximities data sets than to individuals–variables data sets. The ones most used on data sets are the k-means of MacQueen (1967), Isodata from Ball and Hall (1965), and dynamic clouds from Diday (1980). All of the partitioning algorithms can be extended to proximities data sets, but instead of computing the center of gravity, we seek the element of the class whose deviations from the other elements of the class is minimal. This element plays the role of the center of gravity.

6.2. Hierarchical Methods

As with the partitioning methods, hierarchical methds can be performed on proximities data sets. The distances can be computed from individuals–variables data sets or obtained from any reference space. The flowchart of the main procedures is given in Fig. 11.2.

According to the flowchart given in Fig. 11.2, the clustering methods available for proximities data sets are the same as those for individuals–

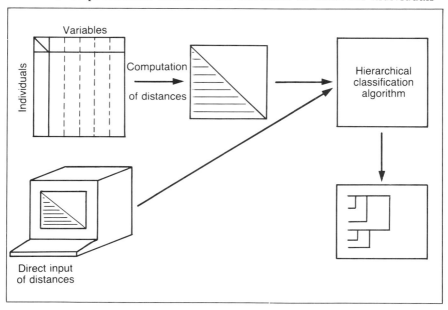

FIGURE 11.2. Flowchart of operations involved in clustering procedures.

variables data sets, which were presented in Chapter 10, since most of them are based on a distance matrix and not on an individuals–variables data set. But, because of the importance of individuals–variables data sets, they were presented in a manner devoted to the classification of individuals–variables data sets. Some of them are excluded from proximities data set clustering, such as any method using an Euclidean framework (center of gravity, methods of variance, etc.). The methods used for proximities data sets are summarized as follows:

> Euclidean proximities $\Rightarrow$ ([CAH.DIST] or [CAHA.DIST];
> data set $\delta_{\text{uim}}, \delta_{\text{usm}}, \delta_{\text{moy}}; d_{ii'}$, no masses);
>
> Non-Euclidean proximities $\Rightarrow$ ([CAH.DIST] or [CAHA.DIST];
> data set any $\delta, d_{ii'}$, mass m_i).

6.3. Case Study: The Family Timetables

Consider again the family timetables data set (cf. Appendix 2, §10, §12). Based on the distances data set, hierarchical clustering with the average criterion was performed, giving the results described in Fig. 11.3. The hierarchical classification highlights six main levels (or subdivisions): the first level (from $N = 55$) distinguished the unemployed women and the other groups (men or women); the second level (from $N = 54$), the employed women and the men; the third level (from $N = 53$), two types of employed women (single and married); the fourth level (from $N = 52$), the men from the USA and the other men.

The reader can explore the graphical tree in more details.

7. Computation of Contributions

The advantage of methods applied to individuals–variables data sets is that contributions can be computed. These can be interpreted in order to highlight the contribution of each element to the variance of an axis or subdivision. The same useful formulas do not exist for the classification and analysis of proximities data sets. This explains why these methods are less frequently used than those involving individuals–variables data sets.

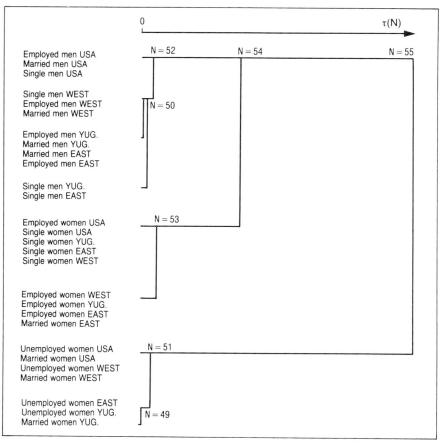

FIGURE 11.3. Family timetables data set. Representation of the hierarchical classification based on the average criterion.

8. Conclusion

• The possibilities for proximities data sets are less than those for individuals–variables data sets because of the absence of simultaneous representations of the sets of individuals and variables.

• The computer programs are available(*cf.* Chapter 12).

Chapter 12

Computer Aspects of Exploratory and Multivariate Data Analysis

1. Place of Exploratory and Multivariate Data Analysis in Statistics

Two parallel paths have existed in data processing; the first one has been devoted to databases and database management, often referred to as information systems and information system management; the second one has been devoted to data analysis and, more generally, exploratory and multivariate data analysis developed by Tukey, Hayashi, and Benzecri around 1960. Nowadays, these two paths can be integrated into the same system, known as a system of database analysis and management.

The differents steps associated with the system are synthesized in Fig. 12.1. This statistical process is considered as a mental tool that can be applied to any statistical study, leading to decision making. In any study, researchers, engineers, and businessmen face the same issues:

- fixing objectives;
- gathering data;
- summarizing by mathematical treatments;
- making decisions and, based on these decisions, fixing new objectives.

These issues provide the real meaning behind the statistical process described in Fig. 12.1, which shows the procedure for an interactive dialogue between a user and his data.

At each point, the analyst can go backwards or forwards according to the previous results, as illustrated in the process associated with 2-D correspondence analysis (Fig. 12.2). What are the actual computer problems in exploratory and multivariate analysis? The only problem is

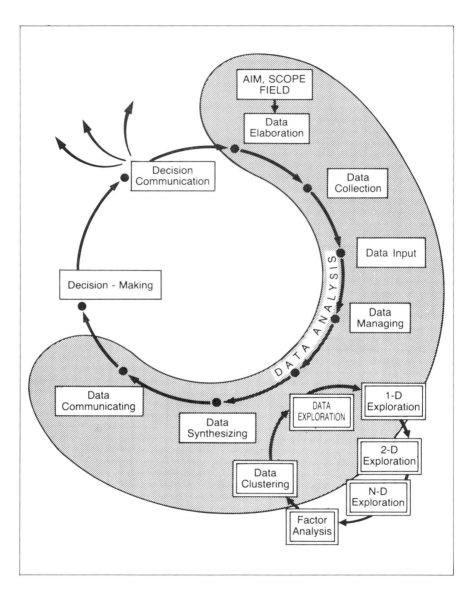

FIGURE 12.1. Flowchart for data analysis processing.

1. Place of Data Analysis in Statistics

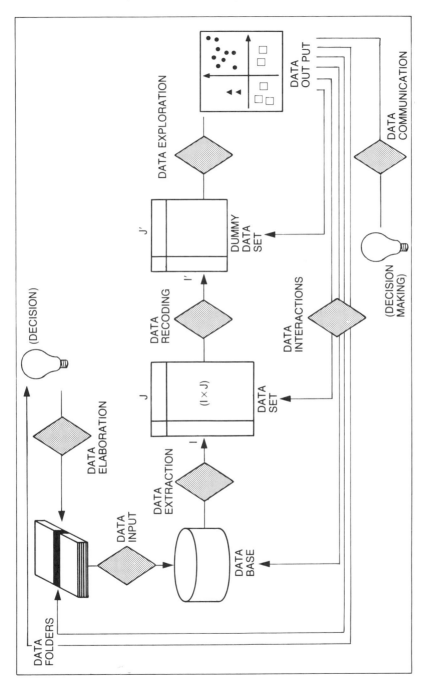

FIGURE 12.2. Data exploration and interactions.

to gather in a single computer system all the functions associated with the management and analysis of statistical data. The method proposed here answers this requirement, providing in addition the users with the means of dialoguing with his data.

2. Basic Features for an Exploratory and Multivariate Data Analysis Software

2.1. Data Sets

The data sets involved are mainly associated with two types of data sets described in previous chapters: individuals–variables data sets and proximities data sets. Surveys time samples, statistical files, databases, questionnaires, opinion polls, measurement campaigns, or experimental results lead to one of these two previous formats. If time is considered as a specific variable, then the data sets are transformed into data cubes or proximities cubes, but this does not change the data processing technique.

2.2. Data Management

Database management systems are generally used, but statistical databases need some other data management functions such as the following:

- creating dummy variables and dummy data sets:
  —forming dummy logical variables from qualitative variables;
  —forming dummy qualitative variables from quantitative variables according to different criteria (equal range, equal frequency, boundaries fixed by the user, etc.);
  —forming dummy logical variables from multiple forms variables;
  —assembling contingency data sets;
  —assembling chronological data sets.
- all other data management functions, such as updating, renaming, merging, sorting, aggregating, checking, coherency testing, etc.

2.4. Contemporary Data Analysis

Here are selected the necessary features for any statistical database analysis:

- 1-D data analysis (*cf.* Chapter 3);
- 2-D data analysis (*cf.* Chapter 4);
- Elementary N-D data analysis (*cf.* Chapter 5);
- Principal components analysis (*cf.* Chapter 7);
- 2-D correspondence analysis (*cf.* Chapter 8);
- N-D correspondence analysis (*cf.* Chapter 9);
- Partitioning and hierarchical clustering (*cf.* Chapter 10);
- Classification and analysis of proximities data sets (*cf.* Chapter 11).

2.5. Conclusions

How does the previous data analysis system differ from all the other statistical packages? First, it is only devoted to "exploratory and multivariate data analysis," while the statistical packages contain generally many other techniques. The data analysis functions are limited, and do not interact with data. Second, many of them do not integrate new developments in data analysis, such as 2-D or N-D correspondence analysis, or hierarchical clustering contributions, because it is rather difficult to introduce any novelty in such packages.

In the next section, we present the different sources enabling the previously described computer system to be built.

3. Data Analysis Libraries

3.1. Introduction

The main sources for building a new computer-aided data multivariate analysis system are from the Department of Statistics of the University of Paris VI, whose head is Benzecri. However, for 1-D to N-D descriptive analyses, the reader is referred to different statistical packages such as STATA, SAS, SPSS, BMDP, SYSTAT, S, STATGRAPHICS, etc. Here, only the sources concerning multivariate data analysis are presented; they are gathered in the data analysis library called DACL

(Data Analysis and Classification Library, from Jambu and Lebeaux, 1983). Before presenting a summary, a historical outline of data analysis is given.

3.2. Historical Outline

The first computer program for correspondence analysis was written by Escofier-Cordier in 1964, based on the correspondence formulas. From 1965 to 1969, Friant and Leroy improved the initial computer program, adding the point plotting associated with 2-D correspondence. In 1969, Lebeaux included the computations of contributions. From 1969 to 1979, different improvements were made by different authors on the diagonalization computer program, and the most circulated version was by Tabet. For proximities data analysis, the first computer program was by Nora in 1968, and it was improved by Jambu in 1976. The classification was known due to the reference text of Sokal and Sneath (1963). In 1968, Benzecri proposed a general algorithm of hierarchical classification with many criteria. The related computer program written by Jambu in 1969 contained many options for aggregation criteria and distances. In 1973, Jambu added the contributions to hierarchical classification, as well as supplementary elements. In 1972, Diday introduced the dynamic clouds algorithm based on an algorithm of moving centers. All these programs were improved year by year by students or researchers, and many versions exist worldwide.

In 1976, for a data analysis summer school, Jambu gathered all the computer programs of the Department of Statistics devoted to exploratory and multivariate data analysis in a library called LTSM. It was finally published in English in 1983 under the name DACL (Data Analysis and Classification Library—Jambu and Lebeaux, 1983).

3.3. Data Analysis and Classification Library—DACL

The actual version of the library contains 38 independent programs, which altogether include more than 300 subroutines written in

3. Data Analysis Libraries

FORTRAN, involving approximately 30,000 statements. The user's guide contains six sections for each program:

Section 1. Aim and Scope;
Section 2. Notation and Formulas;
Section 3. Architecture;
Section 4. Computer Use;
Section 5. Application to a Case Study;
Section 6. Source Listings.

The table of contents associated with DACL is given in Table 12.1 on page 426.

4. Future Prospects

Based on the experience of data analysis over the last twenty years, it is rather difficult to foresee its future development. Nevertheless, some principles can be highlighted:

• Whatever the progress of the computer technology, any "data analysis study" needs human interpretation. It is too soon to gather, explore, and make decisions *automatically*. Some people speak of statistical expert systems, but these can only help in decision making.
• Since the 1960s, needs have changed. Data analysis was initially for researchers in laboratories; it is now considered a standard tool for engineers, teachers, and students. Each category of user does not have the same requirements, and so different data analysis systems are needed.
• The development of computer technology has changed the rules of building computer systems. They have been built by computer scientists and not statisticians, but any data analysis system must be determined by specialists in data analysis in order to obtain a satisfactory and appropriate interactive system.

TABLE 12.1. Table of contents of DACL.

Preliminary description of data sets
1. [DEDOUBLE]—Doubling a score data set
2. [DISJONCT]—Putting qualitative data set in disjunctive form
3. [RECOD1]—Recoding a set of heterogeneous data –1–
4. [RECOD2]—Recoding a set of heterogeneous data –2–
5. [DESCR.CORR]—Usual description of a correspondence data set
6. [CORR.MULT]—Multiple correspondence data set
7. [DESCR.VAR]—Usual description of a qualitative variables data set
8. [HISTO]—Histogram
9. [REORDRE]—Rearranging rows and columns of a data set
10. [REDUIT]—Transformation of a cloud of points
11. [PONDERATION]—Choice of weights for heterogeneous data sets
12. [TRANSPOSE]—Transposition of a data set
13. [DIST.LOGIQUE]—Calculating distances from a logical data set
14. [DISTANCES]—Calculating distances from a quantitative variables data set

Data analysis and classification programs
15. [ANAFAC.CORR]—2-D correspondence analysis
16. [ANAFAC.COMP]—Principal components analysis
17. [ANAFAC.DIST]—Proximities factor analysis
18. [ANAFAC.SOUS.TABLEAUX]—2-D correspondence analysis for a large data set
19. [CAH.DIST.]—Hierarchical classification from proximities data set
20. [CAHA.EUCL1]—Hierarchical classification of individuals–variables data sets
21. [CAHA.INFORM]—Hierarchical classification based on information theory criterion
22. [SQUELETTE]—Tree representation associated with hierarchical classification
23. [DYNAMIC CLOUDS]—Partitioning based on moving centers algorithms
24. [BOULES OPTIMISEES]—Partitioning based on moving balls algorithms
25. [REGR. BOULE]—Regression using factor analysis results
26. [CONTRI.ANAFAC.CORR]—Supplementary analysis of correspondence analysis
27. [CONTRI.ANAFAC.COMP]—Supplementary analysis of principal components analysis
28. [REP.CAH.ANAFAC.]—Graphics with factor analysis and hierarchical classification results
29. [REP.PART.ANAFAC]—Graphics with factor analysis and partitioning results
30. [IMPRIM.CAH.]—Supplementary graphics associated with hierarchical classification
31. [REORDRE.CORR.]—Rearranging rows and columns of data sets according to factor analysis results.
32. [HISTO.COR.CAH.]—Histograms associated with classes of hierarchical classification
33. [CONTRI.CAH.FACT.]—Calculations of contributions between hierarchical classification and factor analysis
34. [CONTRI.CAH.CORR.]—Calculations of contributions associated with correspondence hierarchical classification
35. [CONTRI.CAH.COMP.]—Calculations of contributions associated with Euclidean hierarchical classification
36. [CONTRI.PART.FACT.]—Calculations of mutual contributions between partitioning and factor analysis
37. [COMPLET.CAH.QUADR.]—Classifying supplementary elements on a hierarchical classification
38. [SIM.CAHA.EUCL1.]—Simulation to study the level of hierarchical classification.

Appendix 1. List of Notations

1. General Notations

$I, \{i_1, i_2, \ldots, i_n\}$: set of statistical units, individuals; population

$J, \{j_1, j_2, \ldots, j_p\}$: sets of variables, axes, forms

$I \times J$: product set of I and J

$i, i_1, i_2, \ldots, i_n$: elements of I

$j, j_1, j_2, \ldots, j_p$: elements of J

$(i, j), ij$: indices of $I \times J$

R: space of real numbers

R_I, R_J: Euclidean spaces associated with I, J on R

R^n, R^p: n-, p-dimensional spaces on R

$\emptyset$: empty set

$\mathcal{P}(I)$: set of subsets of I

$\cup$: union of sets

$\cap$: intersection of sets

Δ: symmetric difference of sets

$\in$: is a member of

$\subset$: inclusion of sets

$\leq$: less than or equal

$\geq$: greater than or equal

$\exists$: there exists

$\forall$: for all

$\sum_{i \in I}$: sum of terms indexed by i of I

n_I, Card I: number of elements of I

X': transposed matrix of X

X^{-1}: inverse matrix of X

$d^2(i, i')$: squared distance between i and i'

$\|i - i'\|^2$: squared distance between i and i' relative to the Euclidean norm $\| \ \|$

2. Specific Notations from Chapters 1 to 5

X, Y: statistical variables

X_i: value of X at the unit i of I

p_i: weight associated with the unit i of I

Q: quadratic mean of X

H: harmonic mean

G: geometric mean of X

X_α: α-mean of X

M_0: mode of X

M: median of X

$Q_1, Q_2, \ldots, Q_{k-1}$: k-quantiles of X

Q_1: first quartile

Q_3: third quartile

$D_1, D_2, \ldots, D_9$: deciles of X

$C_1, C_2, \ldots, C_{99}$: centiles of X

R: range of X

IQR: interquartile range of X

MD: mean deviation of X

Δ: mean difference of X

$s, s(X)$: standard deviation X

$s^2, s^2(X)$: variance of X

s_G: geometric deviation of X

Appendix 1. List of Notations

- C_Q: quartile coefficient of X
- V: coefficient of variation
- C_Y: Yule's coefficient of skewness
- C_K: Kelley's coefficient of skewness
- C_P: Pearson's coefficient of skewness
- μ_k: k^{th} centered moment associated with X
- β_1: Pearson's coefficient of skewness
- γ_1: Fisher's coefficient of skewness
- A_K: Kelley's coefficient of kurtosis
- β_2: Pearson's coefficient of kurtosis
- γ_2: Fisher's coefficient of kurtosis
- P_k: k^{th} percentile
- m_k: k^{th} moment of X
- $ij\ (i, j)$: indices of contingency data sets
- n_i: marginal count of i of I
- n_j: marginal count of j of J
- n_{ij}: number of elements in the cell (i, j) of the contingency data set
- f_{ij}: relative frequency of the cell (i, j) of the contingency data set
- f_i: marginal relative frequency of i of I
- f_j: marginal relative frequency of j of J
- χ^2: chi-squared; contingency coefficient; law of χ^2
- ϕ^2: phi-squared; Pearson's coefficient
- C: Cramer's coefficient
- T: Tschuprow's coefficient
- $Y = y_j$: variable X knowing the value y_j of Y
- $\eta^2_{X|y}$: correlation ratio of X knowing the value y of Y
- $r(X, Y), r$: linear coefficient of correlation between the two variables X and Y

3. Specific Notations from Chapters 6 to 9

X_{IJ}, X: individuals–variables data sets

X_{ij}: value of X at the cell (i, j)

(i, j): index of the data set X

n, n_I: number of elements of I

p, n_J: number of elements of J

X'_{IJ}, X': transposed individuals–variables data sets

$\lambda_1, \lambda_2, \ldots, \lambda_\alpha, \ldots$: eigenvalues of XX' and $X'X$

$u_1, u_2, \ldots, u_\alpha, \ldots$: eigenvectors of $X'X$

$v_1, v_2, \ldots, v_\alpha, \ldots$: eigenvectors of XX'

f^j_I: conditional frequency data set associated with I, knowing j; profile of j on I

f^I_J: profiles data set of the elements i of I on J

f^J_I: profiles data set of the elements j of J on I

$\mathcal{P}(I)$: simplex of the probability laws on I

$N(I), N_J(I)$: clouds of points of I set in the Euclidean space R_J

$N(J), N_I(J)$: clouds of points j of J set in the Euclidean space R_I

$d^2(i, i')$: squared distance between the two points i and i' of I

$d^2(f^i_J, f^{i'}_J)$: squared distance between the two points i and i' of I, identified by their profiles f^i_J and $f^{i'}_J$

$\rho^2(i)$: squared distance between the element i of I and the center of gravity of the reference cloud of points $N(I)$; eccentricity of the point i of I

$M^2(I), M^2(N(I)), M^2(N_J(I))$: second central moment of the set I; total variance of the set I

$F_\alpha(i)$: factor coordinate of i relative to the α-axis

$G_\alpha(j)$: factor coordinate of j relative to the α-axis

k_{IJ}: contingency data set; correspondence data set

Appendix 1. List of Notations

k_i: marginal count of i of I

k_j: marginal count of j of J

k_{ij}: element of k_{IJ}; count of (i, j)

k: total count associated with the data set k_{IJ}

k_I: set of marginal counts of i of I

k_J: set of marginal counts of j of J

f_{IJ}: frequency data set associated with $I \times J$

f_I: marginal frequency data set relative to I

f_J: marginal frequency data set relative to J

f_{ij}: element of f_{IJ}; relative frequency of (i, j)

f_i: relative frequency of i

f_j: relative frequency of j

f_j^i: conditional frequency of j knowing i

f_i^j: conditional frequency of i knowing j

f_J^i: conditional frequency data set associated with J knowing i; profile of i on J

f_I^j: conditional frequency data set associated with I knowing j; profile of j on I

α: index of the rank-order of the factor axis

λ_α: eigenvalue of $X'X$ and XX'; absolute contribution of the α-axis to the total variance of $N(I)$; variance of the α-axis

τ_α: relative contribution of the α-axis to the total variance of $N(I)$; rate of variance of the α-axis

$\phi_{\alpha j}$: α-coordinate of the point j of J

$\psi_{\alpha i}$: α-coordinate of the point i of I

$\text{ctr}_\alpha(j)$: relative contribution of the point j of J to the variance of the α-axis: $f_j \phi_{\alpha j}^2 / \lambda_\alpha$

$\text{cor}_\alpha(j)$: relative contribution of the α-axis to the eccentricity of the point j of J: $\phi_{\alpha j}^2 / \rho^2(j)$

$\text{inr}(j)$: relative contribution of the point j to the total variance of $N(J)$: $f_j \rho^2(j) / M^2(N(J))$

$\mathrm{qlt}_s(j)$: quality of representation of the point j in s factor axes

$\mathrm{ctr}_\alpha(i)$: relative contribution of the point i of I to the variance of the α-axis: $f_i \psi_{\alpha i}^2 / \lambda_\alpha$

$\mathrm{cor}_\alpha(i)$: relative contribution of the α-axis to the eccentricity of the point i of I: $\psi_{\alpha i}^2 / \rho^2(i)$

$\mathrm{inr}(i)$: relative contribution of the point i of I to the total variance of $N(I)$: $f_i \rho^2(i) / M^2(N(I))$

$\mathrm{qlt}_s(i)$: quality of representation of the point i of I in s factor axes

qge_s: global quality of representation in s factor axes

Q: set of q questions or categorical variables

q: question or categorical variable; member of Q

J_q: set of forms associated with the question q of Q

n_q: number of forms associated with the question q of Q

b_{JJ}: multiple contingency data set associated with a set of questions or categorical variables; squared matrix associated with J

$b_{jj'}$: element of b_{JJ}

b_j: marginal count of the point j of J

b: total count of $b_{jj'}$ from b_{JJ}

g_{JJ}: correspondence data set associated with k_{IJ}

$g_{jj'}$: element of g_{JJ}

g_j: marginal count of the point j of J

g: total count of $g_{jj'}$ from g_{JJ}

$N(J_q)$: cloud of the forms associated with the question q of Q

$M^2(N(J_q))$: second order central moment of the cloud of points $N(J_q)$

$G_\alpha(q)$: α-coordinate of the point q of Q

$\mathrm{ca}_\alpha(q)$: absolute contribution of the point q of Q to the variance of the α-axis

Appendix 1. List of Notations 433

$\text{ctr}_\alpha(q)$: relative contribution of the point q of Q to the variance of the α-axis

$\text{cor}_\alpha(q)$: relative contribution of the α-axis to the eccentricity of the point q of Q

$\text{inr}(q)$: relative contribution of the point q of Q to the total variance of cloud $N(J)$

$\text{qlt}_s(q)$: quality of representation of the point q of Q in s factor axes.

4. Specific Notations from Chapters 10 and 11

Card I, Card I: dimensions of the sets I, J

$\mathcal{P}(I) - \emptyset$: set of the nonempty subsets of I

Q, $Q(I)$: partition of I

$C(I)$: hierarchical classification of I

a, b, q, s, t: elements of $\mathcal{P}(I) - \emptyset$; classes of elements of I

$\sup_i \{x_i\}$: maximum value of the series $\{\{x_i\}; i \in I\}$

$\inf_i \{x_i\}$: minimum value of the series $\{\{x_i\}: i \in I\}$

Ter $C(I)$: set of terminal classes of $C(I)$, also called terminals of $C(I)$

Nod $C(I)$: set of nonterminal classes of $C(I)$, also called nodes of $C(I)$

Som $C(I)$: set of vertices of $C(I)$

n, N: indices of the nodes of $C(I)$

$v(n), v(N)$: level indices of the nodes n, N of $C(I)$

Suci$(n, C(I))$: set of the successive classes of the class n in $C(I)$

$a(n), a, A(N)$: notation for the eldest of the class n, one of the successive class of n

$b(n), b, b(N)$: notation for the youngest of the class n, the other successive class of n

Card a, Card b: number of elements of the classes a, b,

$N(I), N(J)$: clouds of points associated with I, J

Appendix 1. List of Notations

a_J, b_J: coordinates of the centers of the classes a, b, in the Euclidean space R_J

g_J: coordinates of the center of gravity of the cloud $N(I)$ in the Euclidean space R_J

m_i, $m_{i'}$, $m_{i''}$, ...: masses assigned to the elements i, i', i'', ...

m_a, m_b, m_s, m_t: masses assigned to the classes a, b, s, t

$F_\alpha(q)$, $\psi_\alpha(q)$: α-coordinates of the centers of the classes q of I

$G_\alpha(q)$, $\phi_\alpha(q)$: α-coordinates of the centers of the classes q of J

f_q: marginal frequency of the class q of I

f_q^i: conditional frequency of the class q of I knowing j of J

f_{qj}: relative frequency of the pair (q, j) of $Q \times J$

f_{QJ}: frequency law of $Q \times J$, where Q is a partition of I

$N(Q)$: cloud of the centers of the classes q of $Q(I)$

$M^2(I)$, $M^2(N(I))$: second order central moment of I

$M^2(q)$, $M^2(s)$, $M^2(t)$: second order central moments of the classes q, s, t of I

$M^2(Q)$: second order central moment of the partition Q of I

$M^2\{q, q'\}$: second order central moment of the partition into the two classes q and q' of $Q(I)$

$V^2(I)$, $V^2(N(I))$: variance of I; variance of the cloud $N(I)$

$V^2(Q)$: variance of the partition Q of I

$V^2(q)$, $V^2(s)$, $V^2(t)$: variance of the classes q, s, t of I

$d^2(i, i')$: squared distance between the two elements i and i' of I

$\|i - i'\|^2$: squared distance between the two elements i and i' of I, with respect to the Euclidean norm associated with R_J

$\|\cdot\|_L$: norm associated with the reference space R_J with centre L

$d^2(q, q')$: squared distance between the two centres of gravity of the classes q and q' of I

Appendix 1. List of Notations

$\|q - q'\|^2$: distance squared between the two centres of gravity of the classes q and q' of I with respect to the Euclidean norm associated with R_J

h: index of an algorithmic step

C_h, $C_h(I)$: hierarchical classification on I at the step h

Som C_h, Som $C_h(I)$: vertex of the hierarchical classification $C_h(I)$ at the step h

$\delta(a, b)$: distance between the two classes a and b of I

$\delta_{\text{uim}}(a, b)$: distance between the two classes a and b of I according to the single linkage criterion

$\delta_{\text{usm}}(a, b)$: distance between the two classes a and b of I according to the complete linkage criterion

$\delta_{\text{moy}}(a, b)$: distance between the two classes a and b of I according to the average linkage criterion

$\delta_{\text{med}}(a, b)$: distance between the two classes a and b of I according to the median linkage criterion

$\delta_{\text{bar}}(a, b)$: distance between the two classes a and b of I according to the centroid criterion

$\delta_{\text{mot}}(a, b)$: distance between the two classes a and b of I according to the second order central moment of a partition criterion

$\delta_{\text{var}}(a, b)$: distance between the two classes a and b of I according to the variance of a partition criterion

$\delta_{M^2}(a, b)$: distance between the two classes a and b of I according to the second order moment of the union of classes criterion

$\delta_{V^2}(a, b)$: distance between the two classes a and b of I according to the variance of the union of classes criterion

$\delta_{\text{ang}}(a, b)$: distance between the two classes a and b of I according to the angular similarity criterion

Appendix 1. List of Notations

$\delta^h(a, b)$: distance between the two classes a and b of I at the step h

(s_h, s'_h): pair of classes to be aggregated at the step h

$\text{Num}(n)$: numbering index of the nodes of the hierarchical classification; for the first node, $\text{Num}(n) = \text{Card}\,I + 1$; for the last node, $\text{Num}(n) = 2\,\text{Card}\,I - 1$

$\tau(n)$: rate of variance associated with the node n of $C(I)$

$\tau(Q_n)$: rate of variance of the partition Q established at the level of the node n of $C(I)$

$\rho^2(q)$, $\text{rho}^2(q)$: eccentricity of the class q; distance squared between the center of the class q and the center of gravity

$\text{inr}(q)$: rate of variance of the class q with respect the total variance of the cloud $N(I)$

$\text{cor}_j(q)$: relative contribution of the j-axis to the eccentricity of q of I

$\text{ctr}_j(q)$: relative contribution of the class q of I to the variance of the j-axis

$\text{qlt}_s(q)$: quality of representation of the class q of I in s axes

$\text{dp}(q, q')$: dipole(q, q'); pair of classes q and q' of I

$\text{dp}(a(n), b(n))$: dipole$(a(n), b(n))$; pair of classes $a(n)$ and $b(n)$ of I coming from the node n of $C(I)$

$\text{inr}(\text{dp}(q, q'))$: rate of variance of the dipole $\text{dp}(q, q')$ with respect to the total variance of the cloud $N(I)$

$\text{ctr}_j(\text{dp}(q, q'))$: relative contribution of the dipole $\text{dp}(q, q')$ to the variance of the j-axis

$\text{cor}_j(\text{dp}(q, q'))$: relative contribution of the j-axis to the distance between the two classes q and q' of I

$\text{qlt}_s(\text{dp}(q, q'))$: quality of representation of the dipole $\text{dp}(q, q')$ in s axes

Appendix 1. List of Notations

$v(n, j)$: mutual absolute contribution of the dipole $\mathrm{dp}(a(n), b(n))$ coming from the node n of $C(I)$ and the j-axis to the total variance of the cloud $N(I)$

$\sigma(I)_{JJ}$, $\sigma(I)$: variance–covariance matrix of the variables of J with respect to I

$\sigma(I)_{jj'}$: general term of $\sigma(I)$; covariance of the variables j and j' of J with respect to I

$\sigma(I)^{-1}$, $\sigma(I)_{jj'}^{-1}$: inverse matrix of $\sigma(I)$ and general term (j, j') of the matrix associated with Mahalanobis' metric

$\sigma(Q)$, $\sigma(Q)_{jj'}$: matrix of variances–covariances and covariance of the variables j and j' with respect to the partition $Q(I)$: variance within classes

$\sigma(I - Q)$, $\sigma(I - Q)_{jj'}$: matrix of variances–covariances and covariance of the variables j and j' of J with respect to the classes q of $Q(I)$: variance between classes

Appendix 2. Reference Data Sets

1. Cars Models

A set of 38 cars was studied based on 12 description variables in the view of comparing the cars. Data are given in Table A.1.

2. Marks of Students

A set of 18 students, labelled by E01, E02, ... , E20 (E07 and E17 were omitted), was studied according to their marks in the following subjects: Mathematics, Physics, English, Natural Sciences, Foreign Language, History (*cf.* Table A.2).

3. Statistics of Patents Registration

A set of nine industrial countries was selected for a study on the evolution of patents registration during the years 1980–1986. The data set is divided into two subsets; the first concerns the patents registration in the telecommunications branch (*cf.* Table A.3); the second concerns the patents registration in all of the economic branches mixed. (*cf.* Table A.4)

4. Preferences Given by Students

A set of eighteen students were asked to give their preferences for the following subjects: Mathematics, Physics, English, Natural Sciences, Foreign Language, History. (*cf.* Table A.5.)

TABLE A.1. Car models data set (extract). (From *Graphical Methods for Data Analysis*, by J.M. Chambers, W.S. Cleveland, B. Kleiner, and P.A. Tukey. Copyright © 1983 by Bell Telephone Laboratories Incorporated, Murray Hill, NJ. Reprinted by permission of Wadsworth & Brooks/Cole Advanced Books & Software, Pacific Grove, CA 93950.)

DESCRIPTION VARIABLES

Make & Model	Price $	Mileage mpg	Repair Record 1978	Repair Record 1977	Head-room in	Rear Seat in	Trunk Space cu ft	Weight lbs	Length in	Turn Circle ft	Displacement cu in	Gear Ratio
Chev. Chevette	3299	29	3	3	2.5	26.0	9	2110	163	34	231	2.93
Chev. Impala	5705	16	4	4	4.0	29.5	20	3690	212	43	250	2.56
Chev. Malibu	4504	22	3	3	3.5	28.5	17	3180	193	41	200	2.73
Chev. Monte Carlo	5104	22	2	3	2.0	28.5	16	3220	200	41	200	2.73
Chev. Monza	3667	24	2	2	2.0	25.0	7	2750	179	40	151	2.73
Chev. Nova	3955	19	3	3	3.5	27.0	13	3430	197	43	250	2.56
Datsun 200-SX	6229	23	4	3	1.5	21.0	6	2370	170	35	119	3.89
Datsun 210	4589	35	5	5	2.0	23.5	8	2020	165	32	85	3.70
Datsun 510	5079	24	4	4	2.5	22.0	8	2280	170	34	119	3.54
Datsun 810	8129	21	4	4	2.5	27.0	8	2750	184	38	146	3.55
Dodge Colt	3984	30	5	4	2.0	24.0	8	2120	163	35	98	3.54
Dodge Diplomat	5010	18	2	2	4.0	29.0	17	3600	206	46	318	2.47
Dodge Magnum XE	5886	16	2	2	3.5	26.0	16	3870	216	48	318	2.71
Dodge St. Regis	6342	17	2	2	4.5	28.0	21	3740	220	46	225	2.94
Fiat Strada	4296	21	3	1	2.5	26.5	16	2130	161	36	105	3.37
Ford Fiesta	4389	28	4	—	1.5	26.0	9	1800	147	33	98	3.15
Ford Mustang	4187	21	3	3	2.0	23.0	10	2650	179	42	140	3.08
Honda Accord	5799	25	5	5	3.0	25.5	10	2240	172	36	107	3.05
Honda Civic	4499	28	4	4	2.5	23.5	5	1760	149	34	91	3.30

MODELS OF CARS

Appendix 2. Reference Data Sets

Linc. Continental	11497	12	3	4	3.5	30.5	22	4840	233	51	400	2.47
Linc. Cont Mark V	13594	12	3	4	2.5	28.5	18	4720	230	48	400	2.47
Linc. Versailles	13466	14	3	3	3.5	27.0	15	3830	201	41	302	2.47
AMC Concord	4099	22	3	2	2.5	27.5	11	2930	186	40	121	3.58
AMC Pacer	4749	17	3	1	3.0	25.5	11	3350	173	40	258	2.53
AMC Spirit	3799	22	—	—	3.0	18.5	12	2640	168	35	121	3.08
Audi 5000	9690	17	5	2	3.0	27.0	15	2830	189	37	131	3.20
Audi Fox	6295	23	3	3	2.5	28.0	11	2070	174	36	97	3.70
BMW 320i	9735	25	4	4	2.5	26.0	12	2650	177	34	121	3.64
Buick Century	4816	20	3	3	4.5	29.0	16	3250	196	40	196	2.93
Buick Electra	7827	15	4	4	4.0	31.5	20	4080	222	43	350	2.41
Buick Le Sabre	5788	18	3	4	4.0	30.5	21	3670	218	43	231	2.73
Buick Opel	4453	26	—	—	3.0	24.0	10	2230	170	34	304	2.87
Buick Regal	5189	20	3	3	2.0	28.5	16	3280	200	42	196	2.93
Buick Riviera	10372	16	3	4	3.5	30.0	17	3880	207	43	231	2.93
Buick Skylark	4082	19	3	3	3.5	27.0	13	3400	200	42	231	3.08
Cad. Deville	11385	14	3	3	4.0	31.5	20	4330	221	44	425	2.28
Cad. Eldorado	14500	14	2	2	3.5	30.0	16	3900	204	43	350	2.19
Cad. Seville	15906	21	3	3	3.0	30.0	13	4290	204	45	350	2.24

MODELS OF CARS

TABLE A.2. Marks data set.
SUBJECTS

I \ J	Maths	Physics	English	Natural Science	Foreign Language	History
E01	18	13	2	11	9	7
E02	18	14	2	12	8	6
E03	14	11	6	10	11	9
E04	5	8	15	10	14	12
E05	14	14	6	12	8	6
E06	1	0	19	0	20	20
E08	8	6	12	8	16	14
E09	12	10	8	10	12	10
E10	17	13	3	11	9	7
E11	11	12	9	10	10	8
E12	12	14	8	12	8	6
E13	16	10	4	10	12	10
E14	12	16	8	14	6	4
E15	7	16	13	14	6	4
E16	16	9	4	10	13	11
E18	11	15	9	13	7	5
E19	12	13	8	11	9	7
E20	14	10	6	10	12	10

(STUDENTS)

TABLE A.3. Statistics of patents registration (telecommunications branch).

YEARS

I \ J	1986	1985	1984	1983	1982	1981	1980
USA	986	774	711	591	467	404	258
Japan	653	552	361	307	195	128	43
Germany	405	357	347	251	293	319	208
France	189	158	200	171	153	184	147
Great Britain	204	182	158	137	92	86	67
Italy	31	28	28	21	22	29	15
Holland	64	59	61	64	33	30	11
Sweden	25	19	31	25	15	12	13
Switzerland	23	34	19	30	17	30	15

(TELECOMMUNICATIONS BRANCH)

Appendix 2. Reference Data Sets

TABLE A.4. Statistics of patents registration (all the economic branches mixed).

	J	1986	1985	1984	1983	1982	1981	1980
ECONOMIC BRANCHES MIXED	USA	11126	9726	8969	7486	6243	5820	3574
	Japan	6259	5384	4192	3728	2980	2026	853
	Germany	8710	7892	6604	6360	6002	5747	4087
	France	3304	2891	2797	2735	2252	2371	1634
	Great Britain	3396	3076	2944	2652	2366	2146	1853
	Italy	1180	969	863	748	616	490	206
	Holland	905	779	732	677	545	542	268
	Sweden	843	737	764	659	551	478	302
	Switzerland	1193	1220	1059	1047	912	947	699

TABLE A.5. Preferences data set.

SUBJECTS

	J / I	Maths	Physics	English	Natural Science	Foreign Language	History
STUDENTS	E01	1	2	6	3	4	5
	E02	1	2	6	3	4	5
	E03	1	2	6	4	3	5
	E04	6	5	1	4	2	3
	E05	1	2	6	3	4	5
	E06	4	5	3	6	1	2
	E08	5	6	3	4	1	2
	E09	1	3	6	4	2	5
	E10	1	2	6	3	4	5
	E11	2	1	5	3	4	6
	E12	2	1	4	3	5	6
	E13	1	3	6	4	2	5
	E14	3	1	4	2	5	6
	E15	4	1	3	2	5	6
	E16	1	5	6	4	2	3
	E18	3	1	4	2	5	6
	E19	3	1	4	2	5	6
	E20	1	5	6	3	2	4

5. Responses to a Questionnaire on New Services in Telecommunications

From the survey on the Minitel services three data sets were extracted; these contingency data sets are derived from categorical variables or categorical variables with multiple forms.

(a) Satisfaction, with respect to classes indicating the duration of use. (*cf.* Table A.6.)

TABLE A.6. Contingency data set from the two variables Q8 and Q17 of the survey.

Q17: SATISFACTION VARIABLE

Q8 \ Q17	Very satisfied	Rather satisfied	Rather un-satisfied	Very un-satisfied	Q8
<5 mn	110	449	132	13	704
5–10	83	262	24	1	370
1115	70	170	13	3	256
1630	76	146	9	3	234
3145	31	49	0	0	80
4660	11	39	1	0	51
1–2 h	15	13	0	0	28
2–3 h	12	5	0	0	17
>3 h	4	2	0	0	6
Q17	412	1135	179	20	1746

Q8: CLASSES OF DURATION

(b) Reasons for using the Minitel services, with respect to the main wishes related to the Minitel. (*cf.* Table A.7)

(c) Reasons for using the Minitel services. (*cf.* Table A.8)

6. Financial Data Set

Budget of a company studied according to three branches during the years 1925–1934. (*cf.* Table A.9.)

Appendix 2. Reference Data Sets

TABLE A.7. Contingency data set from the two variables Q12 and Q34; these two variables are with multiple forms.

Q12: REASONS FOR USING THE MINITEL

Q34: MAIN WISHES \ J / I	No response	Practical use	Fastness	To show to people	For learning	Fun	Curiosity	Looking for information
Faster screening	160	12	15	1	5	21	57	22
Faster input	25	0	3	0	6	5	19	6
Colored screen	46	9	8	0	0	16	22	13
Less bulky	87	12	9	2	0	35	69	29
Towards a micro-computer	38	2	1	0	2	10	12	2
Replacing by a micro-computer	8	0	0	0	0	1	4	1
More aesthetic	33	6	7	0	0	12	19	7
User-friendliness input	34	6	6	0	0	18	18	6
Telephone included	10	1	3	0	0	5	11	4
No response	872	114	109	13	3	221	497	157

TABLE A.8. Contingency data set from the qualitative variable Q12 with respect to itself.

Q12: REASONS FOR USING THE MINITEL

I \ J		1	2	3	4	5	6	7
1	For information	1210	73	105	9	3	137	354
2	For curiosity	73	153	24	6	2	16	35
3	For fun	105	24	144	6	1	11	40
4	For learning	9	6	6	16	2	1	4
5	To show to people	3	2	1	2	5	0	2
6	For rapidity	137	16	11	1	0	315	200
7	Practical use	354	35	40	4	2	200	667

TABLE A.9. Financial data set in three components (expressed as percentage)

BRANCHES

I \ J	Goods	Accounts	Credit
1925	37	22	41
1926	57	15	28
1927	38	29	33
1928	61	08	31
1929	50	16	34
1930	64	3	33
1931	68	7	25
1932	69	8	23
1933	67	12	21
1934	74	10	16

(YEARS)

7. Measurements Data Set on Skulls

The problem was to decide how to classify a fossil cranium based on the characteristics given in Table A.10a. To do this, a reference data set was built using data on skulls of dogs and wolves, with a view to applying subsequently discriminant analysis. (*cf.* Table A.10.b)

8. Steel Samples

Two characteristics were studied: the mechanical resistance R, and the carbon grade C. Data are given in Table A.11.

9. Economic Data Set Concerning Investments Abroad

With a view to advising their customers, a group of banks asked some economists to study the risk of their investments abroad; 43 countries were selected according to 15 confidence criteria. Each of the economists gave a mark between 0 (the maximum of risk) and 4 (the minimum of risk). Data are given in Table A.12.

Appendix 2. Reference Data Sets

TABLE A.10.a. Measurements data set.

SKULL VARIABLES

I			CBL	LUJ	WID	LUC	LFM	WFM
	100	▲	129	64	95	17.5	11.2	13.8
	101	▲	154	74	76	20.0	14.2	16.5
	102	▲	170	87	71	17.9	12.3	15.9
	103	▲	188	94	73	19.5	13.3	14.8
	104	▲	161	81	55	17.1	12.1	13.0
	105	▲	164	90	58	17.5	12.7	14.7
	106	▲	203	109	65	20.7	14.0	16.8
	107	▲	178	97	57	17.3	12.8	14.3
	108	▲	212	114	65	20.5	14.3	15.5
	109	▲	221	123	62	21.2	15.2	17.0
	110	▲	183	97	52	19.3	12.9	13.5
	111	▲	212	112	65	19.7	14.2	16.0
	112	▲	220	117	70	19.8	14.3	15.6
	113	▲	216	113	72	20.5	14.4	17.7
	114	▲	216	112	75	19.6	14.0	16.4
	115	▲	205	110	68	20.8	14.1	16.4
WOLF AND DOG SKULLS	116	▲	228	122	78	22.5	14.2	17.8
	117	▲	218	112	65	20.3	13.9	17.0
	118	▲	190	93	78	19.7	13.2	14.0
	119	▲	212	111	73	20.5	13.7	16.6
	120	▲	201	105	70	19.8	14.3	15.9
	121	▲	196	106	67	18.5	12.6	14.2
	122	▲	158	71	71	16.7	12.5	13.3
	123	▲	255	126	86	21.4	15.0	18.0
	124	▲	234	113	83	21.3	14.8	17.0
	125	▲	205	105	70	19.0	12.4	14.9
	126	▲	186	97	62	19.0	13.2	14.2
	127	▲	241	119	87	21.0	14.7	18.3
	128	▲	220	111	88	22.5	15.4	18.0
	129	▲	242	120	85	19.9	15.3	17.6
	200	●	199	105	73	23.4	15.0	19.1
	201	●	227	117	77	25.0	15.3	18.6
	202	●	228	122	82	24.7	15.0	18.5
	203	●	232	123	83	25.3	16.8	15.5
	204	●	231	121	78	23.5	16.5	19.6
	205	●	215	118	74	25.7	15.7	19.0
	206	●	184	100	69	23.3	15.8	19.7
	207	●	175	94	73	22.2	14.8	17.0
	208	●	239	124	77	25.0	16.8	27.0
	209	●	203	109	70	23.3	15.0	18.7
	210	●	226	118	72	26.0	16.0	19.4
	211	●	226	119	77	26.5	16.8	19.3
	001	★	210	103	72	20.5	14.0	16.7

—Biometric analysis of skulls
Data table k_{IJ} associated with skulls (unit: mm)
I: set of skulls; J: set of variables (or descriptors)
▲: code for a dog skull; ●: code for a wolf skull;
★: code for the unknown Jussac skull.

TABLE A.10.b. Keycode for the skull variables.

Key code
Diagram of measurements taken on Jussac skull
(a) calvarium from ventral view
(b) upper left carnivore from occlural view
(c) first upper left molar from occlural view
CBL: caudylo-basal length; LUJ: length of the upper jaw;
WID: width of the upper jaw; LUC: length of the upper
carnivore; LFM: length of the first upper molar;
WFM: width of the first upper molar.

Appendix 2. Reference Data Sets

TABLE A.11. Steel samples data set. C is given in one per 10,000 and R in kg/mm^2.

J \ I	Steel samples															
C	72	55	63	38	10	45	77	67	58	74	65	51	39	27	27	24	32
R	96	82	83	61	39	68	95	92	78	94	88	75	64	53	56	49	57

10. Family Timetables

Twenty-eight groups of population were selected according to the following criteria: sex, marital status, employment status and country; these 28 groups are studied with respect to their activities expressed as "time spent to make (or do an activity)." Ten main classes of activities were retained; the data are given in Table A.13

11. Semantic Field Associated with Colors

The data set represents the association between a set of 11 colors and a set of 89 adjectives. The case represents the number of people who associated a given color and a given adjective. The data are given in Table A.14.

12. Proximities Data Set from the Family Timetables Data Set

From the data set given in Table A.13, distances between groups of population are computed and the results are given in Table A.15.

13. Barataria Data of Grain-Size Measurements

Ninety-eight bottom samples were collected from the Barataria Bay tidal lagoon at the margin of the Mississipi delta, with the objective of evaluating the depositional environment of the lagoon (*cf.* Krumbein and Aberdeen, 1958). Data are given in Table A.16. Only 67 samples fully described were used in the statistical analysis.

TABLE A.12. Confidence marks data sets for the investors abroad, published in *Les cahiers de l'analyse des données* (1976).

COUNTRIES \ VARIABLES	Political Stability	The General Attitude Towards the Investors	Nationalization	Inflation	Balance of Payments	Lateness	Economical Increasing	Convertibility of Currency	Respect of Contacts	Cost of Labor	Services	Communication Services	Marketing Services	Short Term Credit	Long Term Credit
Canada	3.7	2.7	3.5	2.7	2.9	2.9	2.9	3.6	3.7	2.7	3.3	3.4	3.4	3.1	3.1
USA	3.7	3.8	4.0	2.3	1.5	3.2	2.6	3.4	3.7	2.8	3.7	3.9	3.8	3.5	3.5
Mexico	2.9	2.2	2.0	2.7	2.8	2.0	3.0	3.0	2.2	2.1	2.3	2.3	2.5	2.5	2.3
Argentina	1.1	1.7	1.7	0.8	1.0	1.5	1.5	1.0	2.0	2.0	2.4	2.0	2.0	1.6	1.4
Brazil	2.8	2.8	2.6	1.7	2.4	1.9	3.2	2.1	2.2	2.2	2.2	2.1	2.4	2.3	2.1
Chile	0.3	0.0	0.0	0.4	0.4	0.3	0.3	0.4	0.0	0.9	1.5	1.8	1.5	0.7	0.9
Colombia	2.0	2.0	1.7	1.8	1.5	1.4	2.0	1.2	1.8	2.0	1.9	1.9	2.0	2.0	1.7
Peru	1.6	1.2	1.1	1.9	1.5	1.2	1.5	1.0	1.5	1.6	1.8	1.8	1.8	1.3	1.2
Venezuela	2.5	2.1	2.1	2.6	3.3	2.2	2.4	3.2	2.6	2.0	2.6	2.3	2.0	2.4	1.9
Australia	3.4	2.7	3.2	3.0	3.3	3.1	3.0	3.6	3.4	2.5	3.3	3.1	3.2	3.3	3.1
Korea	2.3	2.5	2.8	2.2	2.3	2.2	2.7	2.2	2.2	3.1	2.4	2.4	2.5	1.9	2.8
Indonesia	2.5	2.7	2.2	1.9	1.7	1.6	2.4	1.8	1.8	1.6	1.4	1.6	2.0	1.5	1.5
Japan	3.3	2.5	3.3	2.8	1.4	2.8	3.4	3.6	3.2	3.3	3.5	3.6	3.4	3.2	3.2
Malaya	2.5	2.8	2.5	3.0	2.0	2.2	2.7	3.0	2.6	2.2	2.2	2.2	2.2	2.3	2.1
Philippines	1.4	2.0	2.2	1.4	1.6	1.3	1.6	1.5	1.9	2.0	2.0	2.0	2.1	1.6	1.5
Singapore	3.3	3.3	3.3	3.0	3.8	3.2	2.9	3.3	3.0	2.9	3.0	3.2	3.0	3.2	2.8
Taiwan	2.0	3.2	3.2	2.8	3.5	2.6	2.7	2.4	2.4	2.9	2.7	2.8	2.8	2.4	2.1
India	2.3	1.9	1.9	1.9	1.2	1.2	1.8	1.0	1.4	1.7	2.0	2.0	2.0	1.4	1.4
Iran	2.5	2.3	2.2	2.0	2.2	2.0	2.5	2.4	1.9	2.0	1.9	2.1	1.8	2.1	1.9
Israel	2.9	3.0	2.9	1.7	1.9	2.0	2.5	2.5	3.2	2.7	3.4	3.2	3.1	2.6	2.6
Lebanon	2.0	2.6	2.6	2.5	2.1	2.1	1.9	2.4	2.0	1.9	2.1	2.1	2.2	2.3	2.0
Pakistan	1.2	2.0	1.4	1.0	0.9	1.1	1.2	1.0	1.4	1.4	1.0	1.1	1.4	1.0	0.4
Turkey	1.9	2.1	2.0	1.6	1.5	1.5	2.0	1.4	1.7	2.0	1.9	2.1	1.9	3.4	3.6
Germany	3.2	3.2	3.5	2.6	3.4	2.9	3.0	3.5	3.4	3.2	3.5	3.6	3.4	3.2	3.1
Belgium	3.4	3.6	3.6	2.8	2.8	2.9	2.6	3.5	3.5	3.1	3.3	3.0	3.3	3.4	1.0
Denmark	3.5	3.3	3.3	2.8	2.6	3.0	2.1	3.2	3.3	3.0	3.2	3.2	3.2	2.8	2.5
Spain	2.4	3.0	3.1	2.2	2.6	2.0	2.6	2.4	2.5	2.4	2.3	2.4	2.0	2.2	1.9
France	3.0	2.5	1.0	2.4	2.9	2.2	2.6	2.7	3.0	2.7	3.0	3.0	2.9	2.8	2.7
Greece	2.1	2.9	2.5	2.3	2.1	2.2	2.0	2.1	2.3	2.4	2.2	2.3	2.3	2.8	2.8
Ireland	3.2	3.6	3.4	3.0	2.8	2.8	2.3	3.0	3.1	2.3	2.5	2.6	2.4	2.6	2.4
Italy	2.2	2.6	2.6	2.2	2.5	2.0	2.2	2.7	2.5	2.4	2.8	2.9	2.7	2.8	2.5
Norway	3.2	3.0	3.2	2.9	2.7	2.9	2.5	3.2	3.3	2.8	3.0	3.1	3.2	2.8	2.4
The Netherlands	3.5	3.1	3.4	2.6	2.9	3.0	2.9	3.5	3.5	3.0	3.5	3.6	3.3	3.2	2.8
Portugal	2.6	2.8	3.1	2.5	2.1	1.0	2.2	2.1	2.1	2.0	2.2	2.2	2.0	1.9	1.7
United Kingdom	3.5	3.2	3.3	2.1	2.8	2.9	1.9	3.0	3.6	2.0	3.5	3.6	3.4	3.0	3.6
Sweden	3.2	2.8	2.8	2.4	2.7	2.7	2.3	3.0	3.2	2.8	3.3	3.1	3.2	2.6	2.7
Switzerland	3.9	2.5	3.6	2.7	3.3	3.1	2.7	3.9	3.7	3.0	3.5	3.8	3.4	3.4	3.3
Kenya	2.2	2.5	2.3	2.2	2.3	1.7	2.0	2.2	1.8	1.5	1.5	1.8	1.5	2.0	1.3
Libya	1.4	0.7	0.5	2.3	2.8	0.8	2.5	2.0	1.0	1.2	1.4	1.4	1.1	0.8	0.8
Morocco	1.2	2.0	2.1	2.2	1.9	1.6	1.8	1.8	1.5	1.6	1.6	1.6	1.7	1.5	1.7
Nigeria	2.0	2.2	2.0	1.7	2.2	1.7	2.3	1.5	1.8	1.6	1.6	1.5	1.6	1.8	1.4
Saudi Arabia	1.4	1.4	1.1	1.6	1.2	1.3	1.6	0.8	1.3	1.7	1.7	1.7	1.7	0.9	0.9
South Africa	3.3	3.3	3.1	2.9	3.0	2.8	3.0	3.0	3.2	2.6	3.4	2.9	3.2	3.1	3.1

Appendix 2. Reference Data Sets

This data set. The case contains the number of hours spent by a given group of population on a given activity (data published in *Classification automatique pour l'analyse des données*, Dunod, 1978).

	J	Work	Transport	Household	Children	Shopping	Personal Care	Meals	Sleep	Television	Leisure
Employed men in U.S.A.	EM. USA	610	140	60	10	120	95	115	760	175	315
Employed women in U.S.A.	EW. USA	475	90	250	30	140	120	100	775	115	305
Unemployed women in U.S.A.	UW. USA	10		495	110	170	110	130	785	160	430
Married men in U.S.A.	MM. USA	615	140	65	10	115	90	115	765	180	305
Married women in U.S.A.	MW. USA	179	29	421	87	161	112	119	776	143	373
Single men in U.S.A.	SM. USA	585	115	50		150	105	100	760	150	385
Single women in U.S.A.	SW. USA	482	94	196	18	141	130	96	775	132	336
Employed men in Western countries	EM. WEST	653	100	95	7	57	85	150	808	115	330
Employed women in Western countries	EW. WEST	511	70	307	30	80	95	142	816	87	262
Unemployed women in Western countries	UW. WEST	20	7	568	87	112	90	180	843	125	368
Married men in Western countries	MM. WEST	655	97	97	10	52	85	152	807	122	320
Married women in Western countries	MW. WEST	168	22	528	69	102	83	174	824	119	311
Single men in Western countries	SM. WEST	643	105	72		62	77	140	813	100	388
Single women in Western countries	SW. WEST	429	34	262	14	92	97	147	849	84	392
Employed men in Eastern countries	EM. EAST	650	142	122	22	76	94	100	764	96	334
Employed women in Eastern countries	EW. EAST	578	106	338	42	106	94	92	752	64	228
Unemployed women in Eastern countries	UW. EAST	24	8	594	72	158	92	128	840	86	398
Married men in Eastern countries	MM. EAST	652	133	134	22	68	94	102	763	122	310
Married women in Eastern countries	MW. EAST	436	79	433	60	119	90	107	772	73	231
Single men in Eastern countries	SM. EAST	627	148	68		88	92	86	720	58	463
Single women in Eastern countries	SW. EAST	434	86	297	21	129	102	94	799	58	380
Employed men in Yugoslavia	EM. YUG	630	140	120	15	85	90	105	760	70	365
Employed women in Yugoslavia	EW. YUG	560	105	375	45	90	90	95	745	60	235
Unemployed women in Yugoslavia	UW. YUG	10	10	710	55	145	85	130	815	60	380
Married men in Yugoslavia	MM. YUG	650	145	112	15	85	90	105	760	80	358
Married women in Yugoslavia	MW. YUG	260	52	576	59	116	85	117	775	65	295
Single men in Yugoslavia	SM. YUG	615	125	95		115	90	85	760	40	475
Single women in Yugoslavia	SW. YUG	433	89	318	23	112	96	102	774	45	408

TABLE A.14. Analysis of the semantic field associated with colors. Contingency data set (data published in *Classification automatique pour l'analyse des données*, Dunod, 1978).

I \ J	Blue	Red	Yellow	White	Gray	Pink	Brown	Purple	Black	Orange	Green
Acidic			12							1	4
Agressive		3						1		1	
Anxious		4	1					1	2		
Asiatic			3								
Attractive	1		4							4	4
Autumnal							4				
Foolish				1	3	5					
Burning		4									
Burnt										8	
Misty					7		2		1		
Calm	12					3	4			1	8
Celestial	17		1	6							
Rustic										1	9
Hot		3	1				6	1		6	
Clear	8		2	2	5		4				3
Colored		1						2	1	3	
Comical										2	1
Raw		2	2								
Decorative				1	1			2		3	
Discrete							1	3			
Soft	1			6	2	2	1	3		2	
Hard							3	2			2
Dynamic		1								7	
Ecclesastical				3				11			
Dazzling		2	7	1				1			
Irritable		2				1					
Childish			1		3	13					5
Sunny			7								
Balanced	6						3				
Extensive	3		1					1			
Feminine										3	
Flaming		6	1							2	
Flowery			1			3					
Fragile						6					2
Frank	2	2		1							8
Cold	2			3	2		1				5
Fruity			1							15	
Icy							6				
Harmonious		1					3				
Wintery					2		2				
Immaculate				9							
Joyful	1	1	1			1		1		4	1
Juicy										7	

Appendix 2. Reference Data Sets

Table A.14. (Contd.)
COLORS

I (Adjectives) \ J	Blue	Red	Yellow	White	Gray	Pink	Brown	Purple	Black	Orange	Green
Ugly			1			1	3	1			1
Light	2		1			4				1	
Limpid	4			4							
Smooth				2	5		2				
Faraway	5										
Luminous	3	1	22	3						8	1
Ill		2	1								
Masculine							6				
Mechanical										8	
Finical						12	6				
Dead		2		10	1			7	15		
Mysterious				2	1			4	5		
Natural							14				20
Snowy				6							
Pale				3		8					
Perfumed			1	1		5		4			
Passionate		4								1	
Patriotic	3										
Lost	1		1					2	3		5
Vernal	1					1				1	10
Deep							1	5	7		
Clean	2			5		1					
Pure	2			22							
Refined						1	1	2	3		
Religious								14			
Restful	12										7
Revolutionary		3									
Romantic	2					6					
Round							1			4	
Dirty			2		12	3	6				
Bloody		20								10	
Severe							1	3	6		
Silent									4		
Sinister					2		1		9		
Sober					3			1	4		
Dark	2				7		6	1	9		1
Sonorous			3				2				
Sly			9		2						
Sugary			1			14					
Sad					24		6	8	22		
Empty			6	8	8	5	1	3	4		
Old					4		6				
Keen		18	3							5	
Violent		17							8	2	
Alive	1						1				6
Unstable				2	1	6		3			

TABLE A.15. Proximities data set from the family timetables data set. The distance computed is the correspondence analysis distance.

GROUPS OF POPULATION

	EM. USA	EW. USA	UW. USA	MM. USA	MW. USA	SM. USA	SW. USA	EM. WEST	EW. WEST	UW. WEST	MM. WEST	MW. WEST	SM. WEST	SW. WEST	EM. YUG	EW. YUG	UW. YUG	MM. YUG	MW. YUG	SM. YUG	SW. YUG	EM. EAST	EW. EAST	UW. EAST	MM. EAST	MW. EAST	SM. EAST	SW. EAST
EM. USA	0																											
EW. USA	335	0																										
UW. USA	640	582	0																									
MM. USA	54	372	658	0																								
MW. USA	501	332	176	324	0																							
SM. USA	292	267	589	245	449	0																						
SW. USA	374	122	538	418	382	272	0																					
EM. WEST	354	486	717	337	606	448	546	0																				
EW. WEST	409	389	609	403	485	472	469	240	0																			
UW. WEST	676	616	410	472	433	700	684	561	449	0																		
MM. WEST	360	495	720	341	610	463	555	31	242	559	0																	
MW. WEST	580	559	473	569	441	634	644	447	331	166	444	0																
SM. WEST	399	539	742	389	643	462	596	134	322	591	157	490	0															
SW. WEST	505	454	570	506	492	493	505	311	261	399	324	356	318	0														
EM. YUG	311	387	702	319	564	332	452	305	354	683	326	582	294	440	0													
EW. YUG	412	384	676	413	527	467	490	434	336	660	441	538	474	529	287	0												
UW. YUG	659	546	430	660	408	432	629	595	456	326	606	319	589	413	593	542	0											
MM. YUG	284	382	701	293	542	320	445	298	355	684	317	582	293	446	30	287	602	0										
MW. YUG	499	420	493	497	389	519	529	465	314	431	473	330	487	421	413	282	271	419	0									
SM. YUG	453	468	723	470	608	368	506	472	519	760	498	692	408	496	234	452	608	255	506	0								
SW. YUG	422	338	562	438	409	361	409	404	354	573	426	505	382	358	231	326	414	250	304	230	0							
EM. EAST	261	350	688	268	543	317	413	291	440	330	662	448	545	484	519	289	72	542	288	294	446	0						
EW. EAST	387	333	659	392	501	429	441	440	330	677	305	573	312	447	101	268	413	80	418	314	279	265	0					
UW. EAST	654	525	347	658	348	616	596	609	473	281	618	326	608	392	627	601	158	633	352	638	453	635	389	0				
MM. EAST	227	353	685	227	540	323	415	277	322	669	283	560	322	459	176	277	630	153	428	172	505	328	346	84	0			
MW. EAST	423	348	558	423	417	467	468	440	280	529	445	414	485	461	365	172	418	364	172	256	117	256	442	204	321	0		
SM. EAST	417	463	745	430	625	364	495	415	488	758	440	686	355	479	190	444	642	204	524	117	252	442	664	321	509	0		
SW. EAST																										349	310	0

Appendix 2. Reference Data Sets

TABLE A.16. Barataria data set. The contents are expressed as phi. Only fully described samples are taken into account in statistical analysis. * = Krumbein and Aberdeen's group (Data published in Krumbein and Aberdeen, 1958.)

VARIABLES

I \ J	N^0	*	1	2	3	4	5	6	7	8	9	10
00001	K1	1	0.6	70.2	29.2							
00002	K2	1	1.0	69.9	29.1							
00003	K3	1	0.8	73.7	25.5							
00004	K4	1	0.9	75.3	23.8							
00005	K5	1	0.6	62.5	36.9							
00006	K6	1	1.1	68.8	30.1							
00007	K7	3	0.6	5.9	33.0	24.9	9.4	7.8	5.5	5.4	4.4	3.1
00008	K8	5	1.0	2.3	6.6	16.2	12.0	11.4	13.3	11.0	7.5	18.7
00009	K9	4	1.2	1.6	15.3	38.4	13.0	9.5	5.6	5.3	4.2	5.9
00010	K10	2	9.5	15.8	59.0	6.4	0.9	0.9	1.4	2.3	1.8	
00011	K11	3	0.4	3.9	45.2	24.7	3.7	8.1	3.0	3.8	3.0	4.2
00012	K12	1	5.6	48.4	42.7	3.3						
00013	K13	4	6.3	7.5	25.4	17.2	9.5	6.7	27.4			
00014	K14	2	1.1	16.3	58.7	11.9	3.1	0.7	2.7	1.7	1.4	2.4
00015	K15	3		13.8	39.3	15.4	9.1	4.5	6.4	4.4	3.6	3.5
00016	K16	4	2.3	7.9	23.9	25.5	9.2	7.9	7.7	5.6	4.6	5.2
00017	K17	3	3.0	6.2	30.7	25.7	9.5	7.5	6.4	4.1	3.2	3.7
00018	K19	4	1.0	3.1	15.2	32.0	14.3	10.0	7.2	6.0	4.8	6.4
00019	K20	5	3.2	3.9	10.5	24.1	14.2	15.4	13.5	7.7	5.1	2.4
00020	K21	2	2.4	14.5	53.9	12.2	5.5	1.6	2.5	2.0	1.6	3.6
00021	K22	2	2.2	38.8	42.2	7.9	1.4	1.8	1.0	1.1	1.0	2.6
00022	K23	4		11.5	28.4	19.1	7.3	7.8	4.8	6.4	6.7	6.0
00023	K24	3	2.7	7.4	48.5	19.1	5.3	3.8	3.6	3.6	2.4	3.6
00024	K26	3	1.7	6.8	41.3	16.4	5.6	6.5	6.8	4.8	3.8	6.3
00025	K28	2	1.7	30.4	44.5	11.2	3.0	1.9	2.9	1.5	1.2	1.7
00026	K29	3	4.6	19.2	31.7	16.8	6.4	5.2	5.2	3.7	2.9	4.3
00027	K30	3		12.9	43.2	21.2	6.5	6.8	2.4	3.8	1.8	1.4
00028	K31	2		40.0	32.5	3.8	4.5	6.5	2.7	4.7	3.2	2.1
00029	K32	4	0.8	7.0	31.6	21.1	10.2	9.0	6.3	4.1	3.2	6.7
00030	K36	4		3.4	19.7	25.4	15.7	10.2	9.9	6.9	3.7	5.1
00031	K42	2	1.5	32.2	36.5	12.5	5.1	6.7	5.5			
00032	K43	5	4.4	8.1	8.9	19.9	12.0	11.4	10.8	8.1	5.9	10.5
00033	K44	5	0.5	2.6	7.2	30.0	14.9	12.9	11.2	7.8	6.5	6.4
00034	K45	4	1.4	1.9	14.4	40.2	3.5	8.4	7.1	6.6	5.2	6.3
00035	K47	3	0.7	8.2	27.3	32.7	7.7	5.6	4.6	4.4	4.0	4.8
00036	K48	4	0.4	3.5	18.8	29.5	11.2	10.4	7.5	6.6	4.4	7.5
00037	K49	2	0.3	15.6	54.1	21.3	4.1	2.6	2.0			
00038	K50	2	0.3	24.4	56.0	15.1	4.2					

TABLE A.16. (*Contd.*)

VARIABLES

I \ J	N^0	*	1	2	3	4	5	6	7	8	9	10
00039	K51	2	10.5	29.2	37.3	15.1	4.2	3.7				
00040	K52	2	0.3	13.3	63.5	14.2	4.0	3.4	1.3			
00041	K53	2	1.2	26.9	54.7	11.0	3.9	2.3				
00042	K54	2	0.9	20.4	47.3	17.7	3.3	2.0	3.9	1.8	1.0	1.7
00043	K55	2	0.4	18.0	49.5	12.4	3.3	3.9	4.9	2.7	1.2	3.7
00044	K56	5	0.5	4.1	9.8	27.9	13.5	13.5	7.4	8.3	7.6	7.4
00045	K57	2	0.5	35.6	50.1	7.8	3.9	2.1				
00046	K60	2	1.9	32.5	43.5	11.8	4.0	2.5	1.5	2.3		
00047	K61	3	1.9	11.5	49.5	22.4	5.7	4.5	2.0	2.5		
00048	K62	3	1.0	13.1	33.7	18.3	6.1	5.0	7.7	5.3	5.1	4.7
00049	K65	2	0.7	18.3	53.2	17.2	5.6	5.0				
00050	K70	5	0.7	2.3	5.2	23.2	19.4	14.1	10.1	10.0	8.7	6.3
00051	K72	1	1.1	36.1	58.4	4.4						
00052	K73	1	0.2	21.8	72.8	5.2						
00053	K75	3		18.9	34.4	18.3	6.6	5.6	5.1	3.9	3.2	4.8
00054	K79	3	0.9	13.6	43.9	20.1	7.2	4.8	9.5			
00055	K80	2	0.3	7.4	77.4	9.4	5.5					
00056	K82	3	2.9	15.5	37.0	30.3	5.1	1.9	2.2	5.1		
00057	K83	1	0.8	10.2	79.2	9.8						
00058	K84	1	1.0	16.3	73.8	8.9						
00059	K85	1	1.8	35.7	61.9	0.6						
00060	K86	3	0.9	11.2	39.7	26.8	5.0	3.7	4.6	2.6	1.8	3.9
00061	K87	3	1.6	10.4	43.9	19.4	4.4	3.3	4.3	2.3	5.3	5.1
00062	K88	3		9.5	50.3	16.7	4.9	3.3	6.2	3.7	1.0	4.4
00063	K89	3	4.4	11.0	43.2	21.8	7.1	6.4	6.1			
00064	K91	3	1.3	7.2	45.2	27.8	12.0	6.5				
00065	K93	2	8.6	28.4	26.0	22.5	4.8	6.2	4.5			
00066	K94	3	2.5	17.8	40.3	20.1	4.3	2.5	3.5	3.8	2.5	3.5
00067	K95	3	0.4	5.9	56.5	18.9	2.6	2.3	3.2	1.8	2.6	5.8
00068	K96	2	6.3	17.1	42.4	17.9	9.0	7.3				
00069	K97	3	2.1	16.7	39.6	17.7	8.3	3.5	7.3			
00070	KF18	4	1.4	5.2	22.4	29.3	3.3	6.5	4.6	20.3		
00071	KF25	3	0.9	5.7	41.9	20.1	6.0	5.7	5.0	4.4	3.6	6.2
00072	KF27	3	1.4	7.7	50.2	17.8	2.6	5.4	3.0	3.2	3.6	6.1
00073	KF33	3	1.0	8.1	32.1	23.6	3.0	6.5	6.0	5.0	3.2	6.0
00074	KF34	5	2.8	4.5	7.9	20.8	19.5	16.4	28.1			
00075	KF35	4	0.5	2.1	14.0	37.2	19.9	11.4	6.1	8.8		
00076	KF36	4	0.5	2.0	7.5	37.4	20.8	9.5	22.3			
00077	KF37	5	2.1	2.1	10.7	23.6	15.1	14.0	11.8	20.6		
00078	KF39	2		37.0	45.4	7.3	3.8	3.3	3.8	3.6	2.5	3.3
00079	KF40	3	0.4	4.0	38.2	28.5	6.0	4.3	4.7	4.2	3.0	5.7
00080	KF41	4	1.2	14.7	25.6	17.6	6.8	5.7	8.4	20.0		

(SAMPLES)

Appendix 2. Reference Data Sets

TABLE A.16. (Contd.)
VARIABLES

I	N^0	*	1	2	3	4	5	6	7	8	9	10
00081	KF46	5	0.6	3.6	4.2	17.8	12.4	10.8	9.9	40.7		
00082	KF58	4	3.5	12.6	14.2	28.6	10.1	8.7	22.3			
00083	KF59	2	1.4	40.7	38.9	5.1	3.9					
00084	KF63	5	1.1	2.5	6.5	22.4	17.2	9.3	41.0			
00085	KF64	3	0.4	5.1	31.8	30.3	5.4	7.8	3.0	6.4	5.6	5.2
00086	KF68	3	0.5	5.9	32.2	32.7	4.9	5.4	2.7	5.2	4.6	3.9
00087	KF67	3	1.4	6.2	35.1	27.5	7.3	4.9	3.1	7.6	5.5	2.4
00088	KF68	4	2.3	4.3	11.1	27.8	10.3	9.3	10.5	19.8		
00089	KF69	4	4.2	4.7	14.2	31.7	8.7	8.0	23.5			
00090	KF71	3	1.1	4.9	31.1	41.9	13.9	7.8	3.7	2.4	3.7	3.4
00091	KF74	5	3.4	1.6	4.4	18.0	14.7	15.3	15.1	12.6	10.3	4.6
00092	KF76	4		3.3	19.0	34.7	10.6	9.9	22.5			
00093	KF77	3	7.9	8.5	21.0	19.9	5.9	5.9	6.3			
00094	KF78	3	0.9	6.9	54.0	24.2	5.3	5.3	4.9	3.8	2.1	2.7
00095	KF81	2	0.9	17.5	33.9	26.8	4.0	6.9				
00096	KF90	4		4.1	17.3	31.1	11.4	10.2	9.7	15.9		
00097	KF92	4	0.5	6.3	18.2	28.0	9.1	9.7	9.9	5.6	3.1	9.1
00098	KF98	3	0.3	20.6	55.4	16.6	6.2	6.1	5.5	3.6	3.0	2.7

SAMPLES

14. Quality of Service in the Telephone Network

At France Telecom, the management of quality of service is based on indicators; the synthetized indicator is the "IGQS" (global indicator of quality of service). It is computed from six components: IZAA, EZAA, TSI, VR2, TCR, and TCOM. The meaning of the components are the following:

- IZAA = Intra-urban efficacy;
- EZAA = Extra-urban efficacy;
- TSI = Rate of "out-of-order" telephone equipment;
- VR2 = Rate of restoring telephone equipment;
- TCR = Efficacy at directory inquiries;
- TCOM = Connecting time between two consumers.

These indicators are computed for 21 regions of France (cf. Fig. A.1.) for studying the evolution of the quality. The results for the years 1984–1986 are given in Table A.17a, b, c.

TABLE A.17. Quality of service data sets relative to the telephone network during the years 1984, 1985, 1986. The regions of France are represented in Fig. A.1.

(a) INDICATORS 1984

I \ J	IGQS	IZAA	EZAA	TSI	VR2	TCR	TCOM
Ajaccio	61.4	58.6	55.6	43.3	53.0	31.6	13.4
Amiens	89.1	66.3	66.2	28.0	83.9	68.	18.7
Besançon	95.7	69.5	68.0	19.0	87.5	68.3	15.2
Bordeaux	89.3	67.2	66.1	29.2	84.6	70.6	15.9
Caen	88.0	67.1	66.6	31.5	82.9	74.1	15.6
Châlons	95.8	67.9	68.1	22.5	93.4	78.9	16.6
Clermont	96.0	70.8	69.3	22.5	87.5	75.6	16.6
Dijon	93.4	68.1	68.3	23.2	86.7	74.6	16.1
Lille	91.8	67.1	67.1	28.5	90.2	58.5	16.3
Limoges	83.4	66.7	64.7	38.6	78.9	66.4	15.5
Lyon	93.5	67.4	68.4	21.2	86.1	67.8	17.3
Marseille	87.1	64.0	63.9	29.7	86.3	54.4	16.2
Montpellier	89.3	66.3	64.7	27.6	85.9	60.7	15.5
Nancy	93.8	67.6	68.9	25.4	89.7	73.1	16.6
Nantes	91.8	67.8	67.8	28.9	88.3	74.3	16.7
Orléans	92.7	67.6	67.0	25.7	89.0	70.2	18.5
Poitiers	90.2	67.7	66.8	28.5	84.3	74.2	16.3
Rennes	91.5	68.3	67.9	28.1	85.5	66.2	17.3
Rouen	87.4	68.1	67.8	31.5	78.7	71.3	15.1
Strasbourg	97.8	67.9	70.1	18.8	93.5	82.0	15.8
Toulouse	88.0	67.6	66.9	30.5	80.3	70.2	16.2
Paris	98.2	66.6	68.4	23.2	89.8	54.1	17.1

(b) INDICATORS 1985

I \ J	IGQS	IZAA	EZAA	TSI	VR2	TCR	TCOM
Ajaccio	74.6	61.7	60.2	35.3	67.4	42.1	13.7
Amiens	93.6	66.8	67.6	21.0	87.6	68.1	16.8
Besançon	96.1	68.9	68.3	18.5	88.4	76.9	15.4
Bordeaux	92.2	66.9	67.2	23.9	86.6	68.7	16.3

Appendix 2. Reference Data Sets

TABLE A.17. (*Contd.*)

(b) INDICATORS 1985

	I \ J	IGQS	IZAA	EZAA	TSI	VR2	TCR	TCOM
REGIONS OF FRANCE	Caen	92.0	66.8	67.0	25.7	88.3	72.3	18.0
	Châlons	97.6	68.5	69.3	19.9	93.9	80.7	16.5
	Clermont	97.9	70.8	70.8	19.6	89.1	80.0	16.7
	Dijon	96.1	69.0	69.3	20.2	88.9	80.0	16.6
	Lille	95.0	67.8	68.3	22.0	89.0	69.8	16.3
	Limoges	87.0	66.3	67.4	33.6	81.6	72.2	15.7
	Lyon	95.0	76.4	69.0	20.1	88.4	71.8	17.6
	Marseille	89.1	64.5	64.1	25.9	87.3	53.4	16.6
	Montpellier	91.6	66.5	65.9	23.6	87.2	63.6	16.1
	Nancy	96.0	68.0	69.4	22.2	91.8	78.7	17.2
	Nantes	95.6	68.3	69.4	23.8	91.5	73.8	16.6
	Orléans	94.6	68.2	68.1	23.3	90.4	72.2	16.5
	Poitiers	94.5	68.7	68.0	22.6	88.7	78.4	18.0
	Rennes	94.6	68.4	68.4	22.2	88.5	69.7	16.9
	Rouen	93.4	68.6	68.7	25.4	87.5	72.6	15.8
	Strasbourg	100.5	69.6	72.0	16.8	94.4	86.0	15.5
	Toulouse	90.3	67.6	68.1	27.1	81.8	73.8	16.3
	Paris	99.3	66.8	69.0	20.9	90.0	53.9	17.3

(c) INDICATORS 1986

	I \ J	IGQS	IZAA	EZAA	TSI	VR2	TCR	TCOM
REGIONS OF FRANCE	Ajaccio	77.6	64.0	61.1	34.3	68.5	66.5	15.4
	Amiens	97.3	69.4	70.9	18.7	87.6	69.6	18.7
	Besançon	96.6	69.3	69.2	17.9	87.4	79.3	19.1
	Bordeaux	93.6	68.5	68.6	22.0	84.8	67.8	18.0
	Caen	95.1	68.4	69.0	23.3	90.4	70.9	17.3
	Châlons	98.9	70.2	72.1	19.2	90.6	81.0	17.2
	Clermont	98.1	71.1	72.3	20.1	86.7	82.2	18.1
	Dijon	97.4	69.9	71.5	19.6	87.9	78.5	19.3
	Lille	97.2	69.0	70.2	18.8	90.1	65.5	17.5
	Limoges	88.8	69.6	70.4	31.2	75.6	66.3	19.0
	Lyon	95.8	68.8	69.9	19.3	86.5	73.1	18.4
	Marseille	89.5	65.6	64.8	24.3	84.6	58.6	18.4
	Montpellier	91.2	68.0	67.2	23.8	82.5	71.0	19.3
	Nancy	95.8	69.0	70.5	21.0	87.2	78.2	16.5
	Nantes	98.1	69.3	71.8	20.7	90.9	73.3	18.6
	Orléans	98.2	70.4	72.2	20.1	88.6	73.7	17.4
	Poitiers	96.4	69.7	70.1	20.9	88.3	80.4	19.4
	Rennes	95.5	69.6	69.8	22.0	86.9	77.2	18.5
	Rouen	95.6	70.3	70.1	20.7	86.9	69.2	17.3
	Strasbourg	102.4	71.0	73.8	15.6	94.3	86.4	16.6
	Toulouse	91.3	68.8	69.1	25.0	79.9	72.1	18.2
	Paris	101.0	68.2	70.2	18.4	90.1	52.8	18.2

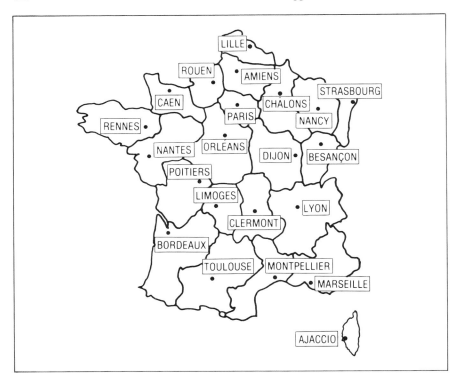

FIGURE A.1. Regions of France identified by their representative town.

15. Crimes Data in the United States of America For 1977

In the United States in 1977 the criminality was studied according to seven crime variables. The statistics were given by state. Data are given in Table A.18.

TABLE A.18. Crime rates per 100,000 population for the year 1977. The states are represented Fig. A.2. (Data reprinted with permission from Springer-Verlag, Young, F., "Visualizir Six-Dimensional Structure with Dynamic Statistical Graphics," in *Chance*, Vol. 2, No. 1, 1989.)

CRIME VARIABLES

	I \ J	Murder	Rape	Robbery	Assault	Burglary	Larceny	Auto Theft
STATES	Alabama	14.2	25.2	96.8	278.3	1135.5	1881.9	280.7
	Alaska	10.8	51.6	96.8	284.0	1331.7	3369.8	753.3
	Arizona	9.5	34.2	138.2	312.3	2346.1	4467.4	439.5
	Arkansas	8.8	27.6	83.2	203.4	972.6	1862.1	183.4
	California	11.5	49.4	287.0	358.0	2139.4	3499.8	663.5

Appendix 2. Reference Data Sets

TABLE A.18. (Contd.)

CRIME VARIABLES

I \ J	Murder	Rape	Robbery	Assault	Burglary	Larceny	Auto Theft
Colorado	6.3	42.0	170.7	292.9	1935.2	3903.2	477.1
Connecticut	4.2	16.8	129.5	131.8	1346.0	2620.7	593.2
Delaware	6.0	24.9	157.0	194.2	1682.6	3678.4	467.0
Florida	10.2	39.6	187.9	449.1	1859.9	3840.5	351.4
Georgia	11.7	31.1	140.5	256.5	1351.1	2170.2	297.9
Hawaii	7.2	25.5	128.0	64.1	1911.5	3920.4	489.4
Idaho	5.5	19.4	39.6	172.5	1050.8	2599.6	237.6
Illinois	9.9	21.8	211.3	209.0	1085.0	2828.5	528.6
Indiana	7.4	26.5	123.2	153.5	1086.2	2498.7	377.4
Iowa	2.3	10.6	41.2	89.8	812.5	2685.1	219.9
Kansas	6.6	22.0	100.7	180.5	1270.4	2739.3	244.3
Kentucky	10.1	19.1	81.1	123.3	872.2	1662.1	245.4
Louisiana	15.5	30.9	142.9	335.5	1165.5	2469.9	337.7
Maine	2.4	13.5	38.7	170.0	1253.1	2350.7	246.9
Maryland	8.0	34.8	292.1	358.9	1400.0	3177.7	428.5
Massachusetts	3.1	20.8	169.1	231.6	1532.2	2311.3	1140.1
Michigan	9.3	38.9	261.9	274.6	1522.7	3159.0	545.5
Minnesota	2.7	19.5	85.9	85.8	1134.7	2559.3	343.1
Mississippi	14.3	19.6	65.7	189.1	915.6	1239.9	144.4
Missouri	9.6	28.3	189.0	233.5	1318.3	2424.2	378.4
Montana	5.4	16.7	39.2	156.8	804.9	2773.2	309.2
Nebraska	3.9	18.1	64.7	112.7	760.0	2316.1	249.1
Nevada	15.8	49.1	323.1	355.0	2453.1	4212.6	559.2
New Hampshire	3.2	10.7	23.2	76.0	1041.7	2343.9	293.4
New Jersey	5.6	21.0	180.4	185.1	1435.8	2774.5	511.5
New Mexico	8.8	39.1	109.6	343.4	1418.7	3008.6	259.5
New York	10.7	29.4	472.6	319.1	1728.0	2782.0	745.8
North Carolina	10.6	17.0	61.3	318.3	1154.1	2037.8	192.1
North Dakota	.9	9.0	13.3	43.8	446.1	1843.0	144.7
Ohio	7.8	27.3	190.5	181.1	1216.0	2696.8	400.4
Oklahoma	8.6	29.2	73.8	205.0	1288.2	2228.1	326.8
Oregon	4.9	39.9	124.1	286.9	1636.4	3506.1	388.9
Pennsylvania	5.6	19.0	130.3	128.0	877.5	1624.1	333.2
Rhode Island	3.6	10.5	86.5	201.0	1489.5	2844.1	791.4
South Carolina	11.9	33.0	105.9	485.3	1613.6	2342.4	245.1
South Dakota	2.0	13.5	17.9	155.7	570.5	1704.4	147.5
Tennessee	10.1	29.7	145.8	203.9	1259.7	1776.5	314.0
Texas	13.3	33.8	152.4	208.2	1603.1	2988.7	397.6
Utah	3.5	20.3	68.8	147.3	1171.6	3004.6	334.5
Vermont	1.4	15.9	30.8	101.2	1348.2	2201.0	265.2
Virginia	9.0	23.3	92.1	165.7	986.2	2521.2	226.7
Washington	4.3	39.6	106.2	224.8	1605.6	3386.9	360.3
West Virginia	6.0	13.2	42.2	90.9	597.4	1341.7	163.3
Wisconsin	2.8	12.9	52.2	63.7	846.9	2614.2	220.7
Wyoming	5.4	21.9	39.7	173.9	811.5	2772.2	282.0

FIGURE A.2. States of the United States of America.

16. Table of percentage points of the χ^2 distribution

This Table is used to test the independence of two categorical variables studied through their associated contingency data set.

PERCENTILE VALUES (χ_p^2)
for
THE CHI-SQUARE DISTRIBUTION
with v degrees of freedom
(shaded area = p)

v	$\chi^2_{0.995}$	$\chi^2_{0.99}$	$\chi^2_{0.975}$	$\chi^2_{0.95}$	$\chi^2_{0.90}$	$\chi^2_{0.75}$	$\chi^2_{0.50}$	$\chi^2_{0.25}$	$\chi^2_{0.10}$	$\chi^2_{0.05}$	$\chi^2_{0.025}$	$\chi^2_{0.01}$	$\chi^2_{0.005}$
1	7.88	6.63	5.02	3.84	2.71	1.32	0.455	0.102	0.0158	0.0039	0.0010	0.0002	0.0000
2	10.6	9.21	7.38	5.99	4.61	2.77	1.39	0.575	0.211	0.103	0.0506	0.0201	0.0100
3	12.8	11.3	9.35	7.81	6.25	4.11	2.37	1.21	0.584	0.352	0.216	0.115	0.072
4	14.9	13.3	11.1	9.49	7.78	5.39	3.36	1.92	1.06	0.711	0.484	0.297	0.207
5	16.7	15.1	12.8	11.1	9.24	6.63	4.35	2.67	1.61	1.15	0.831	0.554	0.412
6	18.5	16.8	14.4	12.6	10.6	7.84	5.35	3.45	2.20	1.64	1.24	0.872	0.676
7	20.3	18.5	16.0	14.1	12.0	9.04	6.35	4.25	2.83	2.17	1.69	1.24	0.989
8	22.0	20.1	17.5	15.5	13.4	10.2	7.34	5.07	3.49	2.73	2.18	1.65	1.34
9	23.6	21.7	19.0	16.9	14.7	11.4	8.34	5.90	4.17	3.33	2.70	2.09	1.73
10	25.2	23.2	20.5	18.3	16.0	12.5	9.34	6.74	4.87	3.94	3.25	2.56	2.16
11	26.8	24.7	21.9	19.7	17.3	13.7	10.3	7.58	5.58	4.57	3.82	3.05	2.60
12	28.3	26.2	23.3	21.0	18.5	14.8	11.3	8.44	6.30	5.23	4.40	3.57	3.07
13	29.8	27.7	24.7	22.4	19.8	16.0	12.3	9.30	7.04	5.89	5.01	4.11	3.57
14	31.3	29.1	26.1	23.7	21.1	17.1	13.3	10.2	7.79	6.57	5.63	4.66	4.07
15	32.8	30.6	27.5	25.0	22.3	18.2	14.3	11.0	8.55	7.26	6.26	5.23	4.60
16	34.3	32.0	28.8	26.3	23.5	19.4	15.3	11.9	9.31	7.96	6.91	5.81	5.14
17	35.7	33.4	30.2	27.6	24.8	20.5	16.3	12.8	10.1	8.67	7.56	6.41	5.70
18	37.2	34.8	31.5	28.9	26.0	21.6	17.3	13.7	10.9	9.39	8.23	7.01	6.26
19	38.6	36.2	32.9	30.1	27.2	22.7	18.3	14.6	11.7	10.1	8.91	7.63	6.84

df													
20	40.0	37.6	34.2	31.4	28.4	23.8	19.3	15.5	12.4	10.9	9.59	8.26	7.43
21	41.4	38.9	35.5	32.7	29.6	24.9	20.3	16.3	13.2	11.6	10.3	8.90	8.03
22	42.8	40.3	36.8	33.9	30.8	26.0	21.3	17.2	14.0	12.3	11.0	9.54	8.64
23	44.2	41.6	38.1	35.2	32.0	27.1	22.3	18.1	14.8	13.1	11.7	10.2	9.26
24	45.6	43.0	39.4	36.4	33.2	28.2	23.3	19.0	15.7	13.8	12.4	10.9	9.89
25	46.9	44.3	40.6	37.7	34.4	29.3	24.3	19.9	16.5	14.6	13.1	11.5	10.5
26	48.3	45.6	41.9	38.9	35.6	30.4	25.3	20.8	17.3	15.4	13.8	12.2	11.2
27	49.6	47.0	43.2	40.1	36.7	31.5	26.3	21.7	18.1	16.2	14.6	12.9	11.8
28	51.0	48.3	44.5	41.3	37.9	32.6	27.3	22.7	18.9	16.9	15.3	13.6	12.5
29	52.3	49.6	45.7	42.6	39.1	33.7	28.3	23.6	19.8	17.7	16.0	14.3	13.1
30	53.7	50.9	47.0	43.8	40.3	34.8	29.3	24.5	20.6	18.5	16.8	15.0	13.8
40	66.8	63.7	59.3	55.8	51.8	45.6	39.3	33.7	29.1	26.5	24.4	22.2	20.7
50	79.5	76.2	71.4	67.5	63.2	56.3	49.3	42.9	37.7	34.8	32.4	29.7	28.0
60	92.0	88.4	83.3	79.1	74.4	67.0	59.3	52.3	46.5	43.2	40.5	37.5	35.5
70	104.2	100.4	95.0	90.5	85.5	77.6	69.3	61.7	55.3	51.7	48.8	45.4	43.3
80	116.3	112.3	106.6	101.9	96.6	88.1	79.3	71.1	64.3	60.4	57.2	53.5	51.2
90	128.3	124.1	118.1	113.1	107.6	98.6	89.3	80.6	73.3	69.1	65.6	61.8	59.2
100	140.2	135.8	129.6	124.3	118.5	109.1	99.3	90.1	82.4	77.9	74.2	70.1	67.3

Source: Catherine M. Thompson, *Table of percentage points of the χ^2 distribution*, Biometrika, Vol. 32 (1941), by permission of the authors and publisher.
Published in Statistics—M. R. Spiegel (1961) McGraw-Hill Book Company—(Schaum's outline series in science).

References

Only references given to students in data analysis courses have been retained and mentioned here. For a study in depth, the reader is referred to the bibliographies of CORMACK (1971), JAMBU (1983), NISHISATO (1986), and of the following reference publications.

AIVASIAN, S., ENUKOV, I., and MECHALKINE, L. (1986). *Eléments de modélisation et traitement primaire des données.* Mir, Moscou.

ANDERBERG, M. R. (1973). *Cluster analysis for applications.* Academic Press, New York.

BALL, G. H., and HALL, D. J. (1965). *Isodata; a novel method of data analysis and pattern classification.* Technical report, Standford Research Institute, Menlo Park, California.

BECKER, R. A., and CHAMBERS, J. M. (1984). *S: An interactive environment for data analysis and graphics.* Wadsworth, Belmont, California.

BENZÉCRI, J.-P. (1969). "Statistical Analysis as a Tool to Make Patterns Emerge from Data," In *Methodologies of pattern recognition* (Watanabe, S., ed.), pp. 35–74, Academic Press, New York.

BENZÉCRI, J.-P. (1972). *La place de l'apriori.* Encyclopedia Universalis, Vol. 17 Organum, Paris.

BENZÉCRI, J.-P., et al. (1973, 1976, 1980a). *L'analyse des données. Tome I. La taxinomie. Tome II. L'analyse des correspondances.* Dunod, Paris.

BASTIN, CH., BENZÉCRI, J.-P., BOURGARIT, CH., and CAZES, P. (1980b). *Pratique de l'analyse des données 2: abrégé théorique et études de cas modèle.* Dunod, Paris.

BENZÉCRI, J.-P., et al. (1981). *Pratique de l'analyse des données 3: linguistique et lexicologie.* Dunod, Paris.

BENZÉCRI, J.-P. (1982). *Histoire et préhistoire de l'analyse des données.* Dunod, Paris.

BENZÉCRI, J.-P., et al. (1984). *Pratique de l'analyse des données en économie.* Dunod, Paris.

BERTIER, P., and BOUROCHE, J. M. (1975). *Analyse des données multidimensionnelles.* Presses Universitaires de France, Paris.

BOCK, H. H. (1973). *Automatische Klassification Theorische und Praktische Methoden zür Gruppierung und Structurierung von Daten.* Van Den Hoecht and Ruprecht, Göttingen.

BRUYNOOGHE, M. (1978). "Classification ascendante hiérarchique de grands ensembles de données—un algorithme fondé sur la construction de voisinages réductibles." *Cahiers de l'analyse des données.* III(1), 35–46.

CALOT, G. (1965). *Cours de statistique descriptive.* Dunod, Paris.

CAZES, P. (1980). "L'analyse de certains tableaux rectangulaires décomposés en blocs: généralisation des propriétés recontrées dans l'étude des correspondances multiples. Partie II. Questionnaire—variantes de codage et nouveaux calculs de contributions." *Cahiers de l'analyse des données.* V(4), 387–403.

CAZES, P. (1982). "Note sur les éléments supplémentaires en analyse des correspondances. I. Pratique et utilisation. II. Tableaux multiples." *Cahiers de l'analyse des données.* VII(1), 9–23; (2), 133–154.

CHAMBERS, J. M. (1977). *Computational methods for data analysis.* Wiley, New York.

CHAMBERS, J. M., CLEVELAND, W. S., KLEINER, B., and TUKEY, P. A. (1983). *Graphical methods for data analysis.* Duxbury Press, Boston.

CORMACK, R. M. (1971). A review of classification." *The Journal of the Royal Statistical Society,* Series A. **134**, (2), 321–367.

RHAM, C. de. "La classification ascendante hiérarchique selon la méthode des voisins réciproques." *Cahiers de l'analyse des données.* V(2), 135–144.

DIDAY, E. (1980). *Optimisation en classification automatique* (2 volumes). Inria, Rocquencourt.

ESCOFIER, B., and PAGES, J. (1988). *Analyses factorielles simples et multiples.* Dunod, Paris.

FENELON, J. P. (1981). *Qu'est-ce-que l'analyse des données.* Lefonen, Paris.

FLAMENBAUM, G., THIERY, J., and BENZÉCRI, J. P. (1979). "Agrégation en boules de rayon fixé et centres optimisés." *Cahiers de l'analyse des données.* IV(3), 137–143.

GNANADESIKAN, R. (1977). *Methods for statistical data analysis of multivariate observations.* Wiley, New York.

GREENACRE, M. (1984). *Theory and applications of correspondence analysis.* Academic Press, London.

HARTIGAN, J. A. (1975). *Clustering algorithms.* Wiley, New York.

HAYASHI, C. (1954). "Multidimensional Quantification I, II." *Proc. Japan Acad.* **30**, 61–65; 165–169.

HAYASHI, C., DIDAY, E., JAMBU, M., and OHSUMI, N. (1988). *Recent developments in clustering and data analysis.* Academic Press, London.

JAMBU, M. (1978). *Classification automatique pour l'analyse des données.* Dunod, Paris.

References

JAMBU, M., and LEBEAUX, M. O. (1983). *Cluster analysis for data analysis.* North-Holland, Amsterdam.

JAMBU, M., STERN, D. and TAN, S. H. (1988). *Etude du suivi régional de la qualité de service par les méthodes d'analyse des données.* Technical Report, Centre National d'Etudes des Telecommunications. Issy-les-Moulineaux.

JARDINE, N. and SIBSON, R. (1971). *Mathematical taxonomy.* Wiley, New York.

JUAN, J. (1982). "Le programme HIVOR de classification hiérarchique selon les voisins réciproques." *Cahiers de l'analyse des données.* **VII**(2), 173-184.

KENDALL, M. G., and STUART, A. (1973). *The advanced theory of statistics,* Vol. I, 1977; Vol. II, 1973; Vol. III, 1976. Griffin, London.

KRUMBEIN, W. C., and ABERDEEN, L. L. (1958). "The sediments of Barataria Bay," *J Sed. Petrol.* **7**(1), 3-17.

KRUSKAL, J. B. (1969a). "Multidimensional scaling by optimizing goodness-of-fit to a non-metric hypothesis." *Psychometrika.* **29**, 1-27.

KRUSKAL, J. B. (1969b). "Non-metric multidimensional scaling: a numerical method." *Psychometrika.* **29**, 115-129.

KRUSKAL, J. B., and WISH, M. (1978). *Multidimensional scaling.* Sage Publications, Beverly Hills, California.

LEBART, L., MORINEAU, A., and FENELON, J. P. (1979). *Traitement des données statistiques.* Dunod, Paris.

LEBART, L., MORINEAU, A., and WARWICK, K. (1984). *Multivariate descriptive statistical analysis.* Wiley, New York.

LERMAN, I. C. (1981). *Classification et analyse ordinale des données.* Dunod, Paris.

MACQUEEN J. (1967). "Some methods of classification and analysis of multivariate observations." In *Proceedings of the 5th Berkeley symposium on mathematical statistics and probability,* 281-297.

MORICE, E., and CHARTIER, F. (1954). *Méthode statistique. Première partie. Elaboration des Statistiqus. Deuxième partie. Analyse Statistique.* Institut National de la Statistique et des Etudes Economiques pour la Métropole et la France d'Outre-Mer, Paris.

NISHISATO, S. (1980). *Analysis of categorical data. Dual scaling and its applications.* University of Toronto Press, Toronto.

NISHISATO, S. (1986). *Quantification of categorical data: a bibliography 1975-1986.* Microstats, Toronto.

ROMEDER, J. M. (1973). *Méthodes et programmes d'analyse discriminante.* Dunod, Paris.

SAPORTA, G. (1990). *Probabilitiés, statistique et analyse des données,* Technip, Paris.

SCHIFFMANN, S. S., REYNOLDS, M. L., and YOUNG, F. W. (1981). *Introduction to multidimensional scaling.* Academic Press, New York.

SHEPARD, R. N. (1974). "Representation of structure in similarity data: problems and prospects." *Psychometrika.* **39**, 373-421.

SOKAL, R. R., and SNEATH, P. H. A. (1973). *Principles of numerical taxonomy.* Freeman, San Francisco.

SPIEGEL, M. R. (1971). *Theory and problems of statistics.* McGraw-Hill, New York.

STERN, D., TAN, S. H., and JAMBU, M. (1988). *Applications de l'analyse des données au traitement d'enquêtes: comportement des usagers non professionnels du MINITEL.* Technical report, Centre National d'Etudes des Telecommunications. Issy-les-Moulineaux.

TEIL, H. A. (1985). *Application and comparison of some ordination and clustering techniques in geology,* Thesis of Geology, Leicester University.

TUKEY, J. W. (1977). *Exploratory data analysis.* Addison-Wesley, Reading, Massachussetts.

VOLLE, M. (1985). *Analyse des données.* Economica, Paris.

YULE, U., and KENDALL, M. C. (1950). *An introduction to the theory of statistics.* Griffin, London.

Author Index

A

Aberdeen, 10, 467
Adanson, 19,
Aivasian, 465
Anderberg, 305, 465

B

Ball, 311, 416, 465
Bastin, 465
Becker, 3, 465
Benzécri, 21, 95, 169, 241, 288, 305, 334, 335, 390, 394, 412, 419, 465, 466
Bertier, 466
Bock, 334, 466
Bourgarit, 466
Bouroche, 466
Bruynooghe, 344, 466
Buffon, 19,

C

Calot, 91, 466
Carroll, 169
Cazes, 241, 465, 466
Chambers, 3, 89, 465, 466
Chartier, 467
Cleveland, 3, 466
Cormack, 305, 466
Cramer, 67
Czekanovsky, 409

D

Diday, 305, 311, 416, 424, 466

E

Enukov, 465
Escofier, 424, 466

F

Fenelon, 466, 467
Fisher, 19, 41, 42
Flamenbaum, 466
Friant, 424

G

Gauss, 222
Gnanadesikan, 466
Greenacre, 95, 241, 466
Guttmann, 224

H

Hall, 311, 416, 465
Hartigan, 305, 334-335, 466
Hayashi, 21, 419, 466
Huyghens, 315

J

Jaccard, 409
Jambu, 95, 241, 322, 349, 424, 466, 467, 468
Jardine, 305, 334, 394, 467
Juan, 467

K

Kelley, 40, 41,
Kendall, 409, 467, 468

Kleiner, 466
Krumbein, 10, 467
Kruskal, 95, 467
Kulcinski, 409

L

Lance, 341
Laplace, 222
Lebart, 95, 241, 467
Lebeaux, 424, 467
Lerman, 305, 334, 467
Leroy, 424
Linné, 19

M

MacQueen, 305, 311, 416, 467
Mahalanobis, 409
Mechalkine, 465
Michener, 409
Morice, 467
Morineau, 467

N

Nishisato, 467

O

Ohsumi, 466

P

Pages, 466
Pearson, 41, 42, 67

R

Rahm, 466
Rao, 409

Reynolds, 467
Rogers, 409
Romeder, 467
Russel, 409

S

Saporta, 467
Schiffman, 95, 467
Shepard, 95, 169, 237, 467
Sibson, 305, 334, 394, 467
Sneath, 305, 334–335, 394, 409, 424, 467
Sokal, 305, 334–335, 394, 409, 424, 467
Spearman, 409
Spiegel, 4, 468
Stern, 467, 468
Stuart, 467

T

Tabet, 424
Tan, 467, 468
Tanimoto, 409
Teil, 468
Thiery, 466
Tschuprov, 67
Tukey, 3, 21, 95, 419, 466, 468

V

Volle, 311, 468

W

Warwick, 467
Williams, 341
Wish, 95

Y

Young, 467
Yule, 28, 30, 40, 468

Subject Index

A

Algorithm
  accelerated alg. of hierarchical classification, 334
  classical alg. of hierarchical classification, 332
  for classifying supplementary elements, 386
  of hierarchical classification, 332
  of moving centers, 311, 316
  [CAH.DIST], 332
  [CAHA.DIST], 343
  [CAHA.EUCL1], 352
  [CAH.EUCL], 341
  [PAR.NC = k], 311
  [PAR.DI = 2R], 316
  [SUP.CAH.DIST], 388
  [SUP.CAH.QUADR], 390
  [SUP.CAH.TER], 386
Analysis
  factor analysis of individuals-variables data sets, 113
  of proximities data sets, 407, 412
  principal components analysis, 125
  ranks factor analysis, 126
  1-D analysis of chronological variables, 53
  1-D analysis of categorical variables, 50, 53
  1-D analysis of quantitative variable, 27
  2-D analysis of categorical variable, 64
  2-D analysis of quantitative variables, 73
  2-D analysis of a categorical and a quantitative variable, 91
  3-D analysis of one categorical and two quantitative variables, 99
  3-D analysis of one quantitative and two variables, 95
  3-D analysis of 3 quantitative variables, 100
  2-D correspondence analysis, 169
  N-D analysis of quantitative variables, 100
  N-D analysis of categorical variables, 107
  N-D correspondence analysis, 241

C

cartography, 162, 394
chart
  bar chart, 52, 56
  cartogram, 61, 112, 161, 400, 404
  hierarchical classification, 307, 310
  component part bar chart, 51
  dipole-by-dipole graphic, 381
  dispersion chart, 73, 100
  dispersion box plot, 48, 56, 97
  factor graphics, 141, 200, 275
  graph-table, 68
  hierarchical representation of a partition, 320
  histogram, 42, 44, 48
  least squares curve, 85
  line graph, 55–56
  multiple bar chart, 61
  multiple scatter diagram, 100, 204
  ogive, 45–46
  pie-chart, 51
  profile diagram, 102
  quality diagram, 104
  regression curve, 80, 82
  scatter diagram, 73, 100
  statistical map (cf. cartogram)
  stereogram, 73, 96
  sun-ray plot, 56, 93, 97, 104
  triangular diagram, 100
center of gravity, 180
centroid, 114
class

class frequency, 42
class limit, 43
class, 306–310
class table, 48
eldest of a class, 309
level index of a class, 308
mass of a class, 313
youngest of a class, 309
classification
　classification of individuals-variables data sets, 305
　classification of proximities data sets, 407
　hierarchical classification, 307
　hierarchical correspondence clustering, 354
　mathematical description of classifications, 306
classifying supplementary elements, 143, 203, 278, 284, 328, 385
cloud of points, 118, 120, 127–128, 177, 182, 184, 221
contributions
　contribution of a dipole, 323, 371
　contribution of a form, 258
　contribution of a question, 258
　contribution to an axis, 181, 194, 197, 218–221
　contribution to the eccentricity, 218–221, 321, 369
　contribution to the variance, 191, 194, 197, 218–221, 256, 260, 371
coordinates
　factor coordinates, 122, 124, 129, 132, 135, 159, 182, 184, 191, 193
　question factor coordinates, 259–260
　supplementary factor coordinates, 150, 206, 280
coefficient
　contingency coefficient, 67
　correlation ratio, 82
　linear correlation coefficient, 87
　of variation, 39
criterion
　angular distance criterion, 340
　average linkage criterion, 335
　complete linkage criterion, 334
　distance between centers of gravity criterion, 335
　hyperplane criterion, 389
　median distance, 336
　single linkage criterion, 334
　variance of a partition, 338

second order central moment of a partition, 337
curve fitting, 75

D

Data
　data exploration, 1, 14, 19
　data collection, 420
　data decision, 14
　data elaboration, 14, 420
　data input, 14, 420
　data management, 14, 422
　data communication, 17
data set
　acceptable data set, 225, 288, 356
　basic data set, 21, 25, 125, 216, 242, 306
　chronological data set, 165
　confusion data set, 234
　contingency data set, 64, 170, 225, 232
　correspondence data set, 170–171, 357
　disjunctive form data set, 243–247, 250
　dispersion data set, 73
　doubled data set, 277
　dummy data set, 227
　factor coordinates data set, 357
　flow data set, 235
　fuzzy data set, 292
　hachured data set, 75, 96
　individuals-variables data set, 25, 306
　logical data set, 226–227
　marginal data set, 172
　marks data set, 160, 226–227, 236
　measurements data set, 161, 226
　multiple contingency data set, 108, 246, 248, 252
　preferences data set, 160, 229, 237
　presence-absence data set, 290
　principal data set, 205
　principal components data set, 356
　profiles data set, 67, 173
　proximities data set, 25, 306, 408
　recoded data set, 242, 248
　questionnaire data set, 229, 242–248
　supplementary data set, 205
Deciles, 33–34
Deviation
　mean deviation, 36
　median deviation, 39
　geometric deviation, 39
　standard deviation, 36

SUBJECT INDEX

Diagonalization, 120, 181-183
Diagram (cf. chart)
Dipole, 323
Distributional equivalence, 179
Distance
  angular distance, 340
  distance in classification, 329, 334, 343
  distance in correspondence analysis, 178, 217
  distance in principal components analysis, 127-128
  distance in proximities analysis, 408-411
  of Kendall, 410
  of Mahalanobis, 409
  of Spearman, 410

E

Eccentricity, 219-220, 369
Eigenvalues, 120, 132, 263
Eigenvectors, 120
Explained points, 191-195
Explicative points, 190-193

F

Factor
  factor analysis, 114
  factor axes, 117, 122, 217
  factor graphics, 141, 200, 275
Forms, 64
Formulas
  contribution formulas, 218-221
  factor analysis formulas, 124
  principal components analysis formulas, 159
  reconstruction formulas, 221
  recursion formulas, 334
  transition formulas, 221
  2-D correspondence analysis formulas, 216
  N-D correspondence analysis formulas, 287
Frequency
  cumulative frequency distributions, 45-46
  frequency curves, 46-47
  frequency distributions, 42, 45, 50, 97
  frequency histogram, 44
  frequency polygon, 44

G

Graphic (cf. chart)

I

Independence, 66
Interpretation
  of a correlation coefficient, 91
  of a correlation ratio, 91
  of a hierarchical classification, 357
  of a partition, 320
  of principal components analysis, 131
  of 2-D correspondence analysis, 189
  of N-D correspondence analysis, 262
Interquartile range, 34

K

Kurtosis, 40

L

Linear adjustment, 113, 118-119
Least squares, 76

M

Mean
  α-mean, 32
  arithmetic mean, 29
  harmonic mean, 31
  geometric mean, 31
  mean deviation, 36
  mean difference, 35
  quadratic mean, 31
Median
  median, 28
  median deviation, 39
Measures
  of association, 67
  of central tendency, 28
  of dispersion, 34
Mode, 33

N

Node, 308

O

Ogive, 45–46

P

Partition, 306
Pattern recognition, 162, 396
Percentiles, 34
Profiles, 174

Q

Quality
  quality of explanation, 199, 265
  quality of representation, 140
Quantiles, 33
Quartiles, 33–34
Quartile deviation, 35
Quartile coefficient of variation, 40
Questionnaire, 229, 242–248

R

Range, 34
Recoding, 227
Reconstruction
  reconstruction of a distance, 131, 188
  reconstruction of a data set, 122, 124, 129, 160, 186
Regression, 75
Rules
  for interpreting factor axes, 134–135, 139, 190, 264
  for selecting explained points, 215
  for selecting explicative points, 158, 214, 286
  for selecting significant classes, 325
  for selecting significant dipoles, 381–382
  for selecting significant variables, 384
  for selecting the number of nodes, 381–382
  for selecting the number of factor axes, 153, 213, 286

S

Simplex of probability laws, 176
Skewness, 40
Statistical map (cf. cartogram)
Statistical series, 27
Statistical set, 24
Statistics, 19
  deductive statistics, 21
  inductive statistics, 21

T

Taxonomy, 19, 162, 394
Terminal of a class, 308
Transition, 174

V

Variable
  categorical variable, 21–22
  chronological variable, 24
  conditional variable, 78
  dummy variable, 24
  logical variable, 23
  marginal variable, 78
  preference variable, 23
  qualitative variable (cf. categorical var.)
  quantitative variable, 22, 242
  with multiple forms, 22
Variance
  of a class, 313–315
  of a cloud of points, 313–315, 376
  of a dipole, 367, 376
  of a factor axis, 193, 218
  of a partition, 313–315
  of a variable, 36